SPECTRAL LINE SHAPES

SPECTRAL LINE SHAPES

18th International Conference on Spectral Line Shapes

Auburn, Alabama 4 – 9 June 2006

EDITORS

Eugene Oks

Michael Pindzola

Auburn University,
Auburn, Alabama

SPONSORING ORGANIZATIONS

U.S. Department of Energy
Alabama Space Grant Consortium
College of Sciences and Mathematics of the Auburn University

AIP
75 Years of Service

Melville, New York, 2006

AIP CONFERENCE PROCEEDINGS ■ VOLUME 874

Editors:

Eugene Oks
Michael Pindzola

Physics Department
206 Allison Lab
Auburn University
Auburn, AL 36849
U.S.A.

E-mail: goks@physics.auburn.edu
pindzola@physics.auburn.edu

L.C. Catalog Card No. 2006936479
ISBN 978-0-7354-0370-3
ISSN 0094-243X

Printed in the United States of America

CONTENTS

NEUTRALS

SYMPOSIUM: SPECTROSCOPIC DIAGNOSTICS FOR MAGNETIC FUSION

ASTROPHYSICS/EARTH ATMOSPHERE

PREFACE

The biennial International Conference on Spectral Line Shapes (ICSLS) is the main international forum for both fundamental and applied works dealing with processes related to the formation of spectral line profiles and with their use for diagnostic purposes. On the fundamental side, the study of the line profiles reveals the underlying atomic and molecular interactions. On the practical side, the line profiles are employed as powerful diagnostic tools for various media, such as neutral gases, technological gas discharges, magnetically-confined plasmas for fusion, laser- and Z-pinch-produced plasmas (for fusion and other purposes), astrophysical plasmas (most importantly, solar plasmas) and planetary atmospheres.

This conference series, which started in 1973 in Paris, France, alternates between Europe and US / Canada. The last two conferences were held in Berkeley, USA (2002) and Paris, France (2004).

The present 18th ICSLS was organized during June 4 – 9, 2006 in Auburn, AL, USA – at the Auburn University. All sessions took place at the Auburn University Hotel and Conference Center. The President of the Auburn University – Dr. E. Richardson – gave an opening address.

The conference was attended by 51 participants representing the following 8 countries (in alphabetical order): Canada, France, Germany, Poland, Russia, Spain, United Kingdom, and USA. They presented 48 papers: 32 invited full-size talks, 16 short talks. There were no posters: all participants were given the opportunity to present their contributions orally. Out of the total 48 papers, 45 of them are collected in this volume.

The papers in the Proceedings are subdivided into the following sections corresponding to the conference sessions:

- Plasmas
- Neutrals
- Symposium "Spectroscopic Diagnostics for Magnetic Fusion"
- Astrophysics/Earth Atmosphere

Most of the members of the International Program Committee provided nominations for speakers. We would like to thank them for the invaluable help.

We are very grateful to the members of the Local Organizing Committee for their assistance. We would like to thank especially:

- Melanie Woods, the Conference Secretary, for being instrumental to numerous tasks in preparing the conference and the Proceedings;
- Phil Forrest, the Technology Advisor, for providing technological help to all participants before and during their presentations;
- Francis Robicheaux for creating and maintaining the conference website;
- Stuart Loch for designing the conference poster;
- Allen Landers for accompanying participants on the tour to Montgomery.

The success of the conference is determined not only by the quality of the presented research, but also by the social events. The pinnacle of the social program was a string quartet concert featuring a prodigious soloist Nadir Khashimov. Being only 15 years old, he is already the prize winner of several prestigious international competitions of violinists. The conference participants were stunned by his virtuosity.

Finally, we would like to thank our sponsors:

- US Department of Energy
- Alabama Space Grant Consortium
- College of Sciences and Mathematics of the Auburn University, and especially the Physics Department.

The 19th ICSLS will occur in Valladolid, Spain in June 2008. Marco Gigosos and Manuel Gonzalez from the University of Valladolid accepted the huge task of organizing the next meeting of the ICSLS.

Eugene Oks and Michael Pindzola
Physics Department
Auburn University

INTERNATIONAL PROGRAM COMMITTEE

Christina Back (USA) --- backca@fusion.gat.com
Alexander Devdariani (Russia) --- snbrn2@yandex.ru
Robert Gamache (USA) --- Robert_Gamache@uml.edu
Volkmar Helbig (Germany) --- helbig@physik.uni-kiel.de ___
John C. Lewis (Canada) --- court@physics.mun.ca
David May (Canada) --- yukon1@pop3.onlink.net
Kenji Ohmori (Japan) --- ohmori@ims.ac.jp
Gillian Peach (U.K.) --- g.peach@ucl.ac.uk
Chantal Stehle (France) --- Chantal.stehle@obspm.fr
Jozef Szudy (Poland) --- szudy@phys.uni.torun.pl
Elisabeth Dilimier (France) --- elisabeth.dalimier@upmc.fr
Milan S. Dimitrijevic (Serbia) --- mdimitrijevic@aob.aob.bg.ac.yu
Michael Hartmann (France) --- jean-michel.hartmann@ppm.u-psud.fr
John Kielkopf (USA) --- kielkopf@louisville.edu
Valery Lisitsa (Russia) --- lisitsa@nfi.kiae.ru
Massimo Moraldi (Italy) --- Massimo.moraldi@fi.infn.it
Eugene Oks (USA) --- goks@physics.auburn.edu
Roland Stamm (France) --- roland.stamm@piim.up.univ-mrs.fr
William Stwalley (USA) --- stwalley@uconnvm.uconn.edu
George Tabisz (Canada) --- tabisz@cc.umanitoba.ca

LOCAL ORGANIZING COMMITTEE

Eugene Oks (Co-Chairman) --- goks@physics.auburn.edu
Michael Pindzola (Co-Chairman) --- pindzola@physics.auburn.edu
Francis Robicheaux--- francisr@physics.auburn.edu
Allen Landers--- landers@physics.auburn.edu
Stuart Loch--- loch@physics.aubur.edu
Phil Forrest (Technology Advisor) ---forrest@physics.auburn.edu
Shahin Abdel-Naby--- abdelsh@auburn.edu
Jesus Hernandez--- hernajv@auburn.edu
Turker Topcu--- topcut1@auburn.edu
Melanie Woods (Secretary)---woodsme@auburn.edu

MINUTES OF THE MEETING OF THE INTERNATIONAL PROGRAM COMMITTEE

The business luncheon/meeting of the International Program Committee occurred on June 6, 2006. The agenda consisted of the following items:

- Financial assistance to needy participants
- Choice of Publishers for Proceedings of the future meetings of the ICSLS
- Committee membership
- Organizers of the next two meetings of the ICSLS

Co-Chairmen of the ICSLS-18 made significant efforts to get grants for the conference. They succeeded in obtaining grants from the US Department of Energy, the Alabama Space Grant Consortium, and the College of Sciences and Mathematics of the Auburn University. Due to these grants, it became possible to provide various kinds of the financial assistance to 13 participants. The organizers of the ICSLS-18 also upheld the rule enacted by the International Committee at the ICSLS-13 in Florence, Italy in 1996 and written in the Minutes of the ICSLS-13: "The organizer of each conference should be granted the registration fee and the hotel expenses for the following conference". In adhering to this rule, Elisabeth Dalimier, who organized ICSLS-17 in Paris in 2004, was granted at the ICSLS-18 the registration fee and the free lodging at the Auburn University Hotel and Conference Center.

John Kielkopf, who was absent due to a family emergency, requested in advance that the Committee should discuss his proposal on the choice of Publishers for Proceedings of the future meetings of the ICSLS. Specifically, he suggested the American Institute of Physics (AIP) to be chosen as the Publishers for Proceedings of all future meetings of the ICSLS – in order to keep the uniformity of the exterior of the volumes. After a thorough discussion of his proposal, the Committee decided that each organizer of the particular meeting of the ICSLS should be given the opportunity to choose the most appropriate Publisher.

The Committee accepted resignations of its long-time members David May and Christina Back. Carlos Iglesias from the LLNL, USA (iglesias1@llnl.gov) and Dionision Bermejo from the CSIC, Spain (dbermejo@iem.cfmac.csic.es) were invited to join the Committee and they accepted (during and after the meeting, respectively).

It has been confirmed that the next 19th ICSLS in June 2008 will be organized by Marco Gigosos and Manuel Gonzalez from the University of Valladolid, Spain. The city of Valladolid has an international airport. The sessions will occur either at the medieval castle (if it would be less than 100 participants) or at the Palace of Congresses (if it would be more than 100 participants). Participating students would be given the opportunity of an inexpensive lodging at the dorm.

John Lewis from the University of Newfoundland (St. John, Canada) offered to organize the 20th ICSLS in 2010. The Committee gratefully accepted his offer.

Eugene Oks

PLASMAS

Emission and Absorption Measurements to Resolve Long-Standing Discrepancies in Spectral Line Broadening

Carlos A. Iglesias

Lawrence Livermore National Laboratory
P.O. Box 808, Livermore, CA 94550, USA

Abstract. It was recently shown that a quantum kinetic theory approach to Stark broadening of spectral lines yields corrections to the standard impact theory that may resolve discrepancies between theoretical and experimental widths of isolated lines. Furthermore, the kinetic theory method predicts, contrary to the standard impact theory, different emission and absorption line widths for plasmas in non-thermal equilibrium. This difference presents new opportunities for experimentally resolving the disagreements in electron-impact line broadening. Quantitative predictions suggest that such experiments are feasible.

Keywords: Collisional broadening, isolated lines.
PACS: 32.70.Jz, 52.25.Os

INTRODUCTION

The Stark broadening of spectral lines is a valuable diagnostic tool as well as a probe to study the interface between atomic and plasma physics. Therefore it is crucial to validate line-broadening theories with experiments. To this end measurements of the *2p-2s* and *3p-3s* line profiles in the Li and Be isoelectronic series of elements ranging from Be to Ne generated valuable data. [1-10] These are isolated lines with small ion broadening that avoid the complicating effects of overlapping lines and Stark level mixing. They also involve simple atomic systems making full quantum mechanical calculations tractable. Consequently, they provide excellent tests for electron collisional line broadening theories.

There is clear evidence indicating discrepancies between experiment and theory for these lines. [1-14] Comparisons of experimental and theoretical line widths are presented in Fig. 1. The figure shows increasing discrepancy between theory and experiment with increasing atomic number for the *3p-3s* Li isoelectronic sequence. Similar results are found for the *3p-3s* Be isoelectronic sequence while factor-of-two errors are reported for the *2p-2s* lines of Li-like ions. These experiments were performed at two independent facilities with distinct plasma sources that obtained consistent line widths, thereby decreasing the probability of large systematic errors. In particular, the experiments at Bochum [2-7] are considerably hotter and denser than those at Belgrade [8-10] as shown in Fig. 2.

CP874, *Spectral Line Shapes: 18th International Conference,*
edited by E. Oks and M. Pindzola

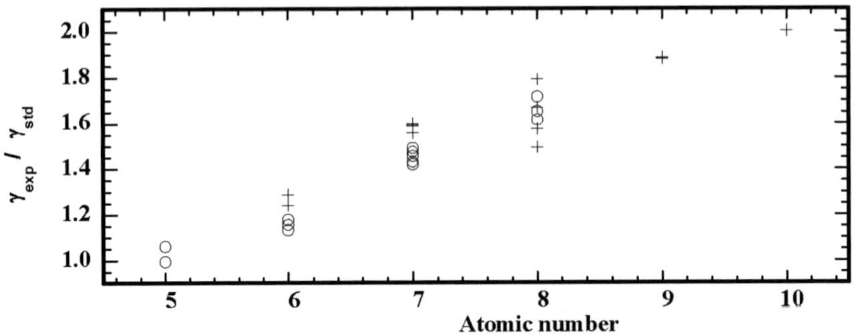

FIGURE 1. Ratio of experimental to theoretical line widths for the *3p-3s* Li isoelectronic sequence. Experimental results are the average of the doublet lines and theory is obtained by interpolating results in Ref. 14: Bochum [2-7] (+) and Belgrade [8-10] (o).

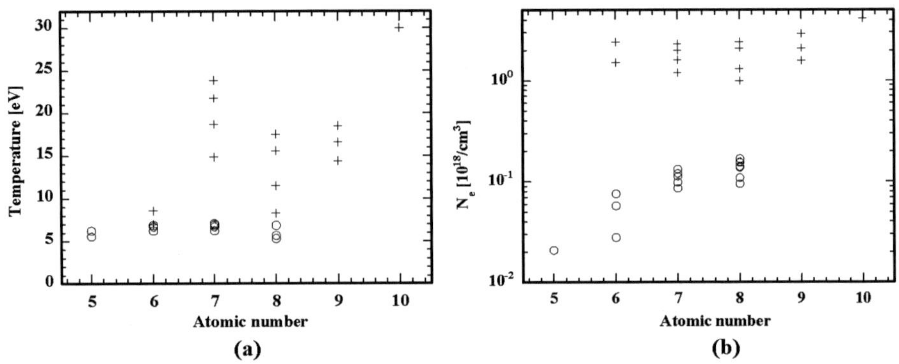

FIGURE 2. Experimental conditions corresponding to Fig. 1: (a) temperatures and (b) free electron densities with Bochum [2-7] (+) and Belgrade [8-10] (o).

On the theory side, recent quantum mechanical calculations of Stark widths [11-14] corroborated earlier theoretical results and agree with the latest version of the non-perturbative semi-classical method. [15] Therefore, the discrepancies are probably not due to numerical inaccuracies. Since broadening of these lines is primarily determined by inelastic binary electron-ion collisions, which can be directly assessed experimentally and crosschecked by comparing calculations, the discrepancies remain a serious concern.

A possible explanation is a failure of the standard impact theory [16] underlying the calculations that is a cornerstone of line broadening. Indeed it was recently shown that a quantum kinetic theory method, [17,18] extended to stationary non-equilibrium conditions, yields corrections to the standard impact theory that may resolve the long-standing discrepancies. [19] These corrections lead to different emission and absorption line widths except for systems in local thermal equilibrium (LTE) when the profiles satisfy the Kirchhoff-Planck relation. [20] The standard impact theory, however, predicts identical emission and absorption line widths independent of conditions, but only as a result of compensating approximations.

The purpose here is to propose an experimental test of line broadening theories by measuring both emission and absorption widths. The techniques for the former have been developed [1-10] while the latter involves measuring the photon transmission. These experiments have two possible outcomes. Firstly, the emission and absorption widths are equal, which the kinetic theory method can only reproduce with LTE populations so that it cannot resolve the discrepancies. [19] The second is different emission and absorption profiles. Then the experiments involve non-LTE radiator level populations, demonstrate limitations of the standard impact theory, and may be explained by the kinetic theory method.

A more fundamental interest is verification of the new width results from the kinetic theory method regardless of whether they explain the existing discrepancies in isolated line widths. If correct, future line width calculations would need to incorporate these corrections.

2. THEORY

The standard impact width for an isolated line connecting upper and lower radiator internal states $|\mu\rangle$ and $|\nu\rangle$ is given by [16]

$$\gamma_{std} = \gamma_{std}^{el} + \gamma_{std}^{in} = \gamma_{std}^{el} + \sum_{\alpha=\mu,\nu} \sum_{J_\beta \neq J_\alpha} \left\langle \sigma_{\alpha\to\beta}^{in} \right\rangle_{cl} \tag{1}$$

where γ_{std}^{el} and γ_{std}^{in} are the elastic and inelastic collision widths from standard impact theory, [16] In order to compare to the kinetic theory results, γ_{std}^{in} was written in terms of the inelastic collision cross-section from internal radiator state $|\alpha\rangle$ with total angular momentum J_α to all radiator states of a level with total angular momentum J_β thermally average over a classical electron gas.

The kinetic theory second-order emission and absorption impact widths, γ_{ems} and γ_{abs} respectively, for the isolated line are given by [19]

$$\left.\begin{matrix} \gamma_{ems} \\ \\ \gamma_{abs} \end{matrix}\right\} = \gamma_{std}^{el} + \sum_{\alpha=\mu,\nu} \sum_{J_\beta \neq J_\alpha} \left\{ \left[e^{\hbar\omega_{\alpha\beta}/T}\, \theta(\omega_{\alpha\beta}) + \theta(\omega_{\beta\alpha}) \right] \begin{pmatrix} P_{\beta\alpha}\delta_{\alpha\mu} + \delta_{\alpha\nu} \\ \\ \delta_{\alpha\mu} + P_{\beta\alpha}\delta_{\alpha\nu} \end{pmatrix} \right\} \left\langle \sigma_{\alpha\to\beta}^{in} \right\rangle_{cl} \tag{2}$$

where

$$\theta(x) = \begin{cases} 1 & x \geq 0 \\ 0 & x < 0 \end{cases} \quad . \tag{3}$$

The relative departure of radiator internal state populations from LTE is described by

$$P_{\beta\alpha} = f_\beta f_\alpha^{-1} e^{\hbar\omega_{\beta\alpha}/T} \tag{4}$$

with

$$\omega_{\beta\alpha} = \omega_\beta - \omega_\alpha \quad , \tag{5}$$

$\hbar\omega_\alpha$ and f_α the energy and population of state $|\alpha\rangle$, respectively, and T the perturbing electron gas temperature in energy units. The widths in Eq. (2) are similar to those from standard impact theory in Eq. (1)but with corrections that depend on radiator internal state populations and detailed-balance property of the perturbing electron gas. Note that

$$P_{\beta\alpha} \xrightarrow[LTE]{} 1, \tag{6}$$

giving equal emission and absorption widths in Eq. (2); thus, satisfying the Kirchhoff-Planck relation. [20] On the other hand, the standard impact theory always has identical emission and absorption widths independent of conditions.

For isolated lines the widths and shifts are much smaller than the energy separation between adjacent radiating levels and the line shape is approximately written as [16]

$$I(\omega) = \pi^{-1} D_{\mu\nu}^2 \left\{ f_\mu \frac{\gamma_{ems}}{(\omega - \omega_o)^2 + \gamma_{ems}^2} + f_\nu \frac{\gamma_{abs}}{(\omega + \omega_o)^2 + \gamma_{abs}^2} \right\} \tag{7}$$

where shifts as well as ion and Doppler broadening were neglected,

$$\omega_o = \omega_\mu - \omega_\nu > 0 \quad , \tag{8}$$

$$D_{\alpha\beta}^2 = \sum_{m_\alpha} \sum_{m_\beta} \left| \vec{d}_{\alpha\beta} \right|^2 \quad , \tag{9}$$

and

$$\vec{d}_{\alpha\beta} = \langle \alpha | \vec{d} | \beta \rangle . \tag{10}$$

Here, $\vec{d}$ is the radiator dipole operator and the sums are over magnetic quantum numbers. The adopted sign convention has $\omega > 0$ ($\omega < 0$) as emission (absorption) by the plasma. Furthermore, all states of a level are assumed degenerate with identical populations so that line shapes are independent of magnetic quantum numbers.

The energy loss and gain by the radiation field to and from the plasma is described by the emission and absorption coefficients. Thus, these coefficients relate experimental quantities to the line shape function and are given by [20]

$$\varepsilon(\omega) = n_R \frac{\omega^4}{3\pi c^3} D_{\mu\nu}^2 f_\mu \Phi_{ems}(\omega) \quad , \tag{11}$$

$$\kappa(\omega) = n_R \frac{4\pi^2 \omega}{3\hbar c} D_{\mu\nu}^2 \left\{ f_\nu \Phi_{abs}(\omega) - f_\mu \Phi_{ems}(\omega) \right\} \quad , \tag{12}$$

where n_R is the radiator number density and the terms with a maximum near $\omega = -\omega_o$ were neglected in the positive-frequency parts of $\varepsilon(\omega)$ and $\kappa(\omega)$. The subtracted term in Eq. (12) accounts for stimulated emission. The line shapes follow from Eq. (7),

$$\Phi_j(\omega)=\frac{\gamma_j/\pi}{(\omega-\omega_o)^2+\gamma_j^2} \quad . \tag{13}$$

In LTE the absorption coefficient simplifies to

$$\kappa_{LTE}(\omega)=n_R\frac{4\pi^2\omega}{3\hbar c}\left(1-e^{-\hbar\omega/T}\right)D_{\mu\nu}^2 f_\nu \Phi_{ems}(\omega) \tag{14}$$

since $f_\mu = e^{-\hbar\omega_{\mu\nu}/T} f_\nu$ and $\gamma_{abs}=\gamma_{ems}$.

Although direct measurement of the emission coefficient is possible, the absorption coefficient typically involves measuring the transmission,

$$T(\omega)=e^{-\ell\kappa(\omega)} \tag{15}$$

where ℓ is the plasma optical path here assumed homogeneous.

3. EXPERIMENTAL DISCUSSION

The results above cannot be readily applied to the experiments [2-10] since radiator level population information is not available. Nevertheless, for the BIII *2p-2s* [4] a reasonable estimate suggests a potential resolution of the width discrepancies due to the corrections from the kinetic theory method. The discussion of the *3p-3s* lines in the Li isoelectronic is more complicated. In these cases rough estimates of the level populations yield emission and absorption widths from Eq. (2). The resulting line profiles are used to demonstrate the feasibility of measuring both the emission intensity and transmission to differentiate between emission and absorption widths.

3.1 BIII *2p-2s* lines

The experiments were performed at Bochum. [4] At the experimental conditions the two-level radiator is an excellent approximation to the BIII 2p-2s transition since the upper and lower levels are well isolated from all other radiator levels. [11] Close-coupling calculations yield for the standard impact widths [11]

$$\gamma_{ex}\approx 2\gamma_{std} \qquad \text{and} \qquad \gamma_{std}^{in}\approx 2\gamma_{std}^{el} \tag{16}$$

with γ_{ex} the experimental width. In addition, [19]

$$\left\langle\sigma_{2s,2p}^{in}\right\rangle_{cl}=3\left\langle\sigma_{2p,2s}^{in}\right\rangle_{cl} \tag{17}$$

so that the emission and absorption widths from Eq. (2) can be written as

$$\left.\begin{matrix}\gamma_{ems}\\ \\ \gamma_{abs}\end{matrix}\right\}=\left\{\begin{matrix}P_{sp}\,e^{\hbar\omega_o/T}+5\\ \\ e^{\hbar\omega_o/T}+3P_{ps}+2\end{matrix}\right\}\frac{\gamma_{std}}{6} \tag{18}$$

and are given in terms of the standard impact theory width, γ_{std}, and the population correction factor $P_{sp} = P_{ps}^{-1}$ where for clarity the principle quantum number $n = 2$ is suppressed.

If the experiments were in LTE, then Eq. (18) yields

$$\gamma_{ems} = \gamma_{abs} = \left(e^{\hbar\omega_o/T} + 5\right)\gamma_{std}/6 \approx 1.1\gamma_{std} \qquad \text{(LTE)} \tag{19}$$

only slightly improving agreement with experiment. It was noted earlier, [11] however, that the BIII *4f-3d* line was missing from the experimental spectrum suggesting that the experimental temperature was lower. Applying this interpretation to the bound electrons (not the perturbing electron gas which is well-characterized by Thomson scattering), then the non-LTE BIII level populations could be described by an effective LTE temperature of say 4 eV. [21] With this estimates Eq. (18) gives

$$\left.\begin{matrix}\gamma_{ems}\\ \gamma_{abs}\end{matrix}\right\} \approx \left\{\begin{matrix}1.6\\ 0.8\end{matrix}\right\}\gamma_{std} \qquad \text{(non-LTE).} \tag{20}$$

Including ion contributions of $\sim 10\%\,\gamma_{ex}$, brings γ_{ems} and γ_{ex} into reasonable agreement. Obviously, the assumed level populations are only a plausible choice, but shows that width corrections from the kinetic theory method can be significant.

Since the absorption width is about half the emission width in Eq. (20), it is interesting to exploit the consequent different frequency dependence of $\varepsilon(\omega)$ and $\kappa(\omega)$. The following assumes that emission lines are optically thin so that the transmission is well approximated by

$$T(\omega) \approx 1 - \ell\kappa(\omega). \tag{21}$$

Figure 3 displays $\varepsilon(\omega)$ and $\kappa(\omega)$ from Eqs. (11) and (12) using γ_{ems} and γ_{abs} from Eq. (20) in the line profiles. The results include Doppler and instrumental broadening. It follows from Fig. 3 that the difference between $\varepsilon(\omega)$ and $\kappa(\omega)$ should be experimentally discernable. This assumes that transmission measurements do not introduce significantly larger experimental errors than emission.

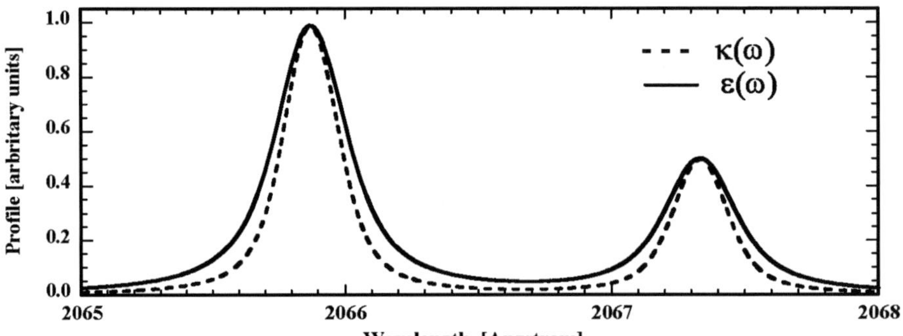

FIGURE 3. Comparison of emission and absorption coefficients at conditions of the BIII *2p-2s* experiment. [4] Profiles are peak normalized and include Doppler as well as instrumental broadening. The latter is a Voigt function with 0.071 Å Lorenztian HWHM and 0.049 Å Gaussian HWHM. [4]

3.2 The Li *3p-3s* isoelectronic sequence

The line width experiments were performed both at Bochum. [2-7] and Belgrade [8-10] for elements raging from B to Ne. A feature of the *3p-3s* lines, mentioned above and displayed in Fig. 1, is the monotonic increase in disagreement along the Li isoelectronic sequence between theory and experiment. This is a disturbing situation since the impact widths of isolated lines from multiply charged ions are dominated by inelastic collisions. [12-14] This is a process for which the convergent close-coupling method used in the standard impact width calculations is considered highly accurate.

Although it is instructive to concentrate on a single experiment (e.g.; the BIII *2p-2s* line attracts attention due to the large discrepancy), ultimately a satisfactory resolution of the discrepancies should address all isolated line width experiments. [1-10] Interestingly, the kinetic theory method may also explain these discrepancies. That is, it is known that the departure of level populations from LTE along an isoelectronic series increases with increasing atomic number. [22] Consequently, the population correction to the widths could explain the trend displayed in Fig. 1.

Estimates for the *3p-3s* lines similar to those in Sect. 3.1 for the *2p-2s* case are more difficult. Firstly, the two-level radiator is no longer valid. On the other hand, at least 90% of the inelastic contribution comes from the dipole-allowed $3s \to 3p$, $3p \to 3s$, and $3p \to 3d$ channels. [14] Consequently, only these three cross-sections are considered below. Secondly, although total inelastic and elastic width contributions for the Li-like *3p-3s* lines are available from convergent close-coupling calculations, [14] Eq. (2) requires individual inelastic channels. Instead of performing new calculations, their relative size is approximated and normalized to the more accurate total inelastic cross-section.

With these approximations the *3p-3s* isolated line widths simplify,

$$\left.\begin{matrix}\gamma_{ems}\\ \\ \gamma_{abs}\end{matrix}\right\} = \left\{\begin{matrix} e^{\hbar\omega_{ps}/T} P_{sp} + \Gamma P_{dp} + 3 \\ \\ e^{\hbar\omega_{ps}/T} + \Gamma + 3P_{ps}\end{matrix}\right\} \frac{\gamma_{std}^{in}}{\Gamma + 4} + \gamma_{std}^{el} \tag{22}$$

where the following

$$\left\langle \sigma_{s\to p}^{in} \right\rangle_{cl} = 3\left\langle \sigma_{p\to s}^{in} \right\rangle_{cl}, \tag{23}$$

valid for a radiator with a single bound electron outside a frozen core [19] and

$$\Gamma = \frac{\left\langle \sigma_{p\to d}^{in} \right\rangle_{cl}}{\left\langle \sigma_{p\to s}^{in} \right\rangle_{cl}} \approx \frac{f_{pd}}{f_{sp}} \left(\frac{\omega_{ps}}{\omega_{dp}} \right)^2 \frac{p(\hbar\omega_{dp}/T)}{p(\hbar\omega_{ps}/T)} e^{\hbar\omega_{ds}/T}, \tag{24}$$

which is approximated by Van Regemorter [23] with $f_{\alpha\beta}$ the absorption oscillator strength connecting the two levels and the function $p(x)$ is available in tabular form. The necessary atomic data to compute Γ is available from experiments. [24] For simplicity, spin was neglected and the principle quantum number $n = 3$ suppressed.

In order to ascertain the effect non-LTE level populations may have on the emission and absorption widths it is assumed that the *3p* and *3d* level population ratio is in LTE; that is, $P_{pd} = 1$. Then P_{sp} is calculated by setting γ_{ems} in Eq. (22) equal to the experimental emission width. Thus,

$$P_{sp} = e^{-\hbar\omega_{ps}/T}\left[\left(\frac{\gamma_{ex} - \gamma_o - \gamma_{std}^{el}}{\gamma_{std}^{in}}\right)(\Gamma + 4) - (\Gamma + 3)\right] \quad . \tag{25}$$

Here, γ_o represents width contributions from mechanisms other than electron collisions (e.g., ion and Zeeman broadening). This correction is typically small [1-10] and is set at $\gamma_o = 0.1\gamma_{ex}$ in the present calculations.

With the above approximations it is possible to estimate the departure from LTE required to explain the line width discrepancies. The assumed level populations can then be used to obtain the ratio of γ_{ems} to γ_{abs} which are displayed in Fig. 4. The figures shows that the difference between these estimates for the kinetic theory emission and absorption widths is larger than between experiment and standard impact theory (see Fig. 1). Consequently, it may be expected that experiments can discriminate between absorption and emission profiles.

Recall, however, that the absorption width measurement is complicated by the stimulated emission term in Eq. (12). As a check, $\varepsilon(\omega)$ and $\kappa(\omega)$ from Eqs. (11) and (12) with γ_{ems} and γ_{abs} from Fig. 4 plus γ_o in the line profiles are plotted in Figs. 5 and 6. The representative cases are for high and low densities. Again, the plots include Doppler and instrumental broadening showing that the different frequency dependence could be experimentally discernable.

Obviously, the procedure employed here to estimate the level populations is *ad hoc*. The ambiguity is caused by having two free parameters (the population ratios of the three $n = 3$ levels) to fit the experimental emission width. Nevertheless, it demonstrates that under non-LTE conditions the difference in emission and absorption widths from the kinetic theory method can be significant.

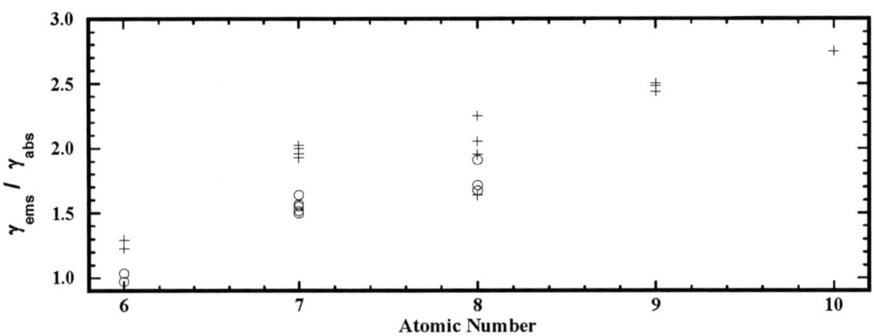

FIGURE 4. Ratio of γ_{ems} to γ_{abs} for the *3p-3s* Li isoelectronic from kinetic theory method sequence computed using non-LTE level populations required to explain width discrepancies.

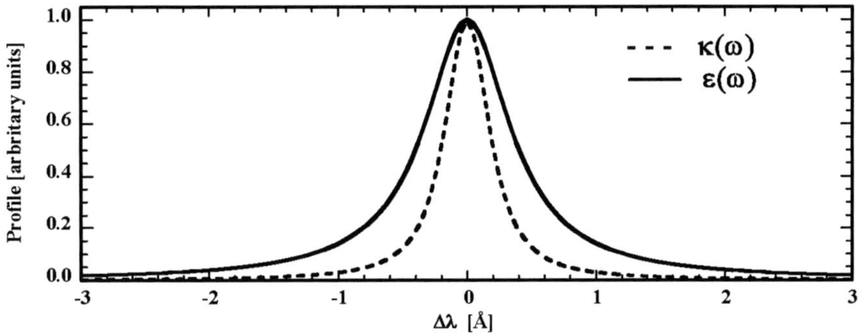

FIGURE 5. Comparison of emission and absorption coefficients near the NV *3p-3s* line at T =23.9eV and $n_e = 2.3\text{x}10^{18}\ \text{cm}^{-3}$. Profiles are peak normalized and include Doppler as well as instrumental broadening. The latter is a Voigt function with 0.25 Å Lorenztian HWHM and 0.125 Å Gaussian HWHM. [2]

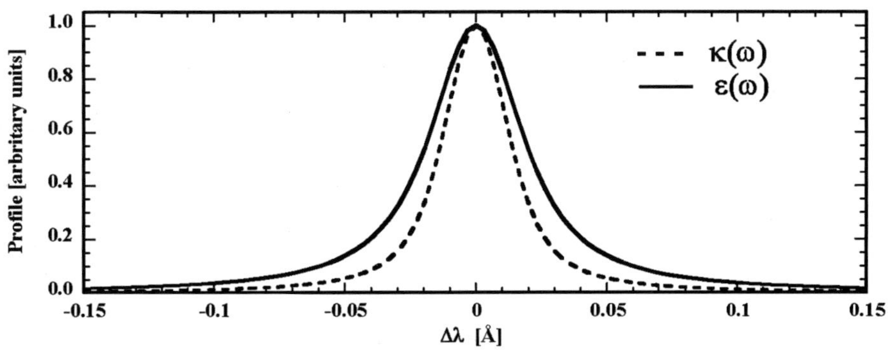

FIGURE 6. Comparison of emission and absorption coefficients at conditions of the NV *3p-3s* line at T =6.4 eV and $n_e = 8.6\text{x}10^{16}\ \text{cm}^{-3}$ Profiles are peak normalized and include Doppler as well as instrumental broadening. The latter is a Voigt function with 0.086 Å Gaussian HWHM. [9].

4. CONCLUSION

A possible explanation of the existing discrepancies [1-14] between experimental and theoretical widths of isolated lines was reviewed. The corrections are due to neglected initial correlations and incomplete treatment of the classical electron gas in the standard impact theory. [16] For isolated lines these approximations introduce significant errors so that a more complete theory is required.

The second-order kinetic theory results [16,17] were extended to stationary non-equilibrium states. It follows that the Rydberg-Ritz principle does not apply to isolated lines involving non-LTE level populations. That is, the line profile is not described by the addition of independent level widths (e.g.; the sum of thermally averaged inelastic cross-sections), but rather a line width that explicitly depends on the radiator states through their steady-state populations. Furthermore, a consistent classical limit of the quantum electron gas introduces corrections to the width. Both corrections are essential in satisfying the Kirchhoff-Planck relation. In second-order theory the corrections enter as multiplicative factors to the thermally averaged inelastic cross-sections allowing for width estimates based on standard impact theory. [19]

A repeat of earlier experiments [1-10] but measuring both photon emission and absorption in the spectral region of the isolated line is proposed. Differences in the energy dependence of the emission and absorption coefficients would substantiate the kinetic theory results, which predict different emission and absorption isolated line profiles for stationary non-equilibrium internal radiator state populations. On the other hand, the standard impact theory satisfies the Kirchhoff-Planck relation but only as a result of compensating approximations and predicts identical emission and absorption line widths independent of the radiator level population distribution.

The proposed resolution of the discrepancies relies on several approximations. The main assumptions are an ideal electron gas and a second-order expansion of the width operator. It is well known [16,17] that the main effect of the plasma interactions is to screen the radiator-electron interaction and should not significantly affect the conclusions. The good agreement between the Coulomb-Born and R-matrix calculations for the BIII *2s-2p* line [11] suggests that using second-order theory to estimate the width corrections is reasonable, at least for that case.

The corrections to the widths can be significant for isolated lines, but are negligible for many other lines. In particular, the radiator level population corrections to line widths are small if the relevant level populations ratios are close to thermal equilibrium even though the levels are not in LTE. For example, for hydrogenic radiators the no quenching approximation [22] is often valid and the population corrections to the width should be negligible.

Finally, the experimental discussion relies on estimates for the radiator internal state populations that were not available from the experiments. The present explanation of the width discrepancies requires non-LTE radiator populations and rough estimates were obtained based on simple arguments. It is also possible to compute the level populations from models. Interestingly, steady-state models calculations [25-27] have yielded only small deviations from unity for the population corrections to the width. Obviously, such results do not support the non-LTE population dependence of the line widths as an explanation to the discrepancies. and the search for alternative explanations based on experimental shortcomings continues. [24] On the other hand, it is possible that steady-state models are not fully describing the experiments. In any case, the radiator state population dependence of isolated line widths is a new theoretical result that can significantly affect line widths. Consequently, it requires experimental verification even if it does not fully explain the existing discrepancies between theoretical and experimental isolated line widths.

ACKNOWLEDGMENTS

Work performed under the auspices of the Department of Energy by Lawrence Livermore National Laboratory under Contract W-7405-ENG-48.

REFERENCES

1 A. Sanchez, M. Blaha, & W. Jones, *Phys.Rev.* **A8**, 774(1973)
2 F. Böttcher, P. Breger, J.D. Hey, & H.-J. Kunze, *Phys.Rev.* **A38**, R2690(1988)
3 S. Glenzer, N.I. Uzelac, & H.-J. Kunze, *Phys.Rev.* **A45**, 8795(1992)
4 S. Glenzer & H.-J. Kunze, *Phys.Rev.* **A53**, 2225(1996)
5 Th. Wrubel, I. Ahmad, S. Büscher, H.-J. Kunze, & S.H. Glenzer, *Phys.Rev.* **E57**, 5972(1998)
6 H. Hegazy, S. Seidel, Th. Wrubel, & H.-J. Kunze, *J.Quant.Spectrosc.Radiat.Transfer* **81**, 221 (2003)
7 S. Glenzer, N.I. Uzelac, & H.-J. Kunze, *Proceeding of the Seventh International Conference on Spectral Line Shapes* (Nova Science Publishers, New York,1993)
8 D. Hadziomerspahic, M. Platisa, N. Konjevic, & M. Popovic, *Z.Physik* **262**, 169(1973)
9 N. Konjevic, *Phys.Reports* **316**, 339(1999)
10 B. Blagojevic, M.V. Popovic, N. Konjevic, & M.S. Dimitrijevic, *J.Quant.Spectrosc.Radiat.Transfer* **61**, 361(1999)
11 H.R. Griem, Yu.V.Ralchenko & I. Bray, *Phys.Rev* **E56**, 7186(1997)
12 Yu.V.Ralchenko, H.R. Griem, I. Bray, & D.V. Fursa, *J.Quant.Spectrosc.Radiat.Transfer* **71**, 595(2001); *Phys.Rev.* **A59**, 1890(1999)
13 H.R. Griem, & Yu.V.Ralchenko, *J.Quant.Spectrosc.Radiat.Transfer* **65**, 287(2000)
14 Yu.V.Ralchenko, H.R. Griem &, I. Bray *J.Quant.Spectrosc.Radiat.Transfer* **81**, 371 (2003).
15 S. Alexiou & R.W. Lee, *J.Quant.Spectrosc.Radiat.Transfer* **99**, 10(2006)
16 M. Baranger, *Phys.Rev.* **111**, 481(1958); *ibid.* **111**, 494(1958); *ibid.* **112**, 855(1958)
17 T. Hussey, J. Dufty, & C. Hooper, *Phys. Rev.* **A12**, 1084(1975); *ibid.* **16**, 1248(1977)
18 J.W. Dufty & D.B. Boercker, *J.Quant.Spectrosc.Radiat.Transfer* **16**, 1065(1976)
19 C.A. Iglesias, *High Energy Density Physics* **1**, 42(2005)
20 D. Mihalas, *Stellar Atmospheres* (Freeman and Co, San Francisco, 1978)
21 LTE calculations show that the BIII 4f-3d line is sufficiently weak only for temperatures less than about 5 eV.
22 H.R. Griem, *Plasma Spectroscopy* (McGraw-Hill, New York, 1964)
23 H. Van Regemorter, *ApJ* **132**, 906(1962)
24 W.L. Wiese, M.W. Smith, & B.M. Glennon, *Atomic Transition Probabilities*, Natl. Bur. Stand. Ref. Data Ser. Natl. Bur. Stand. (U.S.) Circ. No. 4 (GPO, Washington, D.C. 1966), Vol. I
25 H.R. Griem & Yu.V. Ralchenko, *this Volume.*
26 H.K. Hung, 2005 (*private communication*)
27 S.B. Hansen, 2006 (*private communication*)

Are the Discrepancies Between Measured and Calculated Electron Impact Widths of Isolated Ion Lines Due to Non-Equilibrium Level Populations?

Hans R. Griem and Yuri Ralchenko

Institute for Research in Electronics and Applied Physics
University of Maryland, College Park, MD 20742-3511, USA

Abstract. The persisting excess by about a factor of two for measured compared to calculated Stark widths of lines due to transitions between low-lying levels of Li-like ions ranging from B III to Ne VIII would be significantly reduced if the ratios of upper- to lower- level populations were far from local thermodynamic equilibrium (LTE). This possibility was recently discussed using a quantum-kinetic theoretical approach by C. A. Iglesias (instead of Baranger's earlier treatment) and applying LTE validity criteria originally proposed for one-electron atoms or ions. However, these criteria greatly overestimate the critical electron densities required for near-LTE for the ions of interest here, which have no transitions into n=1. Actually, recent NLTE calculations using the NOMAD code indicate steady-state population ratios in the experiments considered here to be within 1% of LTE values in the case of B III 2s-2p and within 5% for Ne VIII 3s-3p transitions, thus invalidating this explanation of the discrepancies. Moreover, our calculations suggest that the electron temperatures in the B III experiment were overestimated by a factor of two. Correcting for this increases the calculated Stark widths by only about 20%. However, since much of the line emission was perhaps from outer layers with lower temperature and higher electron density, this increase may actually be by a factor of 1.6. Macroscopic Doppler effects due to radial expansion could account for the remainder of the discrepancy.

Keywords: electron impact broadening of ion lines
PACS: 52.20.Fs

INTRODUCTION

Over 10 years ago measurements on the well-diagnosed gas-liner pinch discharge at Ruhr University Bochum of the 2s-2p resonance transitions in B III yielded Stark widths [1] almost twice the early five-state close-coupling, quantum-mechanical results [2]. On the other hand, they agreed with some semi-classical calculations of the electron impact broadening [3, 4]. Comparison with a semi-empirical, effective Gaunt factor method [5], however, showed again an even larger disagreement by a factor of 2.15. In fact, a similar disagreement between measurements and quantum-mechanical calculations had already been noticed previously for 2s-2p resonance transitions in another Li-like ion, namely, Be II [6].

CP874, *Spectral Line Shapes: 18th International Conference,*
edited by E. Oks and M. Pindzola

Of even greater interest here is the reported failure to measure Stark widths of these transitions in C IV in the gas-liner pinch [7] at higher densities and temperatures than in the B III experiment.

Meanwhile, improved quantum-mechanical calculations [8] using the convergent close coupling method had confirmed the earlier R-matrix calculations [2], thus supporting the case for a significant overestimate of electron impact widths by semi-classical calculations. This was attributed to the predominance of small L partial waves in the electron-ion scattering and has been confirmed by recent penetrating-collision, semi-classical calculations [9], which also accounted for the entire Coulomb interactions rather than only the dipole interactions. (As a matter of fact, the importance of small L partial waves had already been emphasized by D. E. Roberts [10] in connection with his semi-classical calculations.) The experiment versus theory discrepancy must therefore be considered to be real, motivating the following critical questions especially regarding the B III experiment.

These questions will be discussed in the following sections, first regarding the possibility of non-LTE level populations of the perturbed ions, which according to a quantum kinetic theory approach [11] could resolve the existing discrepancies between theoretical and experimental widths of isolated lines. A second possibility, namely, non-thermal Doppler effects already raised in [8], will then be reviewed in the third section, followed by a summary.

NON-LTE EFFECTS

Implicit assumptions in the standard impact theory [12] are the validity of classical thermal averages and the vanishing of initial correlations between radiators and perturbers. The former assumption is most critical here, as shown in [11] and the preceding paper in these proceedings. Were the relative populations of upper and lower (ground state) and perturbing levels indeed out of LTE, as postulated in [11], the corresponding kinetic theory corrections could indeed be sufficient to remove the long-standing discrepancy between measured and calculated line widths. However, as shown in Table 1, our collisional-radiative NLTE calculations using the NOMAD code [13] yield relative populations very close to LTE (even for $n=3$ levels which contribute only 5% to the calculated widths [8]), thus ruling out the most recent explanation of the discrepancy. (Much larger deviations from LTE were expected in [11] from criteria originally derived for hydrogenic ions [14], for which radiative decay into the $n=1$ state tends to dominate in the rate equations.)

One also notices from the last column in Table 1 that, at the measured electron temperature of 10.6 eV [1], only a rather small fraction of the boron ions is in the B III 2p state. This suggests that outer layers of the compressed plasma may actually contribute significantly to the line-of-sight integral for the observed B III line spectrum. (Our NLTE calculations show that the 2s-2p line intensities increase by a factor of 200 from 10 eV to 5 eV for a given total boron density.) Confirmation of this suggestion would require radial scans and Abel inversions, which were apparently too difficult to do in [1].

kT(eV)	2p	3s	3p	3d	B^{++}	(2p)
3	0.406 (0.407)	0.0005 (0.0006)	0.0010 (0.001)	0.0014 (0.0015)	0.725	0.1508
5	0.903 (0.905)	0.0100 (0.0114)	0.0231 (0.025)	0.0355 (0.0388)	0.252	0.0182
7	1.272 (1.274)	0.0371 (0.0411)	0.0887 (0.098)	0.1399 (0.1554)	0.106	0.0009
10	1.643 (1.647)	0.0942 (0.1070)	0.2412 (0.274)	0.3865 (0.4402)	0.064	0.0001

TABLE 1. B^{++} level populations (relative to the 2s ground state populations) at N_e = 1.8 x $10^{18}cm^{-3}$ from NLTE and (in parentheses) LTE calculations. The last two columns list the fraction of all B^{++} ions and the fraction of those ions in the 2p states.

Instead, one may consider later gas-liner experiments [7, 15] with improved spatial resolution for the Thomson scattering diagnostics. As shown on Fig. 4.6 of [15] at near-peak compression, the electron temperature decreases by about a factor of 2 over a 3 mm increase in radius (and the nominally 0.3% boron impurity density decreases by a factor of about 20), while the electron density increases by nearly 25%. Taken together, these changes would lead to an increase in calculated line widths [8] by a factor of 1.6, thus substantially reducing the discrepancy. However, the plasma conditions in [15] differed from those in [1] at about 50 ns after peak compression in that both the electron density and temperature were higher by a factor of 2, so that a new experiment would clearly be required to confirm all of this. We should also note that the central minimum in the electron density does not persist after peak compression (see Fig. 4.7 of [15]).

We also used the NOMAD code [13] to assess any deviations from LTE in the experiments on 3s-3p transitions in Li-like ions [16], finding that even for Ne VIII the 3s and 3p populations were within 5% of the LTE ratio, i.e., exhibiting only negligible deviations from LTE in the present context. In these experiments Abel inversions had been possible, but the temperature to ionization energy ratios were smaller than in [1] by as much as a factor of 2.

DOPPLER EFFECTS

In an earlier paper [8] with I. Bray, hydrodynamic turbulence and the associated Doppler broadening were suggested as leading to substantial reductions in the Stark widths inferred from the measured profiles. Corresponding corrections to the Thomson scattering profiles calculated according to Evans [17] (but not allowing for the Doppler shifts on the absorption and reemission of the laser photons) are, however, too large to be consistent with the measured profiles [18] (see Fig. 2 there), especially on the lower shoulders. Evidently thermalization of any turbulence is faster than estimated in [8], perhaps due to micro-turbulence. Nevertheless, there may be non-thermal radial motions, as indicated, e.g., by small Doppler shifts of the radially resolved Thomson scattering profiles in [15] (see Fig. 4.5 there). Corresponding Thomson scattering profiles calculated according to [17] and assuming $\delta x = \pm 0.05$ nm Doppler shifts due to expansion (or compression) are shown in Fig. 1. They have a shallower central dip and less steep outer shoulders than the unshifted Evans profile [17] (also shown in Fig. 1), both features possibly causing a slight improvement in fits to observed profiles such as in [18].

Such Doppler shifts were most likely responsible for the splitting by about 0.03 nm of the C IV 2s-2p lines observed in [7]. Because of the longer wavelength, this would scale to almost 0.05 nm for the B III lines, i.e., to over twice the measured width. However, temperatures and densities were lower, and the time of observation later in the earlier experiment [1], suggesting smaller velocities there. We also note that velocities inferred from shifts of the Thomson scattering spectra in [15] range over about 6×10^{6} cm/sec, entirely consistent with the line splitting in [7]. The velocity range chosen for our Fig. 1 amounts to 4×10^{6} cm/sec (assuming that only one of the beams involved in the 90° scattering is Doppler shifted). Therefore, even smaller δx values would still be sufficient to produce significant motional Doppler broadening in [1].

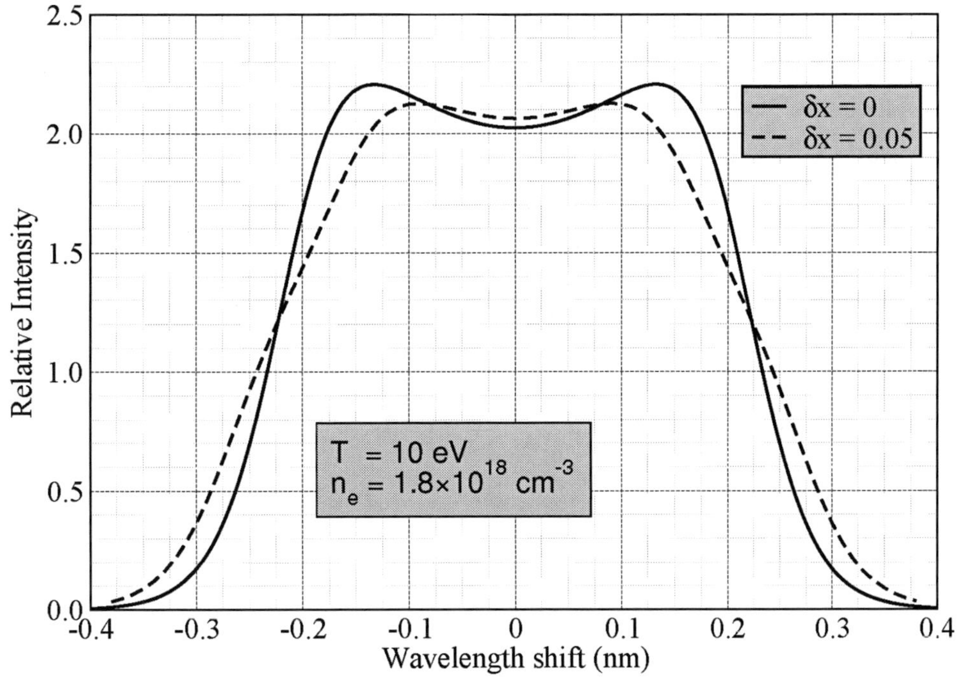

FIGURE 1. Thomson scattering profiles at T_e = 10 eV and N_e= 1.8 x 10^{18}cm^{-3} with and without motional Doppler shifts (see text).

SUMMARY

The suggestion of significant quantum-kinetic theory corrections [11] to Baranger's theory [12] of collisional broadening does not apply to the experiments [1,16] discussed here (nor to more recent experiments on other Li- and Be-like lines, see Refs. 1-9 in [11]), because actual deviations from LTE populations are much smaller than assumed in [11]. However, it did lead us to a re-examination of possible temperature and density variations along the line of sight, which could account for a significant fraction of the discrepancy between measured and calculated widths. The motional Doppler effects estimated here are more likely than not responsible for the remainder of the discrepancy.

ACKNOWLEDGMENTS

We thank our colleague A. W. DeSilva for providing the Thomson scattering profiles.

REFERENCES

1. S. Glenzer and H.-J. Kunze, *Phys. Rev. A* **53**, 2225 (1996).
2. M. J. Seaton, *J. Phys. B* **21**, 3033 (1988).
3. H. R. Griem, *Spectral Line Broadening by Plasmas*, Academic Press, New York, 1974.
4. J. D. Hey and P. Breger, *J. Quant. Spectrosc. Radiat. Transfer* **24**, 349 and 427 (1980); *Spectral Line Shapes*, ed. B. Wende, de Gruyter, Berlin, 1981, p. 191; *S. Afr. J. Phys.* **5**, 111 (1982).
5. M. S. Dimitrijevic and N. Konjević, *J. Quant. Spectrosc. Radiat. Transfer* **24**, 451 (1980).
6. A. Sanchez, M. Blaha, and W. W. Jones, *Phys. Rev. A* **8**, 774 (1973).
7. S. Bűscher, Th. Wrubel, I. Ahmad, and H.-J. Kunze, *Spectral Line Broadening*, ed. R. M. Herrman, AIP Conference Proceedings 467, Melville, New York, 1998, p. 39.
8. H. R. Griem, Yu. V. Ralchenko, and I. Bray, *Phys. Rev. E* **56**, 7186 (1997).
9. S. Alexiou and R. W. Lee, *J. Quant. Spectrosc. Radiat. Transfer* **99**, 10 (2006).
10. D. E. Roberts, Astron. Astrophys. **6**, 1 (1970).
11. C. A. Iglesias, *High Energy Density Physics* **1**, 42 (2005).
12. M. Baranger, *Phys. Rev.* 111, 481 and 494 (1958); *Phys. Rev.* **112**, 855 (1958).
13. Yu. V. Ralchenko and Y. Maron, *J. Quant. Spectrosc. Radiat. Transfer* **71**, 609 (2001).
14. H. R. Griem, *Plasma Spectroscopy*, McGraw-Hill, New York, 1964.
15. Th. Wrubel, Ph.D. thesis, Ruhr University, Bochum, 1999 (unpublished).
16. S. Glenzer, N. I. Uzelac and H.-J. Kunze, Phys. Rev. A **45**, 8795 (1992).
17. D. E. Evans, *Plasma Phys.* **12**, 573 (1970).
18. S. Glenzer, *Atomic Processes in Plasmas*, ed. A. L. Osterheld and W. H. Goldstein, AIP Conference Proceedings 381, Melville, New York, 1996, p. 109.

Stark Widths of Hydrogen Spectral Lines in Plasmas: a Highly-Advanced Non-Simulative Semiclassical Theory and Tables

Eugene Oks

Physics Department, 206 Allison Lab., Auburn University, Auburn, AL 36849, USA

Abstract. Plasma spectroscopy had served as a fertile field for applications of the formalism of Dressed Atomic States in Plasmas (DASP). The theory named in the title of this paper is based primarily on a *generalization* of the formalism of DASP. In its contemporary composition, this theory brings together: A) non-impact description of the indirect coupling of the electron and ion microfields; B) non-binary description of the direct coupling of the electron and ion microfields; C) non-binary non-impact description of the ion-dynamical broadening. The latter produces, in particular, a new counter-intuitive result for the dependence of the Ion-Dynamical Stark Width (IDSW) on the temperature T: it predicts that in distinction to the high-T limit, for which it was known that IDSW ~ $1/T^{1/2}$, in the low-T limit the temperature dependence changes to IDSW ~ $T^{1/4}$. In other words, as the temperature decreases from the high-T limit, the IDSW first increases ~ $1/T^{1/2}$, then reaches a maximum, and then decreases ~ $T^{1/4}$ (this being a non-binary result). A non-binary result on the density dependence of the IDSW is also presented. The theory composed of the above items A), B), C), is used for creating extensive Tables of Stark widths of hydrogen lines in plasmas. By comparison with benchmark experiments, it is shown that these Tables are more accurate than the corresponding tables published previously by other authors.

Keywords: plasma spectroscopy, Stark broadening, lineshapes, dressed atomic states, ion dynamics
PACS: 32.70.Jz, 52.70.Kz, 32.60.+i, 03.65.Sq

THE GENERAL STRUCTURE OF THE NON-SIMULATIVE SEMICLASSSICAL THEORY OF STARK BROADENING OF HYDROGEN/HYDROGENLIKE SPECTRAL LINES IN PLASMAS

Introduction

Plasma spectroscopy had served as a fertile field for applications of the formalism of Dressed Atomic States (DAS) in plasmas [1]. The theory named in the title of this paper is based primarily on a *generalization* of the formalism of DAS in plasmas.

DAS is the formalism initially designed to describe the interaction of a *monochromatic* (or quasi-monochromatic) field – e.g., laser or maser radiation – with gases. Later it was applied for the interaction of a laser or maser radiation with plasmas [1]. The employment of DAS led to the enhancement of the accuracy of analytical calculations and to more robust codes.

CP874, *Spectral Line Shapes: 18th International Conference,*
edited by E. Oks and M. Pindzola

Our *generalization* of DAS is based on using atomic states dressed by a *broad-band* field of plasma electrons and ions [2]. Therefore generalized DAS is a more complicated concept than the usual DAS, where the dressing was due to a monochromatic field.

The employment of the generalized DAS allowed to describe analytically a *coupling* of the electron and ion microfields facilitated by the radiating atom – *indirect coupling* [2]. The indirect coupling *increases with the principal quantum number* n and *with the electron density* N_e. Besides, it *increases as the temperature T decreases.*

The employment of the generalized DAS allowed also obtaining *accurate analytical results for the ion-dynamical broadening.* These results were produced in a *non-binary* description and *without involving the impact approximation* [2].

Our theory takes into account also the Acceleration of (perturbing) Electrons by the Ion Field (AEIF). We describe this phenomenon, representing a *direct coupling* of the electron and ion microfields, in a *non-binary* approach.

Eight Stages in the Design of the Theory

1. *Separation of Static and Dynamic Ions.* For a given value τ of the argument of the correlation function $C(\tau)$, the ion-dynamical part of $C(\tau)$ originates from the collisions, for which the instants of the closest approach t_0 fall within the interval $(-\tau/2,\tau/2)$. The rest of the perturbing ions are considered static. This is consistent with the fact that at $\tau \to \infty$ all ions are dynamic, while at $\tau \to 0$ all ions are static. Thus, to different values of τ correspond different proportions of dynamic and static ions.

2. *Partition of the Hamiltonian.* The Hamiltonian H(t) is broken down in the following four terms:

$$H(t) = H_0 + V^{IS} + V_{\|}(t) + V_{\perp}(t), \quad V_{\|}(t) = V_{\|}^{ID}(t) + V_{\|}^{E}(t), \quad V_{\perp}(t) = V_{\perp}^{ID}(t) + V_{\perp}^{E}(t). \quad (1)$$

Here H_0 is the Hamiltonian of the unperturbed radiating atom/ion (radiator); superscripts IS, ID, and E refer to interactions of the radiator with static ions, dynamic ions, and electrons, respectively; subscripts $\|$ and $\perp$ refer, respectively, to parallel and perpendicular components of the electron and dynamic ion microfields with respect to the direction of the quasistatic ion field at the location of the radiator.

3. *Generalized DAS.* The atomic states of the radiator are dressed by the following time-dependent factor:

$$Q(t) = \exp\{ (i/\hbar)[(H_0 + V^{IS})t + \int_{-\infty}^{t} dt' \, V_{\|}(t')] \}. \quad (2)$$

In other words, the interactions V^{IS} and $V_{\|}(t')$ are taken into account *exactly*, moreover – *analytically*. Then the interaction $V_{\perp}(t)$ is allowed for by Dirac's perturbation theory. The dressing factor Q(t) is controlled by the interaction $V_{\|}(t)$ with a *broadband* field – in distinction to the usual DAS dressed by a monochromatic field (examples of the latter can be found in [1]).

4. *Evolution Operator in the MODIFIED Interaction Representation.* It employs the dressing factor Q(t) from Eq. (2)

$$U(t_1, t_2) = \mathrm{Texp}[-(i/\hbar) \int_{t_1}^{t_2} dt\, Q^*(t)\, V_\perp(t)\, Q(t)], \tag{3}$$

where Texp[...] is the time-ordered exponential.

5. *Calculation of the M-matrix.* The M-matrix that controls the calculation of the correlation function (defined in stage 6) is calculated as follows

$$M(t_1, t_2) = \exp[(i/\hbar) \int_{t_1}^{t_2} dt'\, V_{\|}(t')]\, U(t_1, t_2), \tag{4}$$

where exp[...] is calculated *exactly*, analytically, while $U(t_1, t_2)$ – via the perturbation theory. The impact approximation is *not* used.

6. *Correlation Function.* It is defined by the formula

$$C(\tau) = \mathrm{Tr}\{ \mathbf{d}\cdot\mathbf{d} \exp[-(i/\hbar)(H_0 + V^{IS})\tau]\, \rho <M(\tau/2, - \tau/2)>_{av} \}, \tag{5}$$

where Tr{...} stands for the trace, $<\ldots>_{av}$ denotes the ensemble average.

7. *Lineshape at a fixed quasistatic ion field F.* It is calculated as the Fourier-transform of the correlation function $C(\tau)$.

8. *Averaging over the distribution W(F).* The distribution W(F) of the quasistatic ion field can be produced using, e.g., APEX [3].

Allowance for the Acceleration of Electrons by the Ion Field (AEIF)

1. Simplified, binary description. Perturbing electrons are passing by the system "radiator + the nearest ion" at higher velocities than in the absence of the nearest ion – due to the attraction to the nearest ion. The increase of the electron velocities translates into a decrease of the electron broadening.

This phenomenon can be approximately taken into account by introducing the following reducing factor $\mu(v, \rho)$ into the electron operators $V_{\|}^E(t)$ and $V_\perp^E(t)$

$$\mu(v, \rho) \approx 1 - 1/[1 + R/(2\rho_c) +\rho^2/(2\rho_c^2)], \tag{6}$$

where v is the asymptotic velocity, ρ - the impact parameter, $\rho_c \equiv Z_p e^2/(m_e v^2)$ – the Coulomb radius.

2. Non-Binary Generalization. Instead of the above rough approximation for $\mu(v, \rho)$, we use the following much more accurate expression:

$$\mu(v, \rho) = 1 - 1/\{1 + R/[2\rho_c U(R, N_p)R/(Z_p e)] + \rho^2/(2\rho_c^2)\}. \tag{7}$$

Here the factor $U(R, N_p)R/(Z_p e)$, modulating the potential of the nearest neighbor, is calculated in the spirit of the ion distribution used in the ion sphere models [4, 5]. However, in distinction to the ion sphere models, it represents a smoother form of the potential due to ions: neither the potential nor its derivative has a discontinuity at

$R = R_0 = [3/(4\pi N_p)]^{1/3}$. This amounts to electrons being subjected to a central force due to an "ion of variable charge" – like in one of the latest molecular dynamics simulations [6].

TABLES OF STARK WIDTHS AND THE COMPARISON WITH THE PREVIOUS AUTHORS

Based on the above theory/code, we created Tables of the Stark widths of hydrogen and deuterium lines in plasmas – just as detailed as the Tables published in 1996 by Gigosos and Cardenoso (GC) [7]. Their results are based on rather advanced simulations (independent quasiparticle technique), but they did not allow for all couplings of the electron and ion microfields.

Table 1 compares our results and GC-results with the experimental FWHM of the hydrogen Balmer-α line emitted from *argon* plasmas studied in benchmark experiments by Wiese, Kelleher, Paquette [8] and by Vitel [9] (each group – at a different source). Here W_{exp}(nm) – FWHM measured in the benchmark experiments by Wiese, Kelleher, Paquette [8] (the first three rows) and by Vitel [9] (the last six rows); W_O(nm) – our calculated FWHM, W_{GC}(nm) – FWHM calculated by GC [7].

TABLE 1. **Theoretical and Experimental FWHM of the H_α Line Emitted from *Argon* Plasmas.**

$N_e/(10^{17}cm^{-3})$	$T/(10^3$ K)	W_{exp}(nm)	W_O(nm)	W_{GC}(nm)
0.5	12.5	0.62	0.61	0.66
0.85	12.5	0.80	0.79	0.91
1.0	12.5	0.91	0.90	1.01
6.1	16.0	3.2	3.2	3.5
6.2	17.4	3.5	3.5	3.5
7.4	17.7	3.7	3.8	4.0
8.2	18.0	3.9	4.0	4.3
9.4	18.5	4.1	4.2	4.7
10	18.7	4.4	4.5	4.9

It is seen that our results are in *excellent agreement* with these benchmark experiments (within 3%), while GC-results overestimate the experimental width by up to 15%.

Table 2 compares our results and GC-results (where available) with the experimental FWHM of the hydrogen Balmer-α line emitted from the *helium* plasma of the gas-liner pinch studied by Kunze's group in the most precise [10] out of their two series of similar benchmark experiments.

TABLE 2. Theoretical and Experimental FWHM of the H_α Line Emitted from the *Helium* Plasma of the Gas-Liner Pinch.

$N_e/(10^{18}cm^{-3})$	$T/(10^3\ K)$	$W_{exp}(nm)$	$W_O(nm)$	$W_{GC}(nm)$
0.49	66.9	4.03 ± 0.43	3.6	3.3
0.53	121	4.54 ± 0.38	3.9	
0.68	74.1	4.74 ± 0.65	4.4	
0.99	82.6	5.31 ± 0.65	5.3	5.3
1.35	90.7	6.78 ± 0.25	6.7	
1.96	97.4	8.19 ± 0.56	8.5	
2.54	96.7	9.62 ± 0.95	10.2	8.5

It is seen that our results are in a *very good agreement* also with this benchmark experiment (within 6% compared to the most probable experimental values), while GC-results underestimate the most probable values of the experimental width by up to 18%. (In the corresponding table from our book [2] – Table B.2 – there are two typos: in the last two columns of the last row, it should be 10.2 and 8.5, respectively, rather than 102 and 85.)

Table 3 compares our results and GC-results with the experimental FWHM of the hydrogen Lyman-α line emitted from an *argon* plasma studied in the benchmark experiment by Grützmacher and Wende [11].

TABLE 3. Theoretical and Experimental FWHM of the L_α Line Emitted from the *Argon* Plasma.

$N_e/(10^{17}cm^{-3})$	$T/(10^3\ K)$	$W_{exp}(nm)$	$W_O(nm)$	$W_{GC}(nm)$
1	12.7	0.0175	0.0174	0.0173
2	13.2	0.0255	0.0255	0.0258
3	13.2	0.0320	0.0322	0.0326
4	14.0	0.0363	0.0367	0.0389

Once again, it is seen that our results are in *excellent agreement* with this benchmark experiment (within 1%), while GC-results slightly overestimate the experimental width - by up to 7%.

Now let us briefly compare our results with simulations/theories other than that of GC. Figure 1 shows the comparison of our results and of the results of so-called Conventional Theory (CT) by Griem [12] with the experimental FWHM of the hydrogen Balmer-α line emitted from the *helium* plasma of the gas-liner pinch studied by Kunze's group in the most precise [10] out of their two series of similar benchmark experiments. It is seen that our results are in a *very good agreement* with this benchmark experiment, while the CT *disagrees* with the experiment up to a *factor of two*.

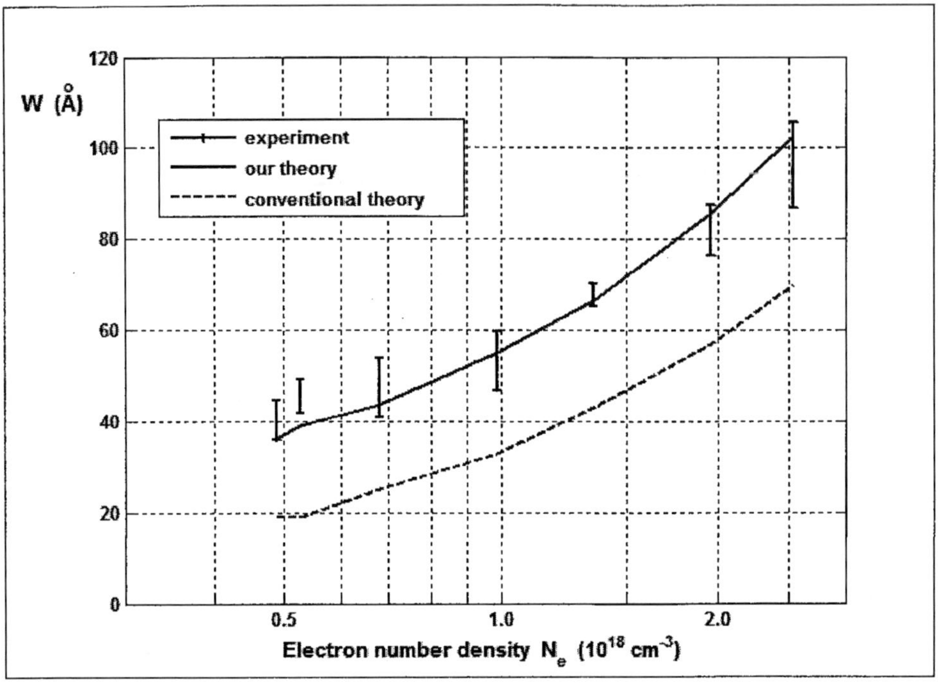

FIGURE 1. FWHM of the H-alpha line of neutral hydrogen emitted from a helium plasma. Vertical bars: benchmark experimental results by Büscher et al. [10] at the gas-liner pinch. Solid curve: our theoretical results. Dotted curve: the conventional theory by Griem [12].

Another example of comparing our results with simulations/theories other than that of GC is presented in Fig. 2. It shows the FWHM of the He II Balmer-alpha line measured in the benchmark experiment by Grützmacher and Johannsen [13] and the corresponding calculations by three theories: by the CT, by the Frequency Fluctuation Model (FFM) as presented in [14], and by our theory. It is seen that: A) the CT *disagrees* with the experiment up to a *factor of 2.7*; B) the FFM is in a much better agreement with the experiment than the CT – it *disagrees by 16-17%*; C) our theory is in the *best agreement* with this benchmark experiment as well.

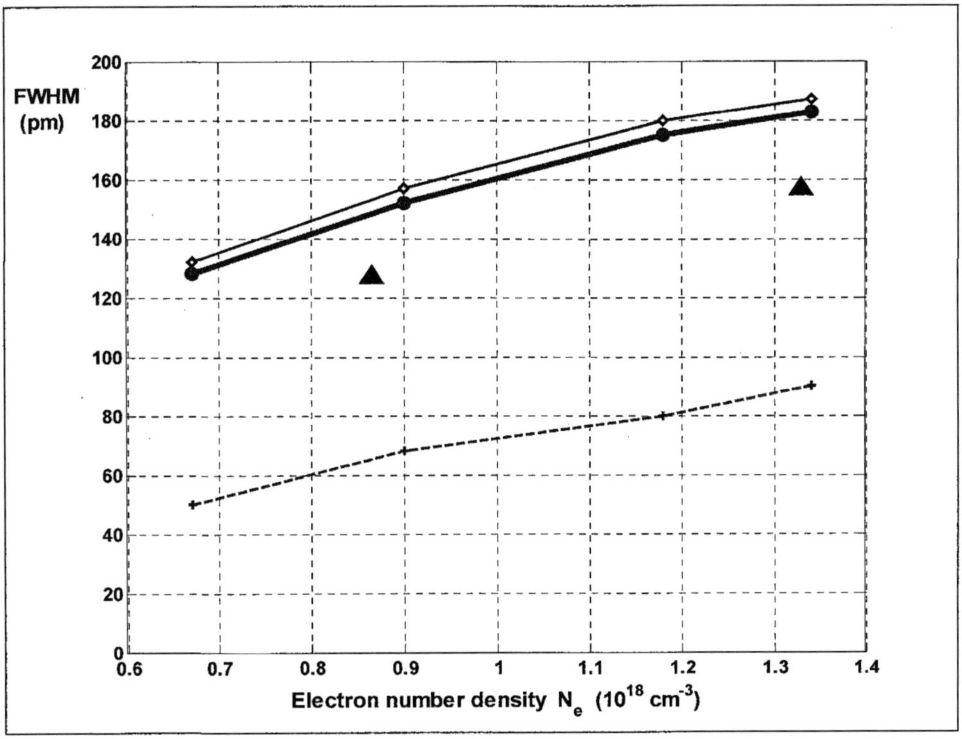

FIGURE 2. FWHM of the He II Balmer-alpha line. Diamonds connected by a thin line: benchmark experimental results by Grützmacher and Johannsen [13]. Bold line: our theory. Dashed line: the conventional theory by Griem [12]. Triangles: the frequency fluctuation model [14].

The rest of the Tables in this paper – Tables 4 - 6 – present our calculated Stark widths (FWHM) of the Balmer-α line of neutral hydrogen and neutral deuterium for the same broad range of plasma parameters as in the tables by GC [7]. Separate Tables are provided for the following values of the reduced mass μ of the pair "radiator – perturbing ion": $\mu = 0.5$ (pure hydrogen plasma), $\mu = 0.8$ (hydrogen in a helium plasma); $\mu = 1$ (either hydrogen in a plasma of heavy ions or a pure deuterium plasma).

In each of the Tables 4 - 6, the first column shows the logarithm of the electron density N_e (cm^{-3}), the base of the logarithm being 10; columns 2 – 7 present our calculated Stark FWHM (nm) corresponding to the values of the temperature $T/(10^3$ K) = 5, 10, 15, 20, 30, and 40, respectively. *The calculated FWHM (in nm) is given in the following notation: for example, 6.4E-2 stands for* 6.4×10^{-2}.

We note that for electron densities $N_e < 10^{16}$ cm^{-3}, our calculated Stark widths for the Balmer-α line practically do not differ from the corresponding results of GC simulations [7]. The difference becomes significant only for $N_e > 10^{16}$ cm^{-3}.

These Tables are presented in our book [2], which contains also the same kind of detailed tables (for the reduced mass $\mu = 0.5$, 0.8, 1, and 2) for some other Balmer lines of neutral hydrogen and neutral deuterium - for the same broad range of plasma parameters as in the tables by GC [7]. We note that in [2] there is a typo in Table B.4 (which corresponds to Table 4 of this paper): in the fifth column of the last row it should be 1.4E1 rather than 1.41.

TABLE 4. Our calculated FWHM (nm) of Balmer-α for the reduced mass μ = 0.5.

$\log(N_e, cm^{-3})$	5000 K	10000 K	15000 K	20000 K	30000 K	40000 K
15.00	6.4E-2	6.7E-2	6.8E-2	6.8E-2	6.6E-2	6.4E-2
15.33	1.0E-1	1.1E-1	1.1E-1	1.1E-1	1.1E-1	1.1E-1
15.67	1.6E-1	1.8E-1	1.8E-1	1.9E-1	1.9E-1	1.9E-1
16.00	2.5E-1	2.8E-1	2.9E-1	3.0E-1	3.1E-1	3.1E-1
16.33	3.8E-1	4.3E-1	4.4E-1	4.5E-1	4.6E-1	4.6E-1
16.67	5.9E-1	6.5E-1	6.7E-1	6.8E-1	7.0E-1	7.2E-1
17.00	9.2E-1	9.9E-1	1.02E0	1.06E0	1.11E0	1.15E0
17.33	1.5E0	1.6E0	1.7E0	1.7E0	1.8E0	1.8E0
17.67	2.5E0	2.6E0	2.7E0	2.8E0	2.9E0	2.9E0
18.00		5.0E0	5.1E0	4.6E0	4.7E0	5.0E0
18.33		9.8E0	9.4E1	8.9E0	9.0E0	9.5E0
18.67			1.3E1	1.4E1	1.4E1	1.4E1

TABLE 5. Our calculated FWHM (nm) of Balmer-α for the reduced mass μ = 0.8.

$\log(N_e, cm^{-3})$	5000 K	10000 K	15000 K	20000 K	30000 K	40000 K
15.00	6.0E-2	6.5E-2	6.7E-2	6.7E-2	6.7E-2	6.6E-2
15.33	9.4E-2	1.04E-1	1.08E-1	1.10E-1	1.12E-1	1.13E-1
15.67	1.5E-1	1.6E-1	1.7E-1	1.8E-1	1.8E-1	1.8E-1
16.00	2.3E-1	2.6E-1	2.7E-1	2.8E-1	2.9E-1	3.0E-1
16.33	3.5E-1	3.8E-1	3.9E-1	4.0E-1	4.1E-1	4.3E-1
16.67	5.6E-1	6.0E-1	6.1E-1	6.2E-1	6.5E-1	6.7E-1
17.00	8.8E-1	9.2E-1	9.4E-1	9.7E-1	1.02E0	1.08E0
17.33	1.5E0	1.5E0	1.5E0	1.6E0	1.6E0	1.7E0
17.67	2.5E0	2.5E0	2.6E0	2.6E0	2.7E0	2.7E0
18.00		4.9E0	4.7E0	4.5E0	4.5E0	4.7E0
18.33		9.6E0	9.2E0	8.7E0	8.7E0	9.1E0
18.67			1.3E1	1.4E1	1.3E1	1.3E1

TABLE 6. Our calculated FWHM (nm) of Balmer-α for the reduced mass μ = 1.0.

$\log(N_e, cm^{-3})$	5000 K	10000 K	15000 K	20000 K	30000 K	40000 K
15.00	5.8E-2	6.3E-2	6.5E-2	6.6E-2	6.7E-2	6.7E-2
15.33	9.1E-2	1.00E-1	1.05E-1	1.08E-1	1.11E-1	1.12E-1
15.67	1.4E-1	1.6E-1	1.7E-1	1.7E-1	1.8E-1	1.8E-1
16.00	2.2E-1	2.5E-1	2.6E-1	2.7E-1	2.8E-1	2.9E-1
16.33	3.4E-1	3.6E-1	3.7E-1	3.8E-1	3.9E-1	4.2E-1
16.67	5.4E-1	5.6E-1	5.8E-1	5.9E-1	6.2E-1	6.5E-1
17.00	8.6E-1	8.9E-1	9.1E-1	9.3E-1	9.9E-1	1.02E0
17.33	1.4E0	1.5E0	1.5E0	1.5E0	1.6E0	1.6E0
17.67	2.4E0	2.5E0	2.5E0	2.5E0	2.6E0	2.6E0
18.00		4.8E0	4.6E0	4.4E0	4.4E0	4.6E0
18.33		9.5E0	9.0E0	8.6E0	8.6E0	8.9E0
18.67			1.3E1	1.4E1	1.3E1	1.3E1

NEW COUNTER-INTUITIVE RESULTS FOR THE TEMPERATURE AND DENSITY DEPENDENCES OF THE ION-DYNAMICAL STARK WIDTH

Ion-Dynamical Part of the Correlation Function

The Tables of the Stark widths presented above and in the book [2] were based on our calculations where *no assumption* was made that the total correlation function $C(\tau)$ could be factorized as the product $C_e(\tau)C_i(\tau)$ of the electronic and ionic correlation functions. If we would make such assumption, then interesting results could be obtained for the temperature dependence of the ion-dynamical Stark width, as follows.

From *first principles*, using the *multiparticle* approach, *without the impact approximation or any artificial cutoffs*, we derived the ion-dynamical part of the correlation function in the form [2]

$$C(\tau) = \exp[-6\tilde{N}_w f(\Omega_w\tau/4)], \qquad f(x) = x\int_0^\infty dy\,[1 - (\sin y)/y]/[y^2(y^2 + 1/x^2)^{1/2}]. \qquad (8)$$

where $f(x)$ is the universal function expressed as the one-fold integral. Here $\tilde{N}_w$ is the number of ions in the sphere of the ionic Weisskopf radius R_w

$$\tilde{N}_{wp} = (4\pi/3)N_pR_{wp}^3, \qquad R_{wp} = 3\,X_{\alpha\beta}\hbar Z_p/(Z_r m_e v_p); \qquad (9)$$

$\Omega_{wi} = v_i/R_{wi}$ is the ion Weisskopf frequency. In Eqs. (8), (9), N_p, Z_p, and v_p are the perturbers density, charge, and velocity, respectively; $X_{\alpha\beta} = [n_\alpha(n_1-n_2)_\alpha - n_\beta(n_1-n_2)_\beta]$ is the usual combination of the parabolic quantum numbers of the Stark sublevels α and β involved in the radiative transition; Z_r is the nuclear charge of the radiator.

Formula (8) is valid for *any* value of $\Omega_w\tau$, *including* $\Omega_w\tau << 1$ (i.e., *"totally" incomplete* collisions). In particular, for $\Omega_w\tau >> 1$, Eq. (8) yields

$$C(\tau) = \exp[-\,\tilde{N}_w\,(\Omega_w\tau/4)\,\ln(\Omega_w\tau/4)], \tag{10}$$

what corresponds to a *quasi-Lorentzian* lineshape. However, for $\Omega_w\tau << 1$ the ion-dynamical part of the correlation function takes a different form:

$$C(\tau) = \exp[-\,(3\pi/32)\,\tilde{N}_w\,(\Omega_w\tau)^2], \tag{11}$$

what corresponds to a *Gaussian* lineshape. This result could be considered as *counterintuitive*: this might be *the first time* when a dynamical Stark broadening by a plasma microfield leads to the Gaussian lineshape.

Thus, for the most interesting case of $\tilde{N}_w >> 1$, the *total* correlation function for the ion broadening (and consequently, the lineshape) has the following *three physically-different regions*: A) *static* region $\Omega_w\tau << 1/\tilde{N}_w^{1/2}$; B) dynamic *Gaussian* region $1/\tilde{N}_w^{1/2} << \Omega_w\tau << 1$; C) dynamic quasi-*Lorentzian* region $1 << \Omega_w\tau$. In the dynamic regions, the correlation function contains both adiabatic and nonadiabatic contributions, but the former predominates over the latter for $\tilde{N}_w >> 1$.

Temperature and Density Dependences of the Ion-Dynamical Stark Width

The general formula (8) allows to easily obtain the dependence of the Ion-Dynamical Stark Width (IDSW) on the temperature T. For the high-T limit, it reproduces the well-known result: IDSW ~ $1/T^{1/2}$. However, for the low-T limit, it predicts the change of the temperature dependence to:

$$\text{IDSW} \sim T^{1/4}. \tag{12}$$

In other words, as the temperature decreases from the high-T limit, the IDSW first increases ~ $1/T^{1/2}$, then reaches a maximum, and then decreases ~ $T^{1/4}$ (this being a *non-binary* result). This is illustrated by Fig. 3.

From the general formula (8) one can easily find also the dependence of the IDSW on the ion density N. In the low density limit $N << N_0$, the dependence is quasi-linear, as expected: IDSW ~ N $\log(N_0/N)$. Here the density N_0 corresponds to $\tilde{N}_w = 1$ (at the fixed temperature), where $\tilde{N}_w$ is defined by Eq. (9).

However, in the high-density limit $N >> N_0$, the dependence becomes really non-linear:

$$\text{IDSW} \sim N^{1/2} \tag{13}$$

(again a *non-binary* result).

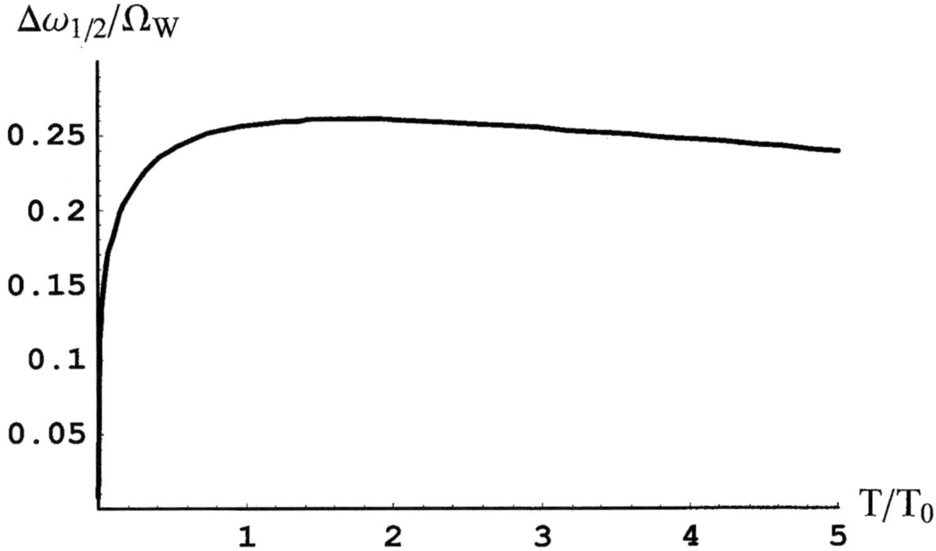

FIGURE 3. Temperature dependence of the ion-dynamical Stark width. Here the temperature T_0 corresponds to $\tilde{N}_w = 1$ (at the fixed electron density), where $\tilde{N}_w$ is defined by Eq. (9).

RESPONSE TO CRITICISM

There are critical remarks in papers [15-17] (the 2nd and 3rd papers mostly "recycle" the remarks from the 1st paper). However, first of all, their *criticism* of all three components of our theory is *limited to the first, approximate versions/drafts* [18-20] of each of these components. Already for this reason, their *criticism has practically no bearing on our current, much more rigorous, highly-advanced theory* - published (partially) in refereed journals [21, 22] (seemingly missed by the authors of [15-17]), presented in book [2] and in this paper, and used in the comparison with the latest high-density experiment [23]. Second, even while discussing the first approximate versions/drafts of the three components of our theory, *these authors made about 30 incorrect and/or irrelevant statements*. Few examples are given below.

Their claim that "the Acceleration of Electrons by the Ion Field (AEIF) is not new" is false. It is the fact that *we were the first to introduce the AEIF* in [20].

Their claim that for the case of $\tilde{N}_w >> 1$ "a significant Ionic Contribution to the Dynamical Stark Width (ICDSW) is not new" is false. It is the fact that *we were the first to introduce the ICDSW* in [19].

Their claim that "the standard theory can reproduce the ICDSW" is false. In reality, *2 assumptions beyond the standard theory* (a.k.a. conventional or Griem's theory [12]) were used by the authors of [15] to reproduce (in part) the ICDSW.

Their claim that "the net result of the Debye screening is the *increase of the Stark width* (by "softening" the AEIF via reducing the potential of perturbing ions) is inaccurate. Actually, the Debye screening also reduces the width of the ion field distribution and thus *decreases the Stark width* – these *two effects essentially cancel each other*.

Alexiou's claim [17] that the well-known "analytical solution by Pfennig-Lisitsa-Sholin is more advanced than our theory" is false. In reality, *Pfennig-Lisitsa-Sholin's solution* [24-26] (progressive in its time) *can treat only one sort of perturbing charges* - either ions or electrons, while *our theory allows analytically for joint effects/couplings* of the electron and ion microfields.

Alexiou [17] stated that the electron-ion coupling is negligible for the parameters of interest to the benchmark experiments at high densities. However, his own simulations [17] contradict to this statement by Alexiou. Indeed, his simulations showed that the electron-ion coupling resulted in a significant narrowing of the H_α line (compared to his other simulations disregarding this coupling): up to 18% for $N_e = 6.2\times10^{18}$ cm^{-3} and T = 0.8617 eV (corresponding to the underwater experiment [27]) or up to 14% for $N_e = 0.49\times10^{18}$ cm^{-3} and T = 5.77 eV (corresponding to the gas-liner pinch experiment [10]).

We can provide the following physical insight to Alexiou's simulations. Direct and indirect couplings of the electron and ion microfields work in the opposite directions for the H_α line. While the direct coupling leads to a narrowing for all lines, the indirect coupling leads to some additional broadening for the H_α line. Therefore, despite their partial compensation for this line, the joint effect of these two phenomena is still a significant narrowing (up to 18% for examples chosen by Alexiou). For some other hydrogen/hydrogen-like lines, both direct and indirect couplings could work in the same direction and produce even more significant narrowing than for the H_α line.

The last but not least: compared to the benchmark experiment by Büscher et al [10] in the gas-liner pinch, *Alexiou's simulations [17] dramatically underestimate the H_α width – by 30% for the lowest density*. Instead of trying to find out what is wrong with his simulations, he suggested that this benchmark experiment is incorrect.

However, first of all, the experiment [10] is commonly accepted as the most reliable of all benchmark experiments at densities $N_e \geq 10^{18}$ cm^{-3}. Second, the primacy of benchmark experiments over theories in physics is another commonly accepted fact. Third, our theory is in a very good agreement with this benchmark experiment, as well as with other benchmark experiments.[1]

---

[1] Griem et al. [15], hereafter GHO, made also other statements contradicting to the facts – on both the experimental and theoretical sides – as we indicate below.

GHO stated that our Generalized Theory (GT) of the Stark broadening [18] was based on the impact approximation for electrons. However, in reality, the GT employed the impact approximation only in its simplified version – to get the message across. In paper [18], the GT was actually developed in the general form (up to Eq. (13)), allowing to obtain results without utilizing the impact approximation. This rigorous "non-impact" GT was actually used in our comparisons with the latest high-density experiments [22, 23], and was presented above and in the book [2].

GHO stated that the GT has a problem because presumably it allows free electrons to penetrate into the excited atomic interior – i.e., allows values of the impact parameter $\rho < R_a$, where $R_a \sim n^2 a_0$ is the size of the hydrogen atom excited to the state of the principal quantum number n. However, in reality one of the advantages of the GT over the Conventional Theory (CT) by Griem [12] is that the GT allows the analytical treatment of free electrons penetrating not inside the atomic interior, but inside the sphere of the electron Weisskopf radius $R_W \sim n^2\hbar/v_{Te}$ (v_{Te} is the electron thermal velocity) - while the CT can handle analytically only collisions having impact parameters $\rho >> R_W$. For the relevant H_α

experiments at the temperature T ~ 1 eV, the volume of the electron Weisskopf sphere is by two orders of magnitude greater than the volume of the excited atom. So, the GT extends the analytical treatment to the surface and to the most of the interior of the electron Weisskopf sphere, leaving out of consideration only a relatively small, practically unimportant volume of the excited atom.

GHO stated that in our calculations incorporating our non-model analytical solution for the multiparticle ion dynamical broadening, we took into account the broadening by the same ions twice: in the quasistatic and dynamical formalisms. This GHO's statement is incorrect: the separation of static and dynamic ions, based on the different ranges of their instants of the closest approach, precedes all other stages in the design of our theory – as presented above and in the book [2].

GHO stated that our non-model analytical solution for the multiparticle ion dynamical broadening used the adiabatic approximation. However, in reality we took into account both adiabatic and non-adiabatic contributions of the perturbing ions – as it was clearly stated in our works [2, 22]. For a particular case, where the number of ions in the sphere of the ion Weisskopf radius is large, the adiabatic contribution turns out to be the dominant one. But this is the consequence of the physics of the ion dynamical broadening and has nothing to do with the "adiabatic approximation" wrongfully attributed to our theory by GHO – no adiabatic approximation was made by us.

GHO's criticism of the AEIF is limited to the first approximate version/draft of the AEIF published in [20]. GHO ignored our later paper [21] on the AEIF, where we eliminated the assumptions/approximations made in [20]. (In particular, we eliminated in [21] the assumption $\rho \ll R_1$, where R_1 is the distance between the emitter and the nearest perturbing ion.) Already for this reason, GHO's criticism of the AEIF has practically no bearing on our current, much more rigorous theory of the AEIF used in the comparison with the latest high-density experiments. Besides, even while limiting the discussion by the first approximate version/draft of the AEIF from [20], GHO made a large number of incorrect statements, as shown below.

While speaking about the draft of the AEIF from [20], GHO stated on p. 983 that presumably we used an incorrect velocity scaling for the electron impact broadening: $1/v^2$ instead of than $1/v$. This GHO's statement on the concept of the AEIF is incorrect. In reality, the situation described in [20] was the following. The frequency of so-called "optical" collisions with electrons (i.e., collisions leading to the Stark broadening) is $\gamma_e = N_e v_\infty \sigma$, where v_∞ is the asymptotic velocity (at infinity) of the incoming electrons and σ is the corresponding cross-section. In this formula, *only the cross-section σ gets affected by the AEIF*. In the CT, this cross-section σ scales as $1/v^2$ rather than as $1/v$. Thus, even in the approximate version/draft of the AEIF [20] we did not use the incorrect velocity scaling ($1/v^2$) for the electron broadening: we used the well-known fact that the cross-section σ scales as $1/v^2$. Seemingly, GHO missed the difference between the electron broadening and the corresponding cross-section.

Instead of the rigorous results on the AEIF from our paper [21] (seemingly missed by GHO), GHO suggested their Eq. (22) for the reducing factor $f_a(\beta)$ as follows:

$$f_a(\beta) = 4x^{-1/2}\int_0^\infty dx(1 + a^2\beta^{1/2}x^{-2}/3)x^2\exp(-x^2), \qquad \text{for } \beta \le \beta_c; \qquad f_a(\beta) = 1, \text{ for } \beta > \beta_c .$$

First, their Eq. (22) is just an estimate – as admitted by GHO. Second, Eq. (22) is based on their misunderstanding of the AEIF discussed above. Third, Eq. (22) contains an arbitrary parameter a, which was not defined by GHO anywhere. Fourth, the reducing factor $f_a(\beta)$ given by their Eq. (22) depends on the scaled velocity x even after the integration over x, what does not make any sense. Fifth, the integral in Eq. (22) with the non-zero second term in the integrand is actually greater than the same integral without second term in the integrand: so, their "reducing" factor $f_a(\beta)$ is actually an "amplifying" factor – contrary to GHO's claim that their Eq. (22) reproduces the AEIF. Even if GHO would now correct possible typographic errors in their Eq. (22), it would still remain a very rough approximation (including the assumption $\rho \ll R_1$ eliminated by us in [21]) that cannot substitute – and less so, benchmark – our rigorous results on the AEIF from [21].

GHO stated that the factorization of the correlation function in the form $C(\tau) = C_i(\tau)C_e(\tau)$, where $C_i(\tau)$ and $C_e(\tau)$ are ionic and electronic correlation functions respectively, "causes an about 10% erroneous increase of FWHM values". We performed calculations without this factorization and found that the difference is unimportant – significantly less than 10% claimed by GHO. Of course, we used our rigorous results for the AEIF, while GHO used very rough, approximate estimates for the AEIF, what might have led GHO to the significant overestimation of the difference. By the way, in paper [23], where we provided the theoretical support for the high-density experiment in a laser-produced hydrogen plasma, our theoretical results were obtained without this factorization.

Trying to bridge a large gap between Griem's theoretical shifts and the experimental shifts, GHO suggest adding to Griem's previous results the ion quadrupole shift. The ion quadrupole shift stands for the ion quadrupole interaction calculated in the first order of the perturbation theory without any coupling with the electron microfield. However, in reality the ion quadrupole shift is identically zero. This was rigorously proven analytically in paper [28]. We emphasize that in [28] it was shown that the ion quadrupole shift is exactly zero regardless of any integration over the microfield distribution: it is zero just as the result of the summation over all Stark components of the spectral line. (So, it does not matter whether or not far wings of the line are taken into account – contrary to the GHO remark.) It was also pointed out in [28] that this result follows from the most fundamental physical laws – the laws of symmetry: specifically, this result is the consequence of the invariance of this physical system with respect to the inversion of the sign of the ion field gradient.

On p. 993 of [15], trying to diminish large discrepancies between the shifts measured in the Bochum gas-liner pinch experiments and the theoretical shifts by Griem, GHO performed an order of magnitude estimate of the $\Delta n \neq 0$ contributions and then invoke again the ion quadrupole shift. However, first of all, such rough estimate of the $\Delta n \neq 0$ contributions cannot really serve as a counterpart for the comparison with the rigorous calculations. Second, the ion-quadrupole shift is identically zero – as already explained above with the reference to paper [28]. Thus, the GHO's estimate on p. 993 of [15] is incorrect and does not justify those discrepancies.

GHO referred to their simulations including the quadratic Stark effect. The quadratic Stark effect results from calculating the ion dipole interaction in the second order of the perturbation theory (PT). However, from the physical point of view, it is inconsistent to add only the quadratic Stark effect without allowing for the ion quadrupole term in the second order of the PT , and for the octupole term in the first order of the PT. Indeed, as early as in 1969 Sholin [29] showed that – due to the degeneracy of the unperturbed states of the hydrogen atom – both the quadrupole term in the second order of the PT and the octupole term in the first order of the PT are of the same order of magnitude as the quadratic Stark effect. So, the lack of those terms in GHO's simulations make them unsuited.

Simulations of shifts, including GHO's simulations, are ill-suited for high-density plasmas of relatively low temperature. First, in such plasmas, *statistical fluctuations* of the simulated distribution should very strongly affect numerical calculations of such subtle effect as the Stark shift. For example, paper [30], having Halenka and Olchawa among the co-authors, stated: "Unfortunately, even after averaging of 10,000 initial perturber configurations, the resulting statistical noise of the imaginary part of the autocorrelation function, Im C(t), did not allow to determine reliable asymmetries and shifts of the profile."

In addition, GHO's simulation technique requires the introduction of some *arbitrary minimal cutoff parameter* R_{min} (minimal impact parameter), as it was emphasized especially clear (several times) in Olchawa's paper [31], as well as in paper [30]. For example, the cutoff chosen by Halenka [32] coincides by the order of magnitude with the electron Weisskopf radius R_W. In distinction, our analytical theory does not require a cutoff at impact parameters comparable to R_W. In the paper on this simulation technique [31] Olchawa, while emphasizing that the line shift in this model *strongly depends* on the choice of R_{min}, wrote that achieving a good fit of calculated H_α shifts to the experimental ones requires the value $R_{min} = 22a_0$, where a_0 is the Bohr radius. Despite this fact, Halenka in his simulations of shifts [32] chose $R_{min} = 13.5a_0$ in one series of simulations and $R_{min} = 5a_0$ in another one - so, he is at odds with the experiments.

It is also worth mentioning that Alexiou (who consulted GHO on paper [15]) wrote [33] that on the issue of shifts, he did not believe that they can be calculated reliably by any sort of simulations. He explained [33]: "The quantity that needs to be integrated over velocities and impact parameters to obtain the shift is a *highly oscillatory function* ... very significant cancellation takes place". No wonder that GHO's simulations yielded a non-zero result for the pure ion quadrupole shift, while this shift has been rigorously proven by analytical calculations to be zero [28].

In our paper [28] we introduced a new source of shifts resulting from a coupling of the ionic and electronic microfields: the quadrupole ionic-electronic shift (QIES). Referring to our paper [28] (which is GHO's Ref. [20]), GHO stated that in papers [8, 10] it was shown that that the QIES is zero. This

GHO's statement is incorrect. In reality, papers [32, 34] did not deal with the QIES. Moreover, there is no publication at all where it would be shown that the QIES is zero.

GHO, made also statements contradicting to the facts on the experimental side – as we show below.

GHO stated that in Flih et al. [22] "the measurements may well have been compromised ... by optical distortions in the walls of the 3 mm inner diameter tube". However, this statement is incorrect for two reasons. First, we have measured the transverse distribution of the continuum at 320 nm at different positions along the flashtube axis, and have obtained a good reproducibility of those profiles, also traducing the good filling of the plasma and its cylindrical symmetry. Second, we have performed the same experiment also with flashtubes of inner diameter 5 mm, filled with an initial pressure of 150 and 200 Torr, the percentage of H_2 being 2 or 3 % [35]. The Stark widths and shifts of the H_α line measured with flashtubes of inner diameter 5 mm are in good agreement compared to those presented in our paper [22] for the same plasma parameters.

GHO stated that in [22] "evidently a constant continuum slope was assumed for subtracting the background emission near H_α". However, this assumption by GHO is inappropriate. In reality, the subtraction of the continuum under the H_α line was done after absorption correction and was fitted by a fifth order polynomial taking into account the continuum curvature observed on the recorded spectra with the low dispersion spectrometer (UFS 200). GHO used the inappropriate assumption of a constant continuum slope also in another argument (in line 5 of their Sect. 5).

Another statement by GHO – that "the line-profile corrections are clearly unreliable ...at least equal or even exceed the Planck intensity at the quoted temperature" – also does not correspond to reality. In Fig. C.1 of book [2] are displayed the experimental radiance spectra obtained in a flashtube of 3 mm inner diameter with a gas mixture Kr (96 %)- H_2 (4 %) at an initial pressure of 400 Torr. The maximum value of the measured optical thickness τ (with the concave mirror placed behind the flashtube [22]) varied between 0.35 and 1.10. Those values are very closed to the ones deduced from the relation $I(\lambda,T)=B(\lambda,T)\,(1-e^{-\tau})$, where T stands for T(0).

While trying to question our diagnostic results [22] obtained from krypton line intensities and from the transverse distributions of continuum intensities, GHO ignored the fact that we have also checked the electron density profiles by a second method – by using He-Ne infrared laser interferometry (type Ashby-Jephcott) at 3.39 μm. We found a good agreement with the diagnostic results obtained by the first method. This fact was explicitly stated in [22].

GHO argued that spurious Kr II lines, which are present at the H_α profiles, have a significant effect. However, in reality their effect is very weak. At the "strongest" plasma parameters condition, where $T(0) = 18080$ K, $N_e(0) = 4.4\text{x}10^{18}$ cm^{-3}, $N_{Kr+}(0) = 4.25\text{x}10^{18}$ cm^{-3} and $N_H(0) = 4.55\text{x}10^{17}$ cm^{-3}, the intensity of the Kr II line at 6420 Å is less than 1 % of the intensity of H_α and for the Kr II line at 6763 Å less than 0.2 %. Taking into account their larger broadening, that explains why their contribution is negligible on the H_α profile at the highest electron densities.

GHO referred to "the apparent lack of any density dependence of the measured widths (64 Å) at the lowest densities for the various fill pressures". If we apply "the classical $N_e^{2/3}$ density scaling" (i.e., a very approximate scaling suggested by GHO) more consistently than GHO did, this scaling would suggest 59, 64, and 69 Å for the $N_e/(10^{18}$ cm$^{-3})$ values of 1.10, 1.25, and 1.40, respectively. However, the values of 59, 64, and 69 Å are within the error bars 64±6 Å clearly indicated by us in [22].

CONCLUSION

Our theory is based primarily on a *generalization* of the formalism of *dressed atomic states*. In its contemporary composition, this theory brings together:

A) *non-impact* description of the *indirect coupling* of the electron and ion microfields;
B) *non-binary* description of the *direct coupling* of the electron and ion microfields;
C) *non-binary non-impact* description of the *ion-dynamical* broadening. The latter produces, in particular, a *new counter-intuitive result for the dependence of the Ion-Dynamical Stark Width (IDSW) on the temperature T and the ion density N.* It predicts that in distinction to the high-T limit, for which it was known that IDSW ~ $1/T^{1/2}$, *in*

the low-T limit the temperature dependence changes to IDSW ~ $T^{1/4}$. In other words, as the temperature decreases from the high-T limit, the IDSW first increases ~ $1/T^{1/2}$, then reaches a maximum, and then decreases ~ $T^{1/4}$ (*non-binary* result). The density dependence in the *non-binary* (high-density) limit becomes nonlinear: IDSW ~ $N^{1/2}$.

We used our theory for creating *extensive Tables of Stark widths* of hydrogen and deuterium lines in plasmas. By comparison with benchmark experiments, it is shown that *these Tables are more accurate* than the corresponding tables published previously by other authors.

REFERENCES

1. E. Oks, *Plasma Spectroscopy: the Influence of Microwave and Laser Fields*, Berlin: Springer, 1995.
2. E. Oks, *Stark Broadening of Hydrogen and Hydrogenlike Spectral Lines in Plasmas: The Physical Insight*, Oxford, United Kingdom: Alpha Science International, 2006.
3. C.A. Iglesias, J. Lebowitz, and D. McGowan, *Phys. Rev. A*, **32**, 1667 (1983).
4. H.R. Griem, *Principles of Plasma Spectroscopy,* Cambridge: Cambridge University Press, 1997.
5. T. Fujimoto, *Plasma Spectroscopy,* Oxford: Clarendon Press, 2004.
6. B. Talin, A. Calisti, and J. Dufty, *Phys Rev. E* **65**, 056406 (2002).
7. M.A. Gigosos and V. Cardenoso, *J. Phys. B*, **29**, 4795 (1996).
8. W.L. Wiese, D.E. Kelleher, and D.R. Paquette, *Phys. Rev. A* **6**, 1132 (1972).
9. Y. Vitel, *J. Phys. B* **20**, 2327 (1987).
10. S. Büscher, Th. Wrubel, S. Ferri, and H.-J. Kunze, *J. Phys. B* **35**, 2889 (2002).
11. K. Grützmacher and B. Wende, *Phys. Rev. A* **11**, 1854 (1975); in *Spectral Line Shapes*, v. 4, Ontario: Windsor, 1978, p. 49.
12. H. R. Griem, *Spectral Line Broadening by Plasmas*, New York: Academic, 1974.
13. K. Grützmacher and U. Johannsen, in *Spectral Line Shapes v. 7*, edited by R. Stamm, B. Talin, New York: Nova Science Publishers, 1993, p. 139; Frühjahrstagung der DPG, Greifswald, 1993.
14. Th. Wrubel, S. Büscher, H.-J. Kunze, and S. Ferri, *J. Phys. B* **34**, 461 (2001).
15. H.R. Griem, J. Halenka, and W. Olchawa: J. Phys. B **38** (2005) 975
16. S. Alexiou, H.R. Griem, J. Halenka, and W. Olchawa, *J.Q.S.R.T.* **99**, 238 (2006).
17. S. Alexiou, *Phys. Rev. E* **71**, 066403 (2005).
18. Ya. Ispolatov and E. Oks, *J.Q.S.R T.* **51** (1994) 129.
19. E. Oks, *Phys. Rev. E* (Rapid Comm.) **60**, R2480 (1999).
20. E. Oks, *J.Q.S.R.T.* **65**, 405 (2000).
21. E. Oks, *J. Phys. B* **35**, 2251 (2002).
22. S.A. Flih, E. Oks, and Y. Vitel, *J. Phys. B* **36**, 283 (2003).
23. Ch.G. Parigger, D.H. Plemmons, and E. Oks, *Appl. Opt.* **42**, 5992 (2003).
24. H. Pfennig, *J.Q.S.R.T.* **12**, 821 (1962).
25. H. Pfennig, *Z. Naturforsch.* **26a**, 1071 (1971).
26. V.S. Lisitsa and G.V. Sholin, *Sov. Phys. JETP* **34**, 484 (1972).
27. A. Escarguel, B. Ferhat, A. Lesage, and J. Richou, *J.Q.S.R.T.* **64**, 353 (2000).
28. E. Oks, *J.Q.S.R.T.* **58**, 821 (1997).
29. G.V. Sholin, *Opt. Spectrosc.* **26**, 275 (1969).
30. T. Wujec, W. Olchawa, J. Halenka, and J. Musielok, *Phys. Rev. E* **66**, 066403 (2002).
31. W. Olchawa, *J.Q.S.R.T.* **74,** 417 (2002).
32. J. Halenka, *Phys. Rev. E* **69**, 028401 (2004).
33. S. Alexiou, in *Spectral Line Shapes*, v. 9, eds. Zoppi M and Ulivi L, AIP Conf. Proc. 386, New York, 1997, p. 79.
34. S. Alexiou, in *Spectral Line Shapes*, v. 12, AIP Conf. Proc. 645, New York, 2002 p. 302; *J.Q.S.R.T.* **81,** 13 (2003).
35. S.A. Flih, *Thèse de l'Université de Paris VI*, 2003.

Temperature Dependence of Stark Broadening Dominated by Strong Collisions

M. A. Gigosos[1], M. Á. González[2] and N. Konjević[3]

[1]Departamento de Óptica, Universidad de Valladolid. 47071 Valladolid (Spain).
[2]Departamento de Física Aplicada, Universidad de Valladolid. 47071 Valladolid (Spain).
[3]Institute of Physics, P.O. box 68, 11081 Belgrade, Serbia and Montenegro.

Abstract. The influence of electron temperature in the broadening of spectral lines dominated by strong collisions has been studied. Computer simulation allows us to study the effects of strong and weak collisions separately. Results shown here are focused on some Sr^+ and Ba^+ resonance lines as examples of lines broadened dominantly by strong collisions. The exact numerical integration of the perturbation process due to the collision with a single particle permits the evaluation of Weisskopf radius. This parameter is usually defined as $r_w \sim 1/v \sim 1/\sqrt{T}$, obtained from Born approximation that is correct for high temperatures. However, at low temperatures the full integration of the collision process permits to test the relationship $r_w \sim 1/T^{1/6}$. This calculation has allowed us to study the influence of temperature on the broadening of the lines dominated by strong collisions. This study has been done in two ways : through a plasma simulation and analyzing the calculated Weisskopf radius for an individual collision. The obtained results show that at low temperatures the width of the line increases for increasing temperature as a consequence of an increase of the number of collisions not compensated by the decrease of Weisskopf radius.

Keywords: Stark broadening, Plasma Diagnosis, Computer Simulation.
PACS: 32.70.Jz 52.70.Kz 52.65.Yy

INTRODUCTION

In order to test the possibility of using electron collision models [1,2] for the evaluation of Stark broadening parameters at very low electron temperatures and, in this way, to test their temperature trend, a numerical computer simulation experiment is performed. This approach is also based on a collision model, but in addition to electrons the ion collisions contribution is estimated as well. The basic difference between the two sets of calculations: the computer simulation method is free from numerous approximations involved in the models described in [1,2].

Let us draw attention to some approximations involved in collision model [2]. As can be seen in [2]: *"the first is that the perturbers do not interact with each other. Each perturber interacts only with the atom, and is otherwise uncorrelated with the motion of the other perturbers. If we are talking about charged perturbers, this is true only of those that are inside the Debye radius, and we may have to introduce later a correction to take into account the mutual screening of the perturbers"* [2]. In this sense, the physical model employed in the calculations [1] and the one used in computer simulations here are identical: perturbers are independent particles that move along straight line trajectories with constant speed according to a Maxwellian

CP874, *Spectral Line Shapes: 18th International Conference*,
edited by E. Oks and M. Pindzola

velocity distribution corresponding to the temperature of thermal equilibrium. The correlation between charges is taken into account approximately using Debye screened fields. These basic assumptions of the collision model are used in [1,2] and in the computer simulation presented here. The difference between the two sets of calculations comes from the different mathematical treatment of these approximations. The calculation in reference [1] uses the so called *"impact approximation"*, which is a mathematical approximation. The simulation, however, does not use those approximations, and the evolution differential equations are fully integrated. In the analytical calculation [1,2] the Born approximation is used for the development of evolution operator (only the dominant term of the perturbative development of the time evolution operator is taken into account), while in the simulation the differential equations giving the emitter evolution are fully integrated. Further, the analytical development assumes that the collisions do not overlap in time, while in the simulation this must appear in a natural way. It must be remembered that the simulation does not consider "collisions", but it uses the electric field due to the complete set of perturbers. In most time of the simulation, there is not a particle, whose electric field clearly dominates. Only occasionally one particle passes very close to the emitter, and, in those cases, the other particles electric field could be neglected. As will be seen later, these interactions, though very rare, will determine the shift and broadening of the spectral lines studied here. Then, differences between the results of both treatments are not due to differences in the emitter or plasma physical model, but of the mathematical treatment of the interaction between both. To be exact, in the problem considered here, the consideration that have a larger influence on the results is the use of Born approximation: in the usual treatment the phase changes due to the collisions are linear with the intensity of the perturbation; in the simulation this linearity does not appear when we are dealing with close collisions.
In addition, the modified semiempirical formulas [3,4] are also used for the evaluation of Sr^+ Stark widths and shifts in a broad temperature range including very low temperatures. These formulas basically use the same theoretical approach [1,2] and the Stark broadening parameters temperature dependence is expected to show the same trend as shown in Fig. 2 of reference [1].

NUMERICAL TREATMENT IN THE SIMULATION

The dipolar emission profile is obtained with the Fourier transform of the emitter dipole moment autocorrelation function [5]:

$$\begin{aligned} I(\omega) &= \frac{1}{\pi}\int_0^\infty dt \ \cos(\omega t) \ \left(C(t)\right) \\ C(t) &= \mathrm{tr}\ \left[\boldsymbol{D}(t)\cdot\boldsymbol{D}(0)\right] \\ \boldsymbol{D}(t) &= U^+(t)\ \boldsymbol{D}(0)\ U(t) \end{aligned} \tag{1}$$

where $\boldsymbol{D}$ is the dipole moment of the transition under study —normalized so that $C(0) = 1$— and $U(t)$ is the time evolution operator of the system, that obeys the Schrödinger equation:

$$i\hbar \frac{d}{dt} U(t) = \left[H_0 + q\boldsymbol{E}(t)\cdot\boldsymbol{R}\right] U(t) \tag{2}$$

H_0 is the unperturbed emitter hamiltonian, $\boldsymbol{E}(t)$ is the electric field sequence undergone by the emitter, and $q\boldsymbol{R}$ is its dipole moment. In our calculations we have not considered the *no-quenching* approximation, so that operator $\boldsymbol{R}$ and $\boldsymbol{D}$ are proportional.

In a computer simulation the perturbers —ions and electrons— behavior is reproduced numerically, and the electric microfield $\boldsymbol{E}(t)$ at the emitter position is calculated. Once this field is evaluated, the equations (2) are solved numerically and by using equation (1) the evolution of dipole moment is obtained. This calculation is repeated a large number of times with a representative set of microfield temporal sequences.

In expression (1) symbols {} mean an average on emitters in the plasma, what in our case means an average on the emitter dipole autocorrelation functions, each of them obtained from a sequence of the perturber microfield $\boldsymbol{E}(t)$. The details how particles are generated in the simulation, including the reinjection technique are described in reference [7].

The correlations between charged particles will be approximated here as in [7,8], using a plasma model of non interacting particles, under thermal equilibrium, with Debye screened fields. For some of the conditions considered in the simulation (low temperature and high density) the plasma coupling parameter is a large number characteristic for non ideal plasma. In these cases it would be necessary to perform a study of the interactions between particles using a molecular dynamics simulation technique. Since the aim of this work was not to provide quantitative data for comparison with experimental results, but to study the influence of temperature within the frame of the physical model used in [1] (independent particles with a screening correction to emulate the correlations), then in order to take into account correlations in the simulations only a screening correction for independent particles is used. In this sense, strongly coupled plasma cases may be considered as an extrapolation. In fact, when one takes into account the correlation between particles the collisions efficiency is reduced and the values of width and shift decrease as well [8]. This effect will contribute to the decrease of width and shift with decreasing temperature.

In the computer simulation, the perturbers movement and the emitter evolution are carried on with discrete *time steps* Δt. These steps are chosen so that the electric field $\boldsymbol{E}(t)$ may be considered static for the time duration of the step. In this case, the solution of the differential equation (2) is:

$$U(t+\Delta t) \simeq \exp\left[-\frac{i}{\hbar}\left(H_0 + q\boldsymbol{E}(t)\cdot\boldsymbol{R}\right)\Delta t\right] U(t) \tag{3}$$

To calculate this exponential, it is necessary to obtain the eigenvalues and the eigenvectors of the hamiltonian. In our calculation, we have used a quick diagonalisation method. Details of this method will be described in a next paper [9]. Calculations have been done considering the states 5s $^2S1/2$, 4d $^2D3/2$, 4d $^2D5/2$, 5p $^2P°1/2$, 5p $^2P°3/2$ and 6s $^2S1/2$ of Sr^+, considering all the dipole transitions between them.

SIMULATION RESULTS

Figure 1 (b) shows a typical example of the dipole autocorrelation function corresponding to the electric field sequence shown in figure 1(a).

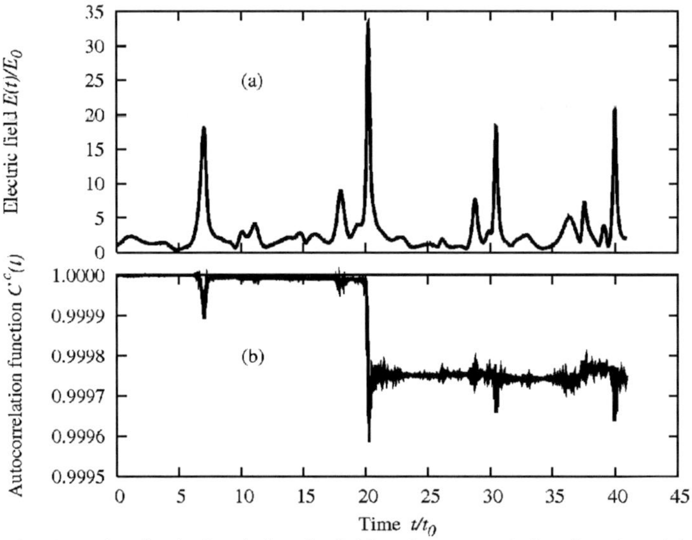

FIGURE 1. An example of calculated electric field and autocorrelation function: (a) a typical time sequence of the electric field modulus; (b) the corresponding emitter dipole autocorrelation function C(t). The function C(t) under weak collisions conditions decreases very slowly. Only when a strong collision occurs a noticeable change in the correlation function is detected. In this example, though the electric field sequence was cancelled except around $t \approx 20\ t_0$, the correlation loss at $t \approx 40\ t_0$ would have been the same. In this figure, $t_0 = r_0/v_0$, $r_0 = (4\pi N_e/3)^{1/3}$, $v_0 = (2kT/m)^{1/2}$, and $E_0 = q_e/(4\pi\varepsilon_0 r_0^2)$.

Under weak field conditions, the correlation loss is very slow. When a strong collision takes place the emitter dipole suddenly looses coherence with a *"step"* function shape. The height of this step depends upon the integral of energy transfer between the perturber and the emitter atom. It should be noticed that when the energy ---expressed as a phase of the evolution operator--- is significant, the relationship between the perturber field and the phase step is not linear. For very strong collisions a monotonous relationship between the phase change and the perturbation magnitude is not necessarily expected, i.e. this is a total coherence breakdown situation. When one considers very weak collisions, the relationship between the phase change and the perturbation energy is linear and the so called impact approximation has physical meaning: non overlapping weak collisions with very short duration. This approximation leads to the width and shift dependence upon electron density and temperature, which after simplification of some details scales the line width with $N_e/\sqrt{T}$ [1,2]. Strong collisions however establish a relationship more similar to $N_e\sqrt{T}$. Namely, very strong collisions give rise to sudden changes of the phase —the *"steps"* — whose magnitude hardly depends on the density and the temperature. The frequency of those collisions, however, is proportional to $N_e\sqrt{T}$. When weak collisions are very weak, their effect is masked by strong collisions, which dominate line broadening. This is exactly what is happening here for the studied transition. Therefore, we can conclude here from the simulations results that the Sr II 421.5 nm line Stark profile is dominated by strong collisions.

Computer simulation permits testing to what extent the strong collisions are the dominant collisions. To achieve this goal, we performed a computer simulation in which the electric microfield sequences were modified according to

$$\boldsymbol{E}'(t) = \begin{cases} \boldsymbol{E}(t) & \text{if } E(t)=|\boldsymbol{E}(t)| \le E_{\max} \\ E_{\max}\dfrac{\boldsymbol{E}(t)}{E(t)} & \text{if } E(t)>E_{max} \end{cases}$$

This means that we truncated the perturber field modulus to a pre-established value $E_{\max}$ but we kept its orientation. The results in table 1 represent several examples obtained in these calculations.

TABLE 1. Electrons only with truncated field

E_{max} (E_0)	FWHM (10^9 s^{-1})	Shift (10^9 s^{-1})	% Collisions
∞	103.7	69.3	0.000
100	5.00	34.0	0.096
50	1.40	23.2	0.290

If the collisions with electric fields larger than one hundred times the normal field strength are eliminated, the total width is reduced to ~5% (see table 1). These collisions correspond only to a 0.096% of the total simulation time. The field larger than 100 E_0 corresponds to collisions with impact parameters lower than 0.1 r_0 what corresponds to 1.33 nm in the considered example.

The complementary process has also been done. For this second case, only the peaks with electric field intensity higher than a given minimum one were considered in the simulation, following the process:

$$\boldsymbol{E}'(t) = \begin{cases} \boldsymbol{E}(t) & \text{if } E(t)=|\boldsymbol{E}(t)| \ge E_{\min} \\ \boldsymbol{0} & \text{if } E(t)<E_{min} \end{cases}$$

Widths and shifts obtained considering only the peaks with intensities higher than $E_{\min}$ are shown in table 2. As can be seen there, a simulation considering only the peaks with intensity higher than 100 E_0 gives the same width and shift than the total simulation. It must be remembered here that the simulation noise is within the 5%.

TABLE 2. Electron field with only peaks

E_{min} (E_0)	FWHM (10^9 s^{-1})	Shift (10^9 s^{-1})
0	103.7	69.3
50	103.6	53.1
100	104.6	45.8

It is well known, that the domain of strong collisions may be characterized by the Weisskopf radius [6], r_w, which is defined as the distance below which the individual collisions are able to produce a change of phase greater than π in the evolution of the emitter dipole. Using this criterion in reference [6] it is estimated that the line broadening due to strong collisions follows a trend proportional to $N_e\sqrt{T}r_w^2$. The numerical treatment employed here permits the evaluation of the Weisskopf radius for our work conditions. In order to achieve this goal the emitter dipole autocorrelation function was calculated for an individual collision between an electron and an emitter atom using a straight trajectory and mean square velocity $v_0=\sqrt{2kT/m_e}$. We have taken as the Weisskopf radius the impact parameter that must be chosen in such a way that the *step* produced in *C(t)* is equal to one —see exp. (\ref{EQ15})—. This means that *C(t)* passes from $C(-\infty)=1$ to $C(+\infty)=0$.

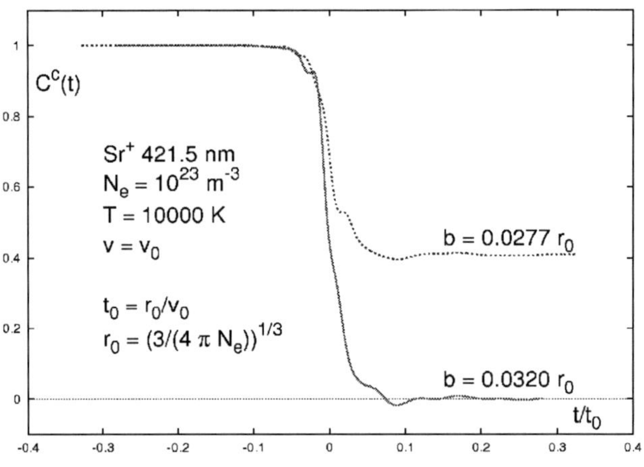

FIGURE 2. Correlation function $C^c(t)$ obtained in two individual collisions simulated for estimating Weisskopf radius for the conditions of electron density and temperature shown in the figure.

Figure 2 shows an example of the process considered for calculating Weisskopf radius integrating the evolution equations for an individual collision. For the example shown in the figure, that corresponds to T = 10000 K, the value considered for the Weisskopf radius is r_w = 0.032 r_0, where r_0 is the mean interparticle distance corresponding to the density pointed out in the figure.

Figure 3 shows calculated values of the Weisskopf radius for two electron densities: the one used here ($N_e = 10^{23}$ m^{-3}), and the lower one used in reference [1] ($N_e = 10^{16}$ m^{-3}).

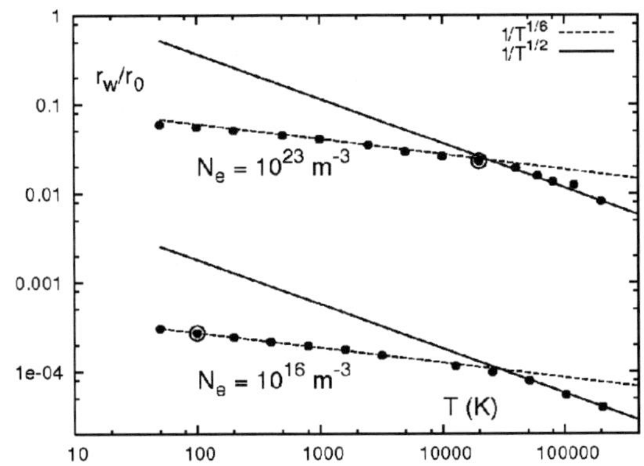

FIGURE 3. Weisskopf radius versus temperature for two electron density conditions. The value of r_w has been fixed as the impact parameter of an individual collision that gives rise to a change equal to one in the emitter dipole autocorrelation function. The values corresponding to the central conditions of the simulation study and those in [1] are marked in this figure with an additional circle. In both cases marked circles are in a region where the dependence of the Weisskopf radius upon temperature is much weaker than $1/\sqrt{T}$. For the low density case this situation is clearer than for the high density one. When Weisskopf radius decreases with a trend weaker than $1/T^{1/4}$ the width induced by strong collisions will have growing trend with temperature. This comes as a consequence of strong collisions frequency, which is proportional to $\sqrt{T}$.

It is important to notice that low density results do not represent results of a linear re-scaling of the high density ones. The Weisskopf radius calculations were performed at the pointed densities and temperatures. It must be taken into account that these calculations were done with a simulation of a single collision, so they are affordable cases in the simulation. In fact, they were obtained considering in the simulation time steps much smaller than those considered in the usual simulation calculations. Both sets of results in figure 3 show temperature dependence that, in the domain analysis, do not correspond to the trend $r_w \sim 1/\sqrt{T}$ which would produce an increase of the spectral line width as temperature decreases. One has to select very high temperatures to find a strongly decreasing trend with the temperature.

It is known that the Weisskopf radius does not depend on electron density and it is only a parameter-indicator when a collision is strong enough to produce a change of phase equal to one. Thus figure 3 shows only that at low density (r_0 is, then, larger) the Weisskopf radius is much smaller than the mean interparticle distance. This means that "strong collisions" are relatively less frequent. Here we must return to figure 1, which shows that the only significant correlation loss for that example is the one induced by the peak at $t \approx 20\ t_0$. Although we would have completely gone without the rest of the electric field, the correlation loss would have been the same. This phenomenon appears when the average collision is very weak. It's more, the weaker the typical collision is, and the more noticeable these effects are. In this sense our extrapolation to low density is completely justified. When density is reduced, the ratio between the Weisskopf radius and the mean interparticle distance becomes smaller. Then, strong collisions are less probable, but, we want to point out that outside the Weisskopf radius collisions have negligible effect in comparison to those of the few strong collisions. If one would expect that the dependence of the line width with temperature be of the type $1/\sqrt{T}$, Weisskopf radius at low densities should recover the trend $1/\sqrt{T}$ at low temperatures. This effect has not been observed in the simulation experiments. The trend $1/\sqrt{T}$ of Weisskopf radius only appears for temperatures over 20000 K. This observation justifies, once more, the extrapolation of our calculations at low densities.

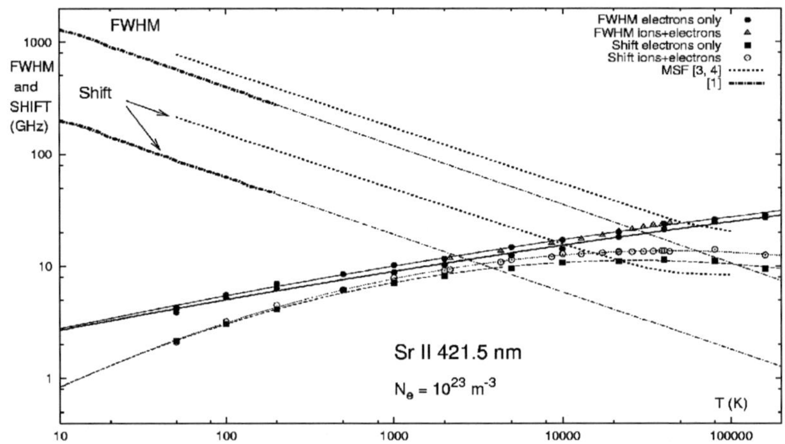

FIGURE 4. Total width at half maximum (FWHM) and shift of the 421.5 nm Sr II line. The simulation results for the joint broadening of electrons and ions are presented with empty triangles, while full circles are used for electrons only broadening. Thin lines show best fit of the simulation results. Results corresponding to the Modified Semiempirical Formulas (MSF [3,4], dotted line) and those appearing in figure 2 of [1] (continuous thick line) are also shown. For these last data, thin line represents temperature extrapolation of data [1] using $1/\sqrt{T}$ relation.

The simulation results are shown in figure 4 together with corresponding data from Modified Semiempirical Formulas [3,4] and calculated ones from figure 2 of reference [1]. For the electron density scaling of all the data the linear dependence of the Stark width and shift is assumed. The same assumption is used to scale up data from Vrinceanu *et al.* [1], corresponding to an electron density of $N_e = 10^{16}$ m^{-3}. These extrapolations are justified in both models, [1] and in our simulations. Since linear dependence of Stark width upon electron density is an important issue here, where data at very different electron densities are compared, a short discussion of this subject will be carried out. It is important to notice that in the treatment of both, very weak and very strong collisions, the dependence of the width upon the electron density is linear. For the first case —very weak collisions— the linearity appears because the perturber field correlation is lost much earlier than the emitter dipole autocorrelation, and the broadening process becomes homogeneous in time. In the second case —very strong collisions— the linearity appears because these collisions never overlap in time —binary approximation is always valid— and two consecutive collisions are statistically independent. In both cases, the simulation results show a linear dependence upon electron concentration for medium densities while at very low densities this dependence can be much easier proven. The proof of the Stark broadening parameters linear dependence upon electron density justifies, a posteriori, our calculations, which were done at much higher density than in [1].

The most important result of the computer simulations is the temperature trend of the line width and shift, which is quite different from the one predicted by the model used in [1]. As expected, the results of Modified Semiempirical Formulas [3,4] show the same trend as shown in [1]. The origin of both approaches is the same and starts from reference [2].

CONCLUSIONS

The aim of this work is not to supply accurate values of Sr^+ line widths and shifts for quantitative comparison with experiment. This would require using Molecular Dynamics simulations of interacting particles with consideration of the emitter charge. It is not expected however that these results would differ qualitatively from those presented here. The most important result of this work is an indication that strong collisions dominate Stark broadening at very low electron temperatures and therefore the theoretical approaches based on weak collisions assumption, see e.g. [2], cannot be applied in this temperature region. Consequently the temperature trend of Stark widths and shifts changes from $1/\sqrt{T}$ used at elevated temperatures to an increasing trend with temperature, which is characteristic of strong collision cases [5]. Our results indicate that width and shift have an increasing dependence with an increase of the temperature and suggest that Stark broadening in ultracold plasmas at relatively low electron densities $N_e < 10^{16}$ m^{-3} is negligible. In favor of this conclusion speaks also an experimental fact: the generation and confinement of ultracold plasmas at very low temperatures would be followed by a huge drift of the optimal laser-plasma interaction wavelength if the Stark shift do no comply temperature dependence of the simulation results. Namely, the wavelengths of Sr II resonance lines are used for laser trapping and cooling of strontium ultracold plasmas. To the authors knowledge such systematic and large shifts of the optimum pumping laser wavelength are not detected even for very low temperatures.

ACKNOWLEDGMENTS

This work was supported by the spanish Ministerio de Ciencia y Tecnología under grant ENE2004-0538/FTN.

REFERENCES

1. D. Vrinceanu, H R Sadeghpour and K Bartschat, *J. Phys. B: At. Mol. Opt. Phys.* **37**, L371-L375 (2004).
2. M. Baranger, *Phys. Rev.* **111**, 481-493 (1958).

3. M. S. Dimitrijević and N. Konjević, *J. Quant. Spectrosc. Radiat. Transfer* **24,** 451-459 (1980).
4. M. S. Dimitrijević and N. V. Kršljanin, *Astron. & Astrophys.* **165,** 269-274 (1986).
5. P. W. Anderson, *Phys. Rev.* **76,** 647-661 (1949).
6. H. Margenau and M Lewis, *Review of Modern Physics* **31,** 569-615 (1959).
7. M. A. Gigosos and V. Cardeñoso, *J. Phys. B: At. Mol. Opt. Phys.* **29,** 4795-4838 (1996).
8. E Dufour, A Calisti, B Talin, M A Gigosos, M Á. González, T. del Río Gaztelurrutia and J. W. Dufty, *Phys. Rev. E* **71,** 066409-1-9 (2005).
9. M. A. Gigosos, M. A. González and N. Konjević, *European Physical Journal D.* (to be published).

Determination of the Local Electric Field Strength in the Cathode Fall of a Low Pressure Gas Discharge Using Doppler-Free Spectroscopy

Volkmar Helbig

Institute of Experimental and Applied Physics, Kiel University, Leibnizstr. 19, D 24098 Kiel, Germany

Abstract. Doppler-free spectra taken in the cathode fall region of a low pressure gas discharge are used to determine the electric field strength. Comparison with Doppler-limited spectra in the same region of the discharge shows good agreement for intermediate fields. The present limitations regarding the minimal field strength values accessible with this technique are discussed.

Keywords: Field strength measurements, Balmer-alpha, Stark effect, Doppler-free spectroscopy
PACS: 32.70.Jz, 33.70.Jg

INTRODUCTION

The Stark effect is of great importance for the diagnostics of terrestrial and extraterrestrial plasmas. The energy levels of a radiating atom will show a shift or in the case of Hydrogen a splitting when affected by a static electric field. If the Stark effect constants are known the electric field strength can be determined from the shift from the zero field position. In case of fluctuating fields the observed lines will show a broadening. Again information about the plasma may be obtained from the measured widths of the lines when the broadening constants are known. In this case the electron density can be determined that is connected to the average value of the fluctuating microfield. In the present investigation we are interested in the absolute value of the electric field strength in the cathode fall region of a low pressure discharge. As field strength probe we use the first line of the Balmer series of Hydrogen because the lines of the Hydrogen atom are the most sensitive with respect to the shift by the Stark effect and therefore promise the highest resolution. Since we are interested in a wide range of electric field strengths ranging from several volts/cm to kilo volts/cm we calculated the splitting of the levels giving rise to the Balmer-alpha line including the fine- structure [1, 2] using standard techniques [3, 4]. The results for the splitting of the levels $n = 2$ and $n = 3$ of Hydrogen are shown in figure 1. For a comparison with the experimental profiles in addition to the splitting the field depending intensities of the single fine-structure components need to be calculated. The results from the calculations for a wide range of field strengths are stored in a matrix that can be read by a least squares fitting routine used to compare experimental line profiles with the calculated ones. There are two sets of experimental data. In a first step Doppler-

CP874, *Spectral Line Shapes: 18th International Conference,*
edited by E. Oks and M. Pindzola

limited profiles were recorded. In this case the Doppler-width is an additional fitting parameter in the mentioned routine. In the second set of data Doppler-free profiles were obtained. In this case the shifts and the intensities are the fitting parameters.

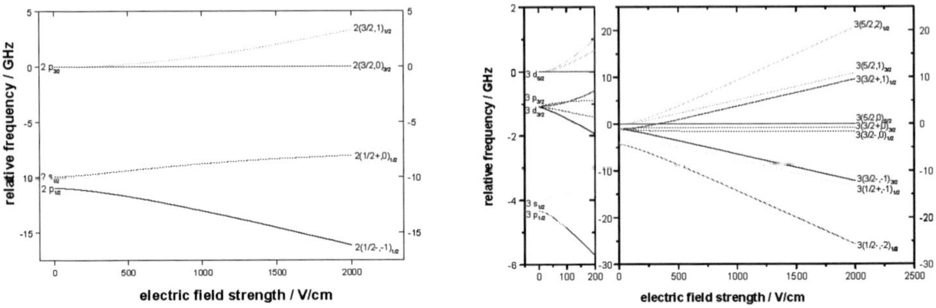

FIGURE 1. Splitting of the levels n=2 and n=3 of Hydrogen with the electric field strength [2].

EXPERIMENTAL SET-UP

The discharge was run in a stainless steel cylinder of about 30 cm in diameter. The cylinder itself served as anode while the cathode was mounted horizontally on a high precision linear drive. The set-up and a print of the discharge are shown in figure 2. The cathode was made from tantalum and the front surface was polished to high

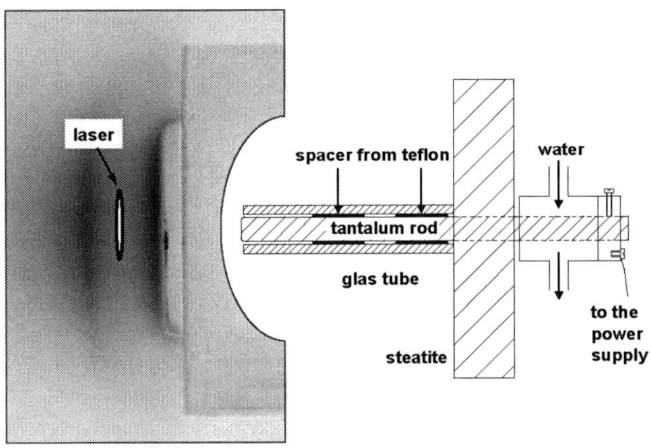

FIGURE 2. The figure shows details of the cathode set-up. The print on the left shows the glow of the discharge in front of the cathode, the glas tube and the front end of the tantalum cathode. The approximate position of the laser beam is also indicated.

quality in order to prevent sparking of the discharge. The discharge was run in Hydrogen at a pressure of p=10 mbar and a current of I=10 mA. Since absorption in the cathode fall region turned out to be too small to obtain useful line profiles optogalvanic spectroscopy [5] was used instead. A homemade Littrow-type diode laser system that could be scanned single mode over a band pass of more than 120 GHz was employed to record the spectra [2]. Using cylindrical lenses the laser beam was focused to a lens-shaped diameter of approximately 50 μm x 2 mm (in figure 2 the aspect ratio is not to scale). In order to investigate the field distribution in front of the cathode it was moved with respect to the laser beam that was at rest. Taking spectra at

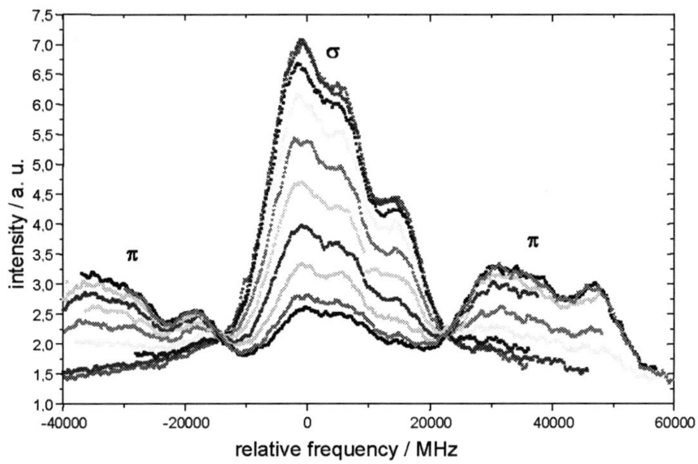

FIGURE 3. Doppler-limited spectra taken in front of the cathode under different angles of polarization. Large σ–intensities correspond to small values of the π–components and vice versa [2].

different angles of polarization as shown in figure 3 we were able to determine the direction of the field in front of the cathode [2] for one distance from the cathode. Repeating this procedure for various distances we mapped the strength and the direction of the electric field in front of the cathode as shown in figure 4. Included in the figure is a print showing the cathode and the laser beam in front of it.

One draw-back of the method described is that the laser probes areas in front of the cathode with different directions of the field. In order to avoid this we changed the set-up as shown in figure 5. The cathode now consists of two parts that are isolated against each other. A central rod and a surrounding tube both in this case made of stainless steel [6] form together the cathode. Both parts of this cathode are shunted separately to the power supply but only from the resistor that feeds the central rod the optogalvanic signal is recorded. This has two advantages: (i) we now get information from a region with a uniform electric field as indicated in the figure (ii) since the laser width matches the diameter of the central part of the cathode we practically probe the

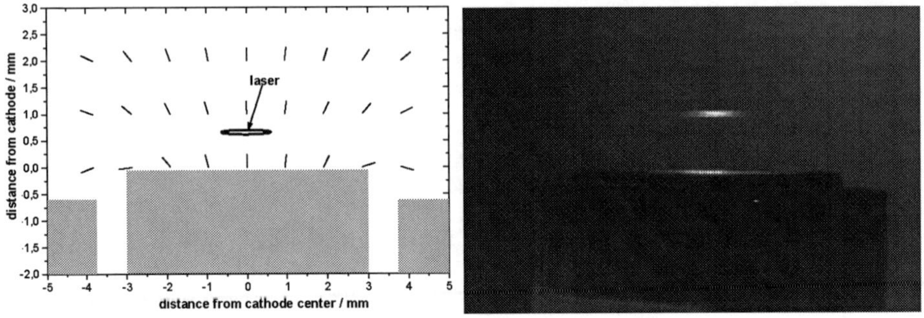

FIGURE 4. The directions of the electric field in front of the cathode on the left and the laser beam in front of the cathode on the right [2, 12].

complete cross-section of that part of the discharge that ends on the central iron rod. This leads to a rather strong optogavanic signal. Details are described in [6]. In figure 5 again the approximate shape of the laser beam cross-section is indicated.

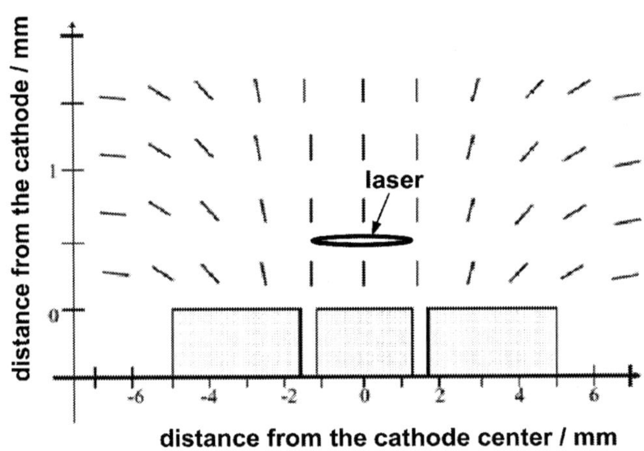

FIGURE 5. Scheme of the splitted cathode and the field strength distribution in front of it. [7]. Only the part of the discharge in front of the central part of the cathode contributes to the optogalvanic signal.

RESULTS FOR THE DOPPLER-LIMITED MEASUREMENTS

The Doppler-limited optogalvanic spectra were fitted to the results of the above mentioned Stark effect calculations. The π-components that are more sensitive to the electric field were used for lower and intermediate field strength values while the σ-components served as field probes for the highest field strengths. Some of these fits for σ-components recorded at different distances from the cathode surface are shown in figure 6. While in the wings the fits are quite satisfactory the fitted profiles show less structure in the center than the experimental line shapes. The same is observed for the π-components. Again a quite accurate field strength value (accuracy about $\pm 2\%$) can be obtained from the outer wings that are formed by the most sensitive components while in the center the experimental profiles exhibit more structure than the theoretical ones. An example for the π-components is given in figure 7.

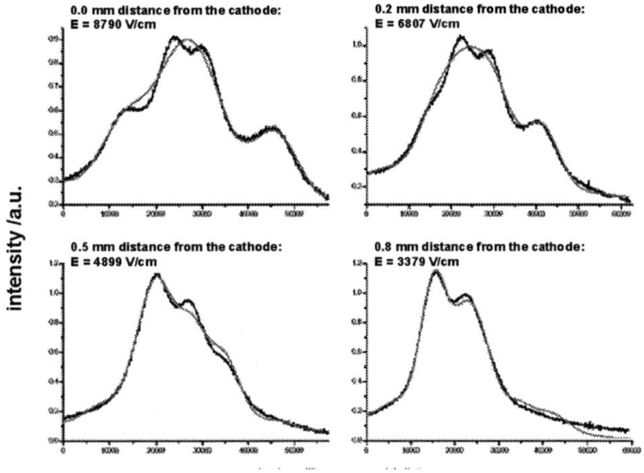

FIGURE 6. Experimental σ-components each with the corresponding least squares fit. The separation from the cathode surface is indicated as well as the obtained electric field strength. The experimental profiles show more structure in the line center than the calculated ones [7].

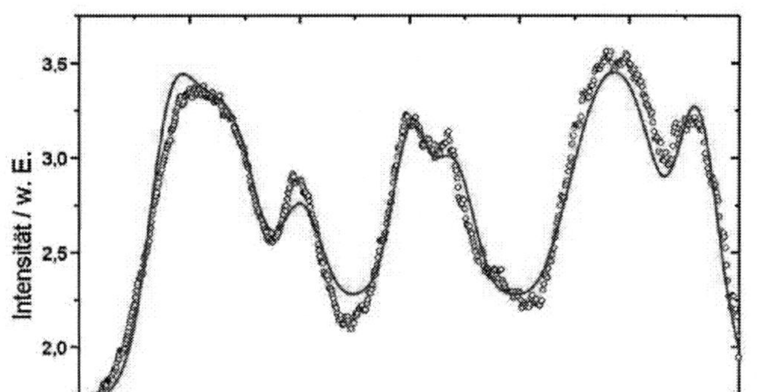

FIGURE 7. Doppler-broadened optogalvanic spectrum of the H_α line [2] recorded in π-polarization. The experimental results are shown together with a least squares fit of calculations including fine structure. From the fit an electric field strength of E = 6330 V/cm is derived with an uncertainty of less than 2%.

The remaining differences between the experimental and the theoretical profiles may have different reasons. We carefully checked the results of our calculations by comparing them with results from a computer code programmed in Meudon [8]. The results were identical and there was no indication whatsoever that an error in the computer code could be the cause for the observed differences. Another possible explanation would be that the neglect of hyperfine structure in our calculations could be the reason that the theoretical profiles show less structure in the line center. In a first step we tried to find an experimental way to give an answer to that question. In figure 8 the hyper-fine splitting of the n=2 levels in Hydrogen and Deuterium are shown. Due to the different spin of the nuclei the splitting is different for both elements. Therefore one would expect differences in the lines profiles taken at identical plasma conditions having their origin in the hyper-fine structure. However, a possible influence of the hyperfine structure is masked by the Doppler-effect that has by far a larger influence on the observed profiles than the hyperfine-structure splitting [9]. An example is given in figure 9.

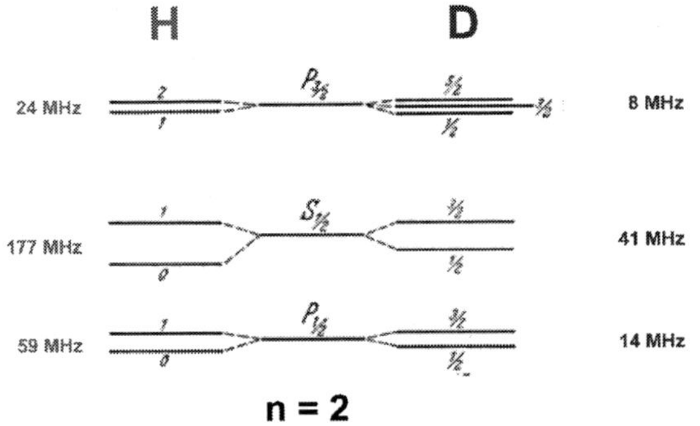

FIGURE 8. Scheme of the hyper-fine splitting of the n=2 levels of Hydrogen and Deuterium (data for Hydrogen from [13], data for Deuterium: s-level from [14], p-levels rough estimate according to [15]).

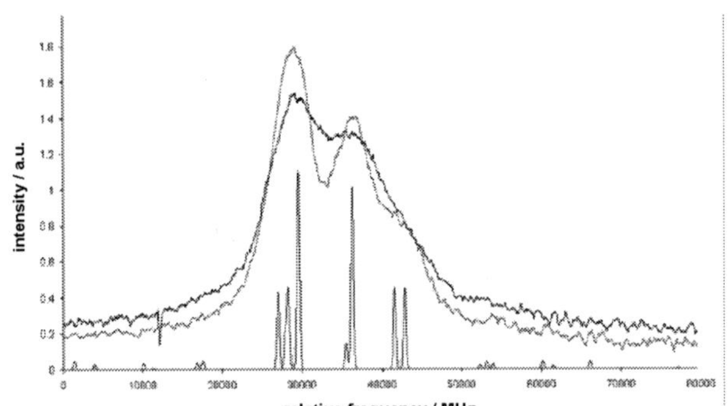

FIGURE 9. Doppler-broadened profiles of H- and D-alpha taken at the same plasma conditions. The Deuterium line shows more structure due to the smaller Doppler effect. The fine structure splitting is indicated at the bottom of the plot [7].

RESULTS FOR THE DOPPLER-FREE MEASUREMENTS

In order to get information about the hyperfine splitting Doppler-free spectra are needed. There are a number of spectroscopic techniques that yield Doppler-free spectra. They are discussed in some detail in [2]. It turned out that the technique suited for spectroscopy in the cathode fall region of the discharge was the intermodulated optogalvanic spectroscopy (IMOGS) first described by [10]. The principle is explained in figure 10. The laser beam is split into two counter propagating beams that probe the region of interest. Both the beams are chopped with different frequencies and

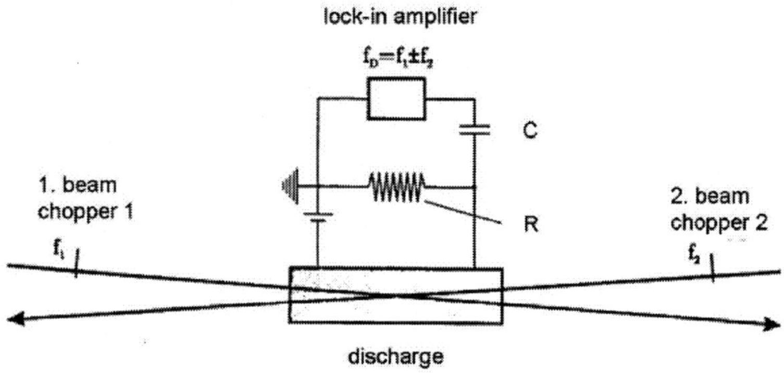

FIGURE 10. Schematic set-up for the Doppler-free measurements. The counter propagating beams are chopped with different frequencies and the signal is recorded on the sum- or the difference-frequency.

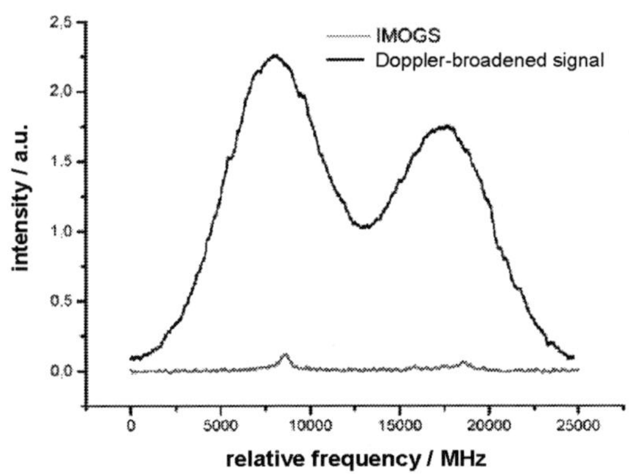

FIGURE 11. Doppler-limited- (upper trace) and Doppler-free- (lower trace) line profile of the Balmer alpha line from the positive column of a Hydrogen reference discharge at identical settings of the amplifier [11].

Doppler-free signal is detected across the resistor with a sensitive lock-in amplifier on the sum- or difference-frequency. The signal is rather weak. Preliminary tests in the positive column of a Hydrogen reference discharge showed, that the Doppler-free signal is only of the order of 1 % of the original optogalvanic signal. In figure 11 a comparison of a Doppler-limited and a Doppler-free signal taken at the same setting of the amplifier in the positive column of the reference discharge is shown [11]. The situation is even worse in the cathode fall because of the limited population density of the n=2 level. Therefore the spectra become quite noisy. This limits the accuracy of the present measurements. The signal to noise ratio depends critically on the laser power. Therefore attempts have been made to amplify the output radiation of the diode laser spectrometer with the help of an Argon-ion-laser pumped dye-jet. Though a small increase of the laser power has been observed [12] the amplification was not large enough to improve the quality of the signal sufficiently. Therefore we have to live for the moment with Doppler-free spectra as shown in figure 12 taken from [12] where Balmer alpha profiles from two different discharges are displayed. In the upper plot the profile from a discharge in Deuterium and in the lower plot a corresponding H-alpha line profile are shown. In both cases the distance of the laser beam from the

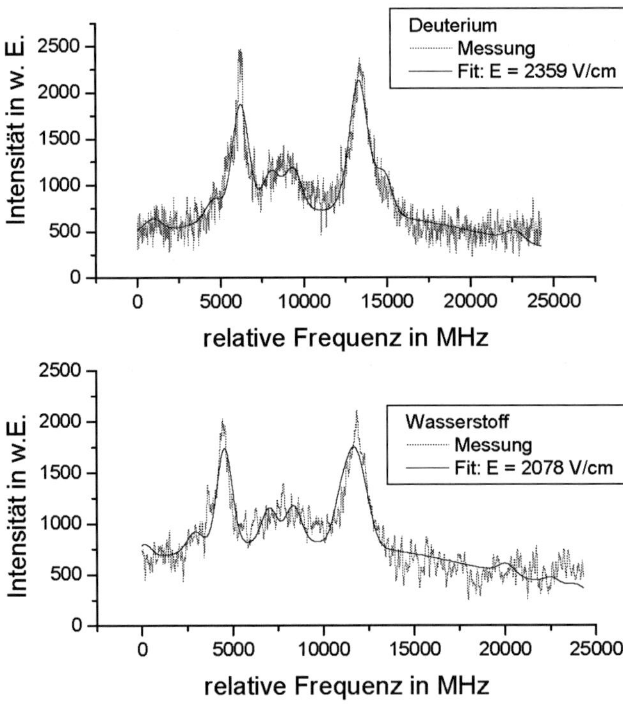

FIGURE 12. Doppler-free spectra of D-alpha (upper plot) and H-alpha (lower plot) together with a least squares fit to the theoretical profiles. The field strength obtained from the fit is indicated in the insert [12].

cathode was the same as were all other discharge parameters with exception of the gas filling. The experimental profiles are given together with the best fit from our fitting routine (solid line). The derived field strength values differ slightly because the discharge is run in different gases. Due to the poor signal to noise ratio only the position of strongest components can be determined with some trust. However, these components are fairly insensitive to the field as is shown in figure 13. The more sensitive components are those with low intensity and their position is difficult to determine as long as we don't have a better signal to noise ratio. In figure 14 the shift of the most sensitive component is shown.

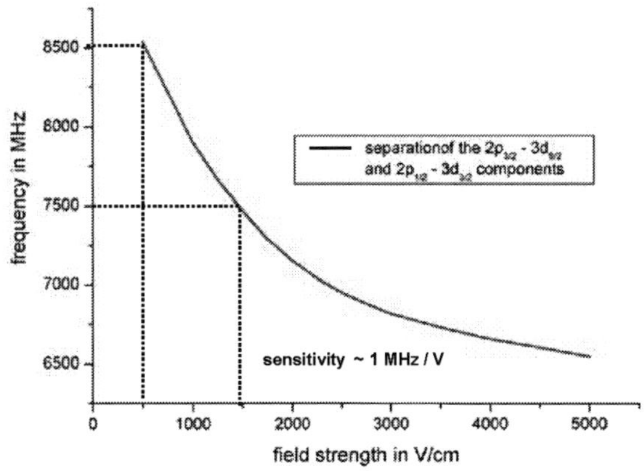

FIGURE 13. Difference of the position of the two strongest components plotted versus field strength. For the lower field strength values a sensitivity of about 1 MHz/V can be read from the figure [12].

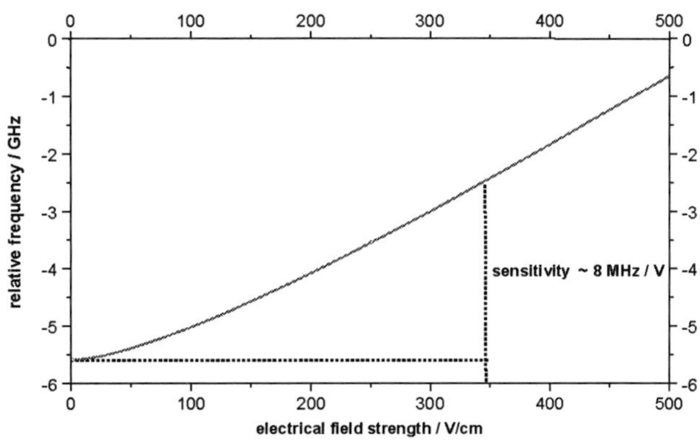

FIGURE 14. Shift of the 2 $s_{1/2}$ - 3 $p_{1/2}$ component plotted versus field strength. This component is the one that is the most sensitive of all fine structure components. For field strength values smaller than 350 V/cm it can be resolved from the other components and a sensitivity of roughly 8 MHz/V can be taken from the plot [12].

CONCLUSIONS

We have shown that Doppler-free measurements are desirable if one aims for the determination of low electric field strength values. We present first Doppler-free spectra from the cathode fall region of a glow discharge [12]. Even though the signal to noise ratio is not yet satisfying the spectra prove that the method works in principle and that the lower limit for the measurement of electric field strength by means of the Stark splitting of H_α finally may be pushed down from typically kV/cm in the Doppler-limited case to maybe several tenth of V/cm for the Doppler-free measurements if sufficient laser power is available.

REFERENCES

1. J. Neumann, Diploma thesis, Kiel University, 2000.
2. T. Rieper, Ph.D. thesis, Kiel University, 2000.
3. G. Lüders, *Ann. d. Physik* **6** (8), 301-321 (1951).
4. G. Lüders, *Z. Naturforschg.* **5a**, 608-611 (1950).
5. J.E.M. Goldsmith and J.E. Lawler, *Contemp. Physics* **22**, 235 (1981).
6. Z. Altug, Diploma thesis, Kiel University, 2001.
7. S. Herzog, Diploma thesis, Kiel University, 2003.
8. A. Demura, C. Stehle, Z. Altug, D. Esch, T. Rieper, and V. Helbig, in *Spectral Line Shapes*, Vol. 12, edited by C. Back, AIP Conference Proceedings V. 645, N.Y., 2002, pp. 79-85.
9. D. Esch, S. Herzog, Zhou Li and V. Helbig, Verhandl. DPG 4/2004, P 3.14, 2004.
10. J.E. Lawler and D.A. Doughty, *Phys. Rev. Lett.* **42**, 16 (1979).
11. M. Amin, Diploma thesis, Kiel University, 2004.
12. D. Esch, Ph.D. thesis, Kiel University, 2006.
13. S. Bashkin and J.O. Stoner, Atomic Energy Levels And Grotrian Diagrams, Vol. I, North-Holland Publishing Company, Amsterdam, 1975.
14. N. Kolachevsky, P. Fendel, S.G. Karshenboim, and T.W. Hänsch, *Phys. Rev.* A **70**, 062503, 2004.
15. H. A. Bethe and E. E. Salpeter, Quantum Mechanics Of One- And Two-Electron Atoms, Academic Press Inc., New York, 1957.

Stark Profiles of Spectral Lines Corresponding to the Transitions to Rydberg Levels of Argon: Possibilities for Diagnostics of Weak Electric Fields in Plasmas

V.P. Gavrilenko

Center for Surface and Vacuum Research, Federal Agency on Technical Regulating and Metrology, Novatorov street 40/1, Moscow 119421, Russia

Abstract. Quantitative description of the Stark effect on Rydberg levels of argon atoms is given. Theoretical dependences of the shift and splitting of both allowed and forbidden spectral lines of argon atoms on the magnitude of the electric field are obtained These dependences can be used for laser-aided measurements of weak electric fields in low temperature plasmas.

Keywords: Stark effect, Rydberg atoms, measurements of electric fields.
PACS: 52.70.-m, 52.70.Kz, 32.60.+i, 32.70.Jz.

INTRODUCTION

At present laser spectroscopy of Rydberg levels of argon atoms is widely used for diagnostics of electric fields (EFs) in plasmas (see, e. g., [1-6]). For EF measurements, two main schemes of excitation in argon atoms are used. A feature of the first scheme [1-3] is a single-photon excitation from the metastable level $4s[3/2]_2$ to Rydberg $n'f$ levels ($n' >> 1$). Within the second scheme [4-6], a two-step excitation from the $4s[3/2]_J$ level to Rydberg levels $n'd$ and $(n'+2)s[3/2]_J$ is used, and laser-induced fluorescence-dip spectroscopy [7] is employed to detect the Stark spectra (see Fig. 1). As an intermediate level, the np level is used, where $n = 4$ or 5. Experiments performed with the use of this scheme (see [4-6]) revealed a very complicated structure of Stark spectra of argon.

In the present work, we give a quantitative description of the Stark effect for argon atoms within the excitation scheme employed in Refs. [4-6]. The results, obtained in the present work, can be used for laser-aided measurements of weak electric fields in plasmas.

THEORETICAL CALCULATIONS AND RESULTS

The wave functions and the positions of energy levels of Rydberg states of argon atoms are determined from the following Schrödinger equation:

$$H\psi = \varepsilon\psi , \quad H = H_0 + ezF . \tag{1}$$

CP874, *Spectral Line Shapes: 18th International Conference,*
edited by E. Oks and M. Pindzola

Here H_0 is the unperturbed Hamiltonian of an argon atom, and ezF is the operator of the electric dipole interaction of the argon atom with the electric field $\vec{F}$. It is assumed that the z-axis is chosen parallel to the direction of the vector $\vec{F}$.

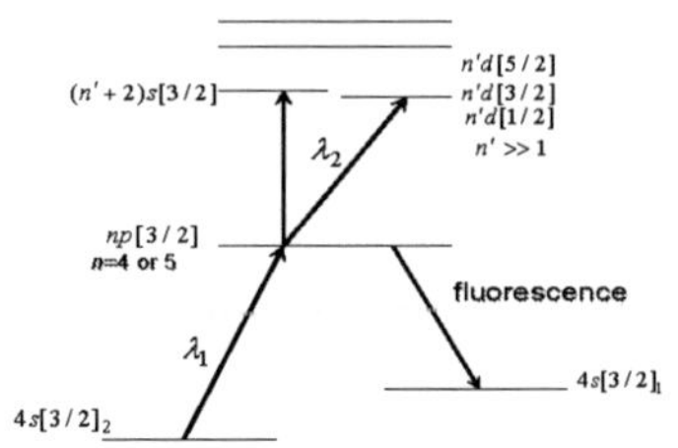

FIGURE 1. Partial scheme of energy levels of an argon atom. Two-step excitation scheme is used to populate Rydberg levels of argon. It is assumed that the Stark spectrum is recorded using laser-induced fluorescence-dip spectroscopy.

We solved Eq. (1) by performing numerical diagonalization of the matrix $H = H_0 + ezF$. We assume that the vector of polarization of the laser radiation having the wavelength λ_2 (cf. Fig. 1) is parallel to the vector $\vec{F}$. In our numerical code that calculates the Stark spectra corresponding to the excitation of the transitions to Rydberg levels n' of argon, the following Rydberg levels of argon are taken into account: $n'l$ $(l = 2,3,...,n'-1)$, $n'-1,l$ $(l = 2,3,...,n'-2)$, $n'+1,l$ $(l = 0,1,2,...,n')$, $(n'+2)s$, $(n'+2)p$, and $(n'+3)s$.

Theoretical Stark spectra of argon atoms calculated for the excitation to levels of $n' = 25$ are shown in Fig. 2. For rather weak electric field F=70 V/cm, the spectrum consists of three intense allowed lines corresponding to the transitions from the $np[3/2]$ level to the levels $25d[3/2]_2$, $25d[5/2]_J$ $(J = 2;3)$, $27s[3/2]_2$. Two weak forbidden lines corresponding to the transitions to the $26p$ and $27p$ levels can also be seen. For sufficiently strong electric field, the Stark spectra of argon consist of many components. The positions of most of these components depend strongly on the strength of the electric field.

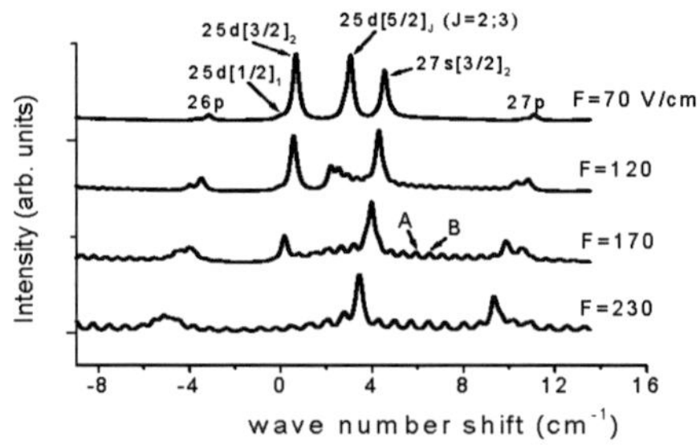

FIGURE 2. Theoretical Stark spectra of argon atoms for the excitation scheme shown in Fig. 1.

For measurements of moderate electric fields, it is convenient to use the separation between the closely spaced spectral lines that are indicated in Fig. 2 by letters **A** and **B.** When the electric field is weak and the separation between these lines cannot be measured accurately, the strength of the electric field can be deduced from the Stark shift of the allowed spectral line corresponding to the transition $np[3/2] \rightarrow 27s[3/2]_2$, as well as of the forbidden spectral lines corresponding to the transitions $np[3/2] \rightarrow 26p$ and $np[3/2] \rightarrow 27p$, where n=4 or 5. Figure 3 (curves 1-3) shows the theoretical dependence of the Stark shift of these lines on the electric field strength. The curve 4 in Fig. 3 shows the experimental dependence of the shift of the line $4p[3/2] \rightarrow 26p$ obtained in Ref. [4]. One can see that there is a reasonable agreement between the experimental and theoretical dependencies of the Stark shift of the line $4p[3/2] \rightarrow 26p$ on the electric field strength.

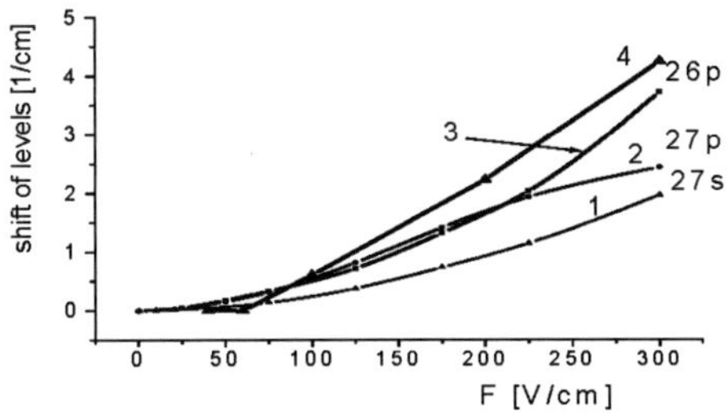

FIGURE 3. Stark shift of spectral lines corresponding to the transitions $np[3/2] \rightarrow 27s[3/2]_2$, $np[3/2] \rightarrow 27p$, and $np[3/2] \rightarrow 26p$, respectively, versus the magnitude of the electric field F. Curves 1 ($np[3/2] \rightarrow 27s[3/2]_2$), 2 ($np[3/2] \rightarrow 27p$) and 3 ($np[3/2] \rightarrow 26p$) show the theoretical dependences of the shift on F, n being either 4, or 5. Curve 4 shows the experimental shift of the transition $4p[3/2] \rightarrow 26p$ against the magnitude of the electric field F, obtained in Ref. [4].

ACKNOWLEDGMENTS

This work was partially supported by the Russian Foundation for Basic Research (project No 06-02-17011).

REFERENCES

1. J.B. Kim, K. Kawamura, Y.W. Choi, M.D. Bowden, K. Muraoka, and V. Helbig, *IEEE Trans. Plasma Sci.* **26,** 1556 (1998).
2. V.P. Gavrilenko, H.J. Kim, T. Ikutake, J.B. Kim, Y.W. Choi, M.D. Bowden, and K. Muraoka, *Phys. Rev. E* **62,** 7201 (2000).
3. V.P. Gavrilenko, H.J. Kim, T. Ikutake, J.B. Kim, M.D. Bowden, and K. Muraoka, *Phys. Rev. E* **63,** 047401 (2001).
4. K. Takizawa, K. Sasaki, and K. Kadota, *Jpn. J. Appl. Phys.* **41,** L1285 (2002).
5. K. Takizawa, K. Sasaki, and A. Kono, *Appl. Phys. Lett.* **84,** 185 (2004).
6. E.V. Barnat and G.A. Hebner, *Appl. Phys. Lett.* **85,** 3393 (2004).
7. U. Czarnetzki, D. Luggenhölscher, and H.F. Döbele, *Phys. Rev. Lett.* **81,** 4592 (1998).

FFM Applications to Dense and Warm Hydrogen Plasma Study

A. Calisti[1], L.A. Bureyeva[2], V.S. Lisitsa[3], D. Shuvaev[3] and B. Talin[1]

[1] *Laboratoire de Physique des Interactions Ioniques et Moléculaires, CNRS UMR6633, Université de Provence, centre St Jérôme, case 232, 13397 Marseille cedex 20.*
[2] *Institute of Spectroscopy, Troitsk, Moscow region, 142190, Russia*
[3] *Russian Research Center "Kurchatov Institute", Moscow, 123182, Russia*

Abstract. A study of hydrogen lines emitted in dense and low temperature plasmas is presented. A transition from impact to quasi-static broadening for electrons is analyzed with the help of the Frequency Fluctuation Model (FFM). Electron broadening of Balmer series lines is studied for different densities and temperatures spanning a wide domain from impact to quasi-static limit. It is shown that electronic broadening makes a transition from impact to quasi-static limit depending on plasma conditions and principal quantum number. Even for the Balmer alpha line, at a density equals 10^{18} cm^{-3} and a temperature equals 1 eV, this transition occurs both in the wings and the core of the line.

Keywords: Plasma, Energy Level Populations.
PACS: 03.65.Sq, 02.60.Cb

CALCULATION METHOD

The starting points of calculations are static profiles for electron and ion broadening given by the equation[1]:

$$J(\omega) = \sum_{i,f} \int dFW(F) J_{i,f}(\omega, F) j_i(F), \tag{1}$$

where $J_{i,f}(\omega, F)$ is the line shape at the fixed value of electric field F, $W(F)$ is static electric field distribution function, $j(F)$ is a dilution factor due to the field ionization effect of atomic states given by the equation:

$$j_i(F) = A_{i,f} / [A_{i,f} + \Gamma_i(F)], \tag{2}$$

where $A_{i,f}$ are radiative decay rates between initial (i) and finite (f) atomic energy states, $\Gamma_i(F)$ are ionization decay rates in plasma electric field F, given by the equation (see [2] and references there)

$$\Gamma_i = \left(\frac{4}{Fn^3}\right)^{2n_2+m+1} \exp[3(n_1 - n_2) - \frac{2}{3Fn^3}] / n^3 n_2!(n_2 + m)!, \tag{3}$$

where n, n_1, n_2, m are principle and parabolic quantum numbers of the upper level. The sum in eq.(1) runs over quantum numbers of initial and finite atomic states. It can be connected with a sum over emitted frequencies with account of the relationship

CP874, *Spectral Line Shapes: 18th International Conference,*
edited by E. Oks and M. Pindzola

$$\omega_{i,f} = \frac{3}{2} F[n(n_1 - n_2) - n'(n_1' = n_2')] \qquad (4)$$

Here *n, n'* corresponds to the upper and lower levels.

The total (normalized to the total radiative intensity) intensities $I_{i,f}(\omega)$ of radiative transitions follow from eq.(1) by its multiplication by corresponding inverse radiative transition probabilities.

The static electric field distribution function in eq.(1) is obtained by molecular dynamics (MD) method, see [3,4]. The FFM shape is obtained by an introduction of electric field jumping frequency $\nu = N^{1/3} v$ (N, v being the electric charges density and particle thermal velocity) resulting in intensities exchange between different spectral domains of static line shape.

The transition from impact to static broadening is determined by the ratio g of typical static Stark energy levels splitting $CN^{2/3}$ to the jumping frequency $\nu = N^{1/3} v$. Substituting the typical value of Stark constant $C \propto n^2$ one arrives to the criteria

$$g = N(n^2 \hbar / Mv)^3, \qquad (5)$$

where $\hbar$ is Plank constant, M is a mass of plasma particles (ions or electrons). When $g >> 1$ the dominant broadening mechanism is static whereas for $g << 1$ the broadening mechanism is impact one.

The FFM takes into account a smooth transition from static to impact line shapes with account of the intermediate case $g \cong 1$. For dense plasma conditions considered below the value of the parameter g is typically large for ion broadening whereas it changes from small to large values for electron broadening. The estimations of a specific transition value of the parameter g (5) can be obtained from direct comparison of the static and impact values of line widths.

The resulting spectral line shape taking into account both ion's and electrons effects can be obtained by a direct convolution of FFM ion's and electron's line shapes.

RESULTS AND DISCUSSION

The FFM calculations of initial members of hydrogen Balmer series were carried out for plasma parameters: densities $N = 10^{18}$ - 10^{19} cm^{-3} and temperatures $T = 1$ eV. Fig.1 presents H_α and H_β line shapes calculated in the frame of standard impact approximation and FFM.

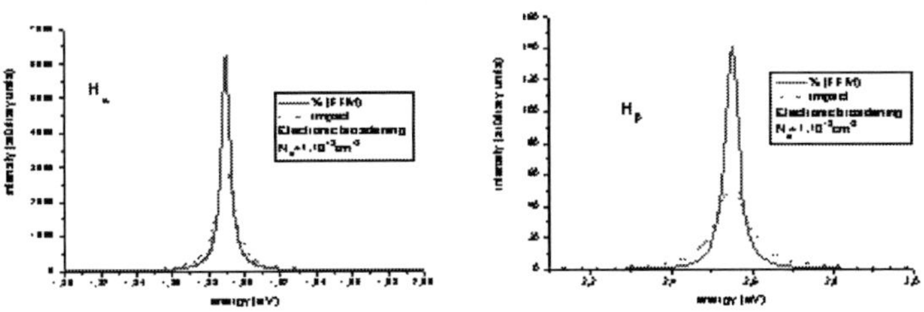

FIGURE 1. H_α and H_β line profiles with impact (dotted) and FFM (solid) electron broadening ($N_e = 10^{18} cm^{-3}$, $T_e = 1$ eV).

One can see an essential change of the shapes widths due to the transition of electron broadening from impact to quasistatic regime. It was studied the dependence of spectral line electron's widths for Balmer series as function of principle quantum numbers n for plasma densities 10^{18} and $10^{19}\ cm^{-3}$. One can see that the dependence follows much smaller power low (closer to static dependence n^2) as compared with impact theory (n^4). In order to compare the correct transition of FFM theory to the impact results for high temperatures the comparison of widths scalings in the frame FFM and impact theories was made for $T=80\ eV$ and low density $10^{16} cm^{-3}$. It was seen a good correspondence between both theories in this case.

The comparison of experimental data with three types of theory, namely impact, molecular dynamic simulation and FFM was realized for H_α line as a function of temperature. One can see a good correspondence between FFM and molecular dynamics. Note that the case under consideration is the most worse results in FFM theory.

The effect of ionization (dilution factor eqs. (2,3)) on spectral line intensity distribution is presented in Fig. 2 Balmer lines H_α, H_γ lines at plasma densities 10^{18}, $10^{19}\ cm^{-3}$.

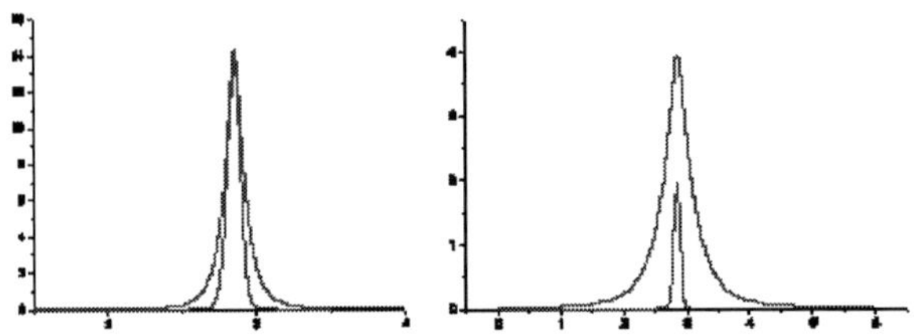

FIGURE 2. The effect of dilution factor on H_γ line profile at electron density 10^{18} cm^{-3} (left) and 10^{19} cm^{-3} (right) and temperature 1 eV

The strong effect on spectral line profiles is especially evident for large plasma densities. It is clear however that higher lines can disappeared in this case under large continuum radiation. It is an interesting point in $H\alpha$ line profile that the ionization effect results in an increase of the line intensity near its center. The effect can be understood taking into account different dilution effect for central (unshifted) and shifted components of the line.

The general conclusion from the consideration above is that the transition of electron broadening from impact to static limit can change the spectral line shapes dramatically for dense plasma conditions. Two effects are responsible for the change: modification of electron broadening and plasma electric field ionization effect. Both effects can change the problem of interrelation between discrete and continuum spectra in dense plasmas.

ACKNOWLEDGMENTS

The work is supported by project INTAS 03-546348 and RFBR 06-02-16614.

REFERENCES

1. H.R. Griem, *Spectral line shapes by plasmas*, N.Y.: Mc Grow Hill, 1982.
2. L.A.Bureyeva, V.S. Lisitsa, *A Perturbed Atom,* Gordon and Breach, 2002.
3. B. Talin, A. Calisti and J.W. Dufty, *Phys. Rev.* **E 65**, 056406 (2002).
4. E. Dufour, A. Calisti, B. Talin, et al., *Phys. Rev.* **E 71**, 066409 (2005).

Line Shapes and Emission Properties in Colliding Plasmas

F.B. Rosmej[1], R. Schott[2], E. Dalimier[2], D. Riley[3], A. Delserieys[3], O. Renner[4], E. Krousky[4], V.S. Lisitsa[5]

[1] *PIIM, Université de Provence, UMR 6633, Centre de St. Jérôme, 13397 Marseille Cedex 20, France*
[2] *Université de Paris VI et LULI, Paris et Palaiseau, France*
[3] *Queens University, Belfast, BT7 1NN, UK*
[4] *Institute of Physics, Prague, Czech Republic*
[5] *Russian Research Center Kurchatov, 123182 Moscow, Russia*

Abstract. Single (Al) and double (Al/Mg) foils have been irradiated with high energy laser pulses to study the radiative properties during the interaction of highly ionized plasma jets with gases and solids. X-ray microscopy with two-dimensional curved Bragg crystals was shown to provide a detailed topological structure with 15 μm resolution and a 5 cm large field of view. The realization of simultaneously high spectral resolution enabled to identify not only resonance and intercombination lines but even single satellite transitions. Gas pressure dependent emission features for resolved line transitions inside the jets of H-, He- and Li-like aluminum ions was discovered when the jet moves into the gas. In the case of two-foils, accumulation features near the non-irradiated target have been observed. At large distances from irradiated single foils, anomalous high (3 orders of magnitude) intensities of Li-like intercombination transitions from double excited states (satellite lines) have been identified in the front and rear side plasma. Charge exchange driven cascading in autoionizing states is proposed to explain the experimental findings.

Keywords: laser produced plasma jets, colliding plasmas, x-ray spectroscopy, dielectronic satellites, intercombination transitions, charge exchange.
PACS: 32.70.Fw, 32.30.Rj, 34.70.+e

INTRODUCTION

Interpenetration of counter-propagating plasmas is of fundamental interest in plasmas physics as it occurs widely in astrophysics and in modeling of various astrophysical situations using intense lasers [1-4], inertial confinement fusion research [5], x-ray laser research [6], magnetic fusion [7, 8] and pinching plasmas [9]. The prediction of the shock formation and the temperature evolution in plasmas is difficult, thus several spectroscopic experiments have been designed to obtain simulation independent information about plasma conditions [10-12]. As plasma collisions are three-dimensional and include an interaction of at least two different materials (see Figure 1), the numerical simulations are extremely difficult. The fluid simulations have to resolve and to follow the contact surface of the interacting plasmas and their eventual mixing. The kinetic and particle-in-cell calculations have to include the

CP874, *Spectral Line Shapes: 18th International Conference,*
edited by E. Oks and M. Pindzola

particle collisions and to cover large temporal and spatial scales. For these reasons the development of new methods capable to independently characterize the topology of colliding plasmas are urgently requested.

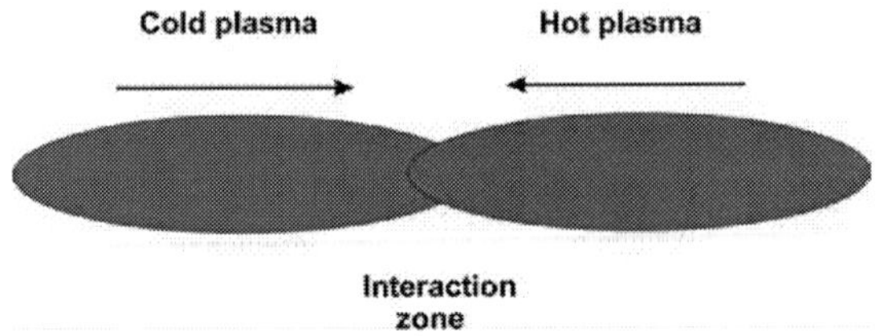

FIGURE 1. Counter propagating plasmas

In the present work high resolution (spatial and spectral) x-ray microscopy of plasma jets realized with two-dimensional curved Bragg crystals to study the collision processes of gases and plasmas. The simultaneously achieved high spectral and spatial resolution enabled us for the first time to identify resonance and intercombination line transitions from single and doubly excited states, characterize their novel diagnostic potential and record a spectrally resolved topological structure of counter-propagating plasmas.

EXPERIMENTAL SETUP

The experiment has been carried out at the PALS kilojoule laser facility [13] in the framework of the LASERLAB access program [14]. The schematic experimental setup is depicted in Figure 2. Aluminium foils with 0.8 μm thickness have been irradiated at normal incidence with the iodine PALS laser at 3ω (λ = 0.438 nm) using laser intensities between 10^{14} and 10^{16} W/cm^2. The focal spot diameter was about 80 μm and the delivered laser energy varied between 5 and 250 J. The laser pulse length was 0.25-0.3 ns. Target gas pressures (air) have been varied between 10^{-4} and 1 mbar.

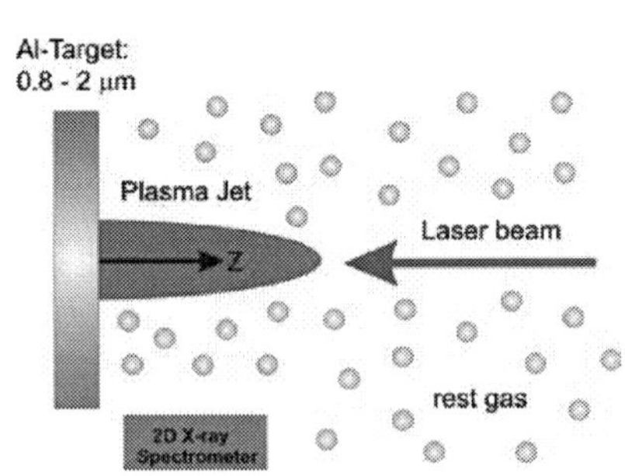

FIGURE 2. Schematic experimental setup

High resolution x-ray microscopy has been realized with a spherical bent mica crystal [15]. The distance between target and crystal was 33 cm with a central Bragg angle of 56.1° (central wavelengths λ_0 = 5.515 Å). The x-ray film (Kodak Industrex CX) was covered with two 1 μm thick mylar foils covered on each side with 0.1 μm Al to protect the film from the visible light. The crystal was covered with a 1 μm thick polypropylene foil to avoid debris damage. X-ray films have been processed with a drum scanner (EUROCORE, 10.000 dpi) with 3000 dpi providing a spatial resolution of 22.4 μm per scanned pixel. The optical thickness of the x-ray film D is related to the scanner electrical values I_{16bit} according $D = 5.0508 \cdot 10^{-5} \cdot I_{16bits} - 0.21213$, the relation to intensity values I is according $D = \log(I/I_0)$, I_0 is a threshold intensity.

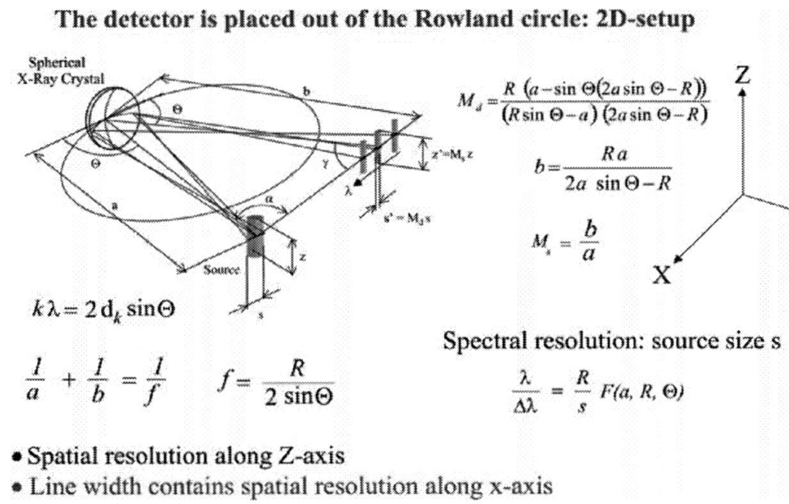

FIGURE 3. High resolution x-ray imaging with 2-dimensional curved Bragg crystals. Placing the detector out of the Rowland circle, spatial resolution is achieved along the Z-axis and also along the x-axis (wavelengths axis).

Spatial resolution is along the Z-direction (target normal in Figure 2). If the detector is placed out of the Rowland circle, Figure 3, additional spatial information along the X-axis is obtained. In this case the X-coordinate coincides with the wavelength-coordinate and leads to a decrease of the spectral resolution $\lambda/\Delta\lambda$. However, the defocusing can be adjusted in a manner to keep still high spectral resolution to resolve single line transitions while containing information about another spatial coordinate (e.g., radial). This additional coordinate can provide, e.g., information about the angular distributions of plasma jets [16].

The crystal was centered to the target surface to provide x-ray imaging simultaneously from the front and rear side of the irradiated target, Figure 4a. A demagnification factor of $M_S = 2.65$ allowed a large geometrical window of 1-2 cm on either side of the target.

This particular arrangement permitted us to observe, simultaneously, the Al H-like Lyman-series np $^2P_{1/2,3/2}$ – 1s $^2S_{1/2}$ in the 3rd order of reflection and the Al He-like resonance line W = 1s2p $^1P_1$ – 1s^2 $^1S_0$, intercombination line Y = 1s2p $^3P_1$ – 1s^2 $^1S_0$ with its associated Li-like satellite transitions 1s2l2l' – 1s^{2}2l'' in 2nd order. Figure 4 b shows, that a detailed topological high resolution x-ray image is obtained which shows at the same time a high spectral resolution (e.g., the Lyman series shows distinct transitions up to principal quantum numbers of about 13).

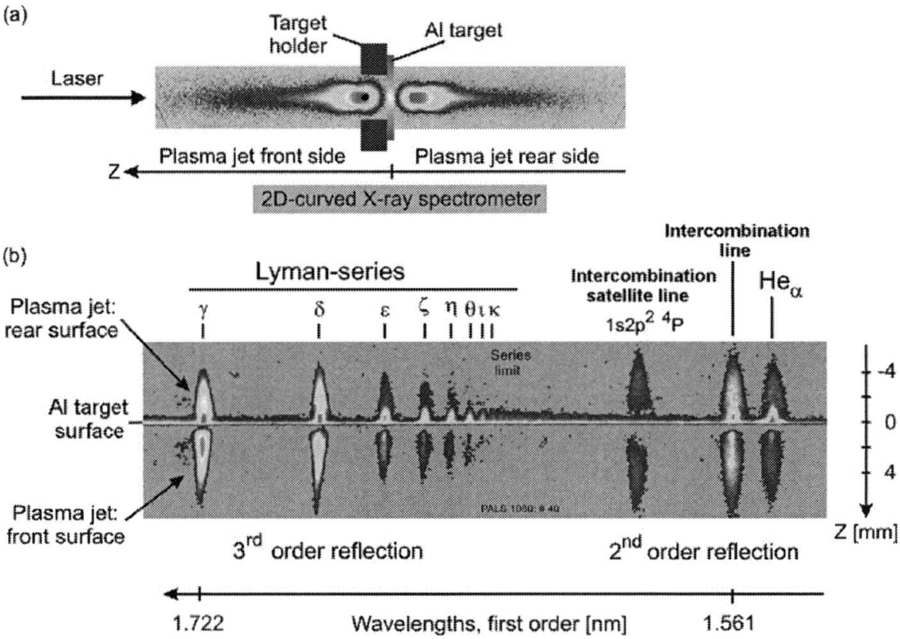

FIGURE 4. a) Scheme of the experimental setup, b) high resolution x-ray microscopy of the spectrally resolved plasma jets of the rear and front side interacting with air (p = 0.19 mbar). Distinct emission structures are visible which show large differences between the various spectrally resolved line transitions.

INTERACTION OF PLASMA JETS WITH GASES

Anomalous intensities (3 orders) of dielectronic satellite lines

Figure 4b shows the spectrally resolved x-ray image of Al emission taken with the gas pressure of p = 0.19 mbar in the interaction chamber. The laser energy of 23 J provided an averaged intensity of about $2 \cdot 10^{15}$ W/cm^2 on the target. The image shows strong front side (positive Z-direction) and also rear side emission (negative Z-direction) which indicates an effective burn through. The figure shows the jets visualizing the emission of the Al H-like Lyman-series: Ly_γ until the continuum (taken in 3rd order of Bragg reflection), the He-like resonance line He_α and the He-like intercombination line (taken in 2nd order of Bragg reflection). At the left hand side of the He-like intercombination line, a very bright and spectrally rather broad emission structure is visible. Its experimentally determined wavelength is 7.926 Å (using the He-like resonance and intercombination lines as reference lines). Atomic structure calculations carried out with the MZ-code [17] show that no strong Li-like satellite transitions are located at this wavelengths. However, several extremely weak intercombination satellite transitions do have transitions near 7.926 Å. Table I. depicts the detailed transitions and corresponding atomic data.

Table I. Atomic data (spontaneous transition probability A, autoionizing rate Γ, wavelength λ and satellite intensity factor Q) of the He-like resonance line $1s2p\ ^1P_1 - 1s^2\ ^1S_0$, intercombination line $1s2p\ ^3P_1 - 1s^2\ ^1S_0$ and intercombination satellite lines $1s2p(^3P)2p\ ^4P - 1s^22p\ ^2P$ of Li-like Al ions. For reference, the strongest Li-like satellite transition $1s2p^2(^1D)\ ^2D_{5/2} - 1s^22p\ ^2P_{3/2}$ is also listed.

transition	A [s^{-1}]	Γ [s^{-1}]	λ [10^{-10} m]	Q [s^{-1}]
$W = He_\alpha = 1s2p\ ^1P_1 - 1s^2\ ^1S_0$	$2.78\cdot10^{13}$	-	7.7573	-
$Y = 1s2p\ ^3P_1 - 1s^2\ ^1S_0$	$7.62\cdot10^{10}$	-	7.8070	-
$1s2p(^3P)2p\ ^4P_{3/2} - 1s^22p\ ^2P_{1/2}$	$1.23\cdot10^{8}$	$2.54\cdot10^{10}$	7.9247	$3.00\cdot10^{8}$
$1s2p(^3P)2p\ ^4P_{1/2} - 1s^22p\ ^2P_{1/2}$	$1.17\cdot10^{10}$	$1.80\cdot10^{9}$	7.9263	$3.04\cdot10^{9}$
$1s2p(^3P)2p\ ^4P_{5/2} - 1s^22p\ ^2P_{3/2}$	$2.61\cdot10^{10}$	$3.80\cdot10^{11}$	7.9265	$1.47\cdot10^{11}$
$1s2p(^3P)2p\ ^4P_{3/2} - 1s^22p\ ^2P_{3/2}$	$1.61\cdot10^{10}$	$2.54\cdot10^{10}$	7.9283	$3.93\cdot10^{10}$
$j = 1s2p^2(^1D)\ ^2D_{5/2} - 1s^22p\ ^2P_{3/2}$	$1.31\cdot10^{13}$	$1.86\cdot10^{14}$	7.8764	$7.32\cdot10^{13}$

The satellite intensity factor Q is defined as

$$Q_{ji,k} = \frac{g_j A_{ji} \Gamma_{j,k}}{\sum_i A_{ji} + \sum_k \Gamma_{j,k}} \tag{1}$$

where A_{ji} is the spontaneous transition probability for a bound-bound transition $j \rightarrow i$, g_j is the statistical weight and $\Gamma_{j,k}$ is the autoionising rate of level j into channel k. The dielectronic recombination rate for a satellite line which originates from dielectronic capture is given by

$$\langle DR \rangle_{ji,k} = 1.656 \cdot 10^{-22} \frac{Q_{ji,k}}{g_k} \frac{\exp(-E_s / kT_e)}{(kT_e)^{3/2}} \quad [cm^3 s^{-1}] \tag{2}$$

g_k is the statistical weight of the level before capture, kT_e is the electron temperature [eV] and E_s is the dielectronic capture energy [eV]. As can be seen from Table I. , the satellite intensity of the intercombination satellite transitions is 2-3 orders of magnitude smaller than that of the strongest satellite transition j. This is in agreement with the experimental spectra observed near the target surface (Figure 5a), however, at larger target distances the intensity rise to the level of the resonance line (Figure 5b) is in strong contradiction to the Q-factors.

In order to explore this discrepancy we have performed experiments at different gas pressures. Figures 5a,b show spectra obtained at p=0.19 mbar whereas Figures 5c,d have been obtained at $p=1.5\cdot10^{-4}$ mbar; note, that no smoothing procedure has been applied in the wavelength direction, i.e., the horizontal resolution equals 1 px (corresponding to $\delta\lambda$ = 0.151 mÅ). Near the target surface, the vertical or spatial resolution is 2 px (=45 μm), at large target distances 15 px (=337 μm) was chosen. For low target pressures (Figures 5c,d) no emission at λ = 7.926Å is visible, however, its intensity at large target distances gradually rises with higher pressure p (Figure 5b).

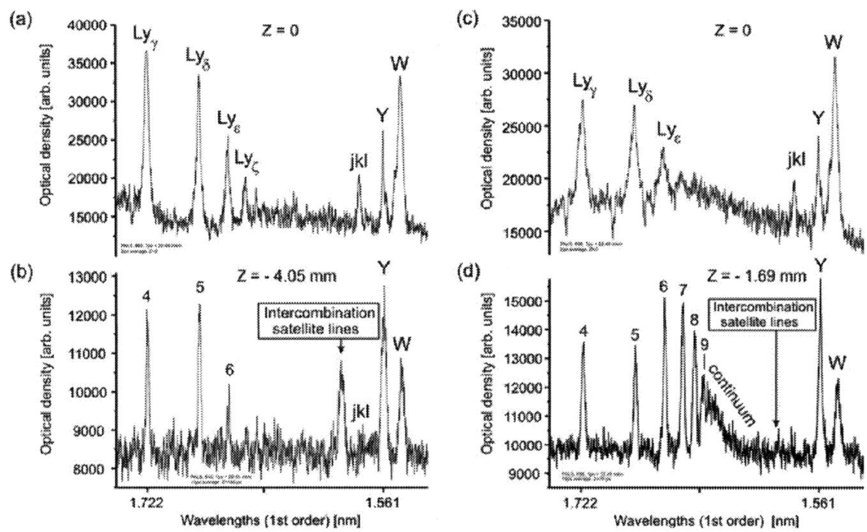

FIGURE 5. Single shot space resolved x-ray spectra for different target chamber gas pressures p and target distances Z: a) $p = 1.9\cdot10^{-1}$ mbar, Z = 0 mm, b) $p = 1.9\cdot10^{-1}$ mbar, Z = -4.05 mm, c) $p = 1.5\cdot10^{-4}$ mbar, Z = 0 mm, d) $p = 1.5\cdot10^{-4}$ mbar, Z=-1.69 mm.

Figure 6 explores the detailed spatial dependence of various emission lines: He-like intercombination line (indicated as $\Delta S=1$ line, Figure 6a,b), Li-like intercombination satellite lines (indicated as $\Delta S=1$ satellite line, Figure 6d) and the H-like Ly_γ (Figure 6e) and Ly_ε (Figure 6f). Three important observations are made: first, at high pressures, the $\Delta S=1$ line shows additional emission maxima A and E (Figure 6a), second, these emission maxima are spatially correlated with the emission maxima of the $\Delta S=1$ satellite line (Figure 6c,d) and, third, Ly_γ shows similar emission maxima at high pressures (Figure 6e). For Ly_ε, however, these maxima are absent (Figure 6f). The feature indicated as "C" corresponds to the Bremsstrahlung emission at the target surface and the maximum indicated as feature "B" is due to the hot dense plasma emission near the target surface. It is important to note that the feature "B" shows up for all observed lines except the Li-like intercombination satellite transitions, their emission is absent up to almost 1 mm target distance (see Figure 6d).

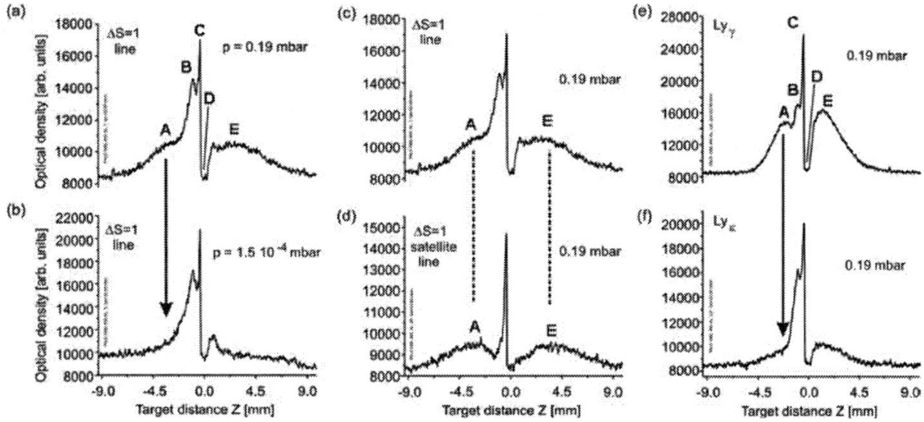

FIGURE 6. Spatial intensity variation for different lines and gas pressures: a) and c) ΔS=1-line, p=1.9·10^{-1} mbar, b) ΔS=1-line, p=1.5·10^{-4} mbar, d) ΔS=1-satellite lines, p=1.9·10^{-1} mbar, e) H-like Ly_γ and Ly_ε, p=1.9·10^{-1} mbar. The intensity minimum indicated as feature "D" is an artefact due to the blend of the line of sight from the target holder.

Charge exchange driven radiative properties of colliding plasmas

Charge exchange driven cascading in the autoionizing quadruplet system 1s2p[^{3}P]nl ^{4}L is proposed to explain the experimental findings. The principle mechanism is schematically explained in Figure 7. We start with Figure 7a considering cascading processes in the He-like singlet and triplet system. The process starts from the H-like 1s $^2S_{1/2}$-state. The decay cannot effectively be transformed via photon emission in the triplet system because the transition probabilities of the intercombination transitions 1snp $^3P_1$ - 1s^2 $^1S_0$ is very small. All population received by the states 1snl ^{3}L cascades, therefore, down to the states 1s2l ^{3}L and then, having no other possibility, decays via the Y-line. At low gas pressures, the relevant atomic process is the well known radiative recombination which manifests itself in a spatially decaying wing (for large target distances, Figure 6b). Figures 6e,f show, that the population channel at large target distances is selective with respect to the main quantum number n. This is a typical feature for charge exchange processes and denies the generally accepted radiative recombination interpretation. The x-ray emission spectra depicted in Figure 5a,b support this consideration. At large target distances, the emission from the n=5 line is stronger than those from the n=4 line (Figure 5b), whereas near the target surface, a continuous decrease of the line intensity with increasing quantum number n is observed (Figure 5a). Near the target surface, the hot plasma drives the usual collisional-radiative processes leading to a continuous decrease of the intensity with increasing quantum numbers n and to the emission maximum B (Figure 6). At large target distances, however, the electron temperature is too low to provide effective collisional excitation and radiative recombination and charge exchange recombination are the main excitation channels. The comparison between Figures 6a,b demonstrates that charge exchange recombination driven cascading in the triplet system is the most effective mechanism. The simple picture of the cascading processes (Figure 7) is in agreement with detailed collisional-radiative simulations [18] which have shown that charge exchange initiated cascading drives a strong emission of the intercombination and forbidden components from the 1s2l ^3L states (note that in high density plasma the forbidden lines 1s2p $^3P_2$ - 1s^2 $^1S_0$ and 1s2s $^3S_1$ - 1s^2 $^1S_0$ are not observable for light elements).

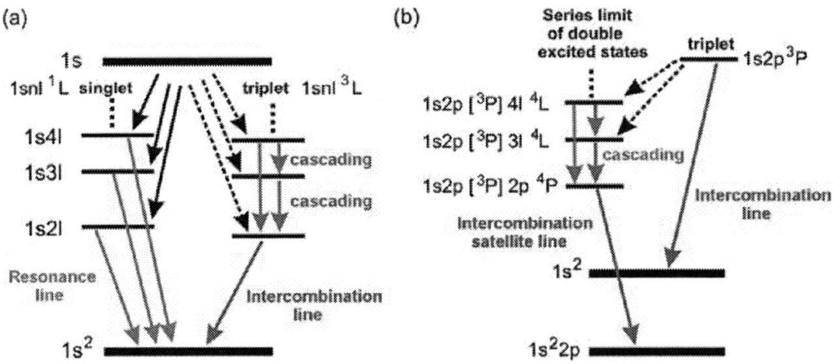

FIGURE 7. Cascading in single and double excited levels: a) He-like singlet and triplet levels 1snl ^{3}L, b) Li-like double excited quadruplet levels 1s2lnl' ^{4}L.

We now suggest a charge exchange driven chain reaction of the type

1. 1s + N_2 → 1s2l ^{3}L + N_2^+ and 1s2l ^{3}L → 1s^2 $^1S_0$ + (ΔS=1-line)

2. 1s2p ^{3}P + N_2 → 1s2p[^{3}P]nl ^{4}L + N_2^+ and 1s2p[^{3}P]nl ^{4}L → 1s2p[^{3}P]2p ^{4}P and 1s2p[^{3}P]2p ^{4}P → 1s^{2}2p ^{2}P + (ΔS=1-satellite lines)

The chain reaction "2" can only be effective, if the charge exchange recombination rate <Cx> is faster or at least of the same order than the radiative decay rate of the intercombination line:

$$\langle Cx\rangle \approx n_{gas} V_{Jet} \sigma_{Cx} \geq \overline{A(1s2l\ ^3L)} \approx 2\cdot 10^{10}\ s^{-1} \qquad (3)$$

Taking $n_{Gas} = 10^{16}$ cm^{-3}, $V_{Jet} = 3\cdot 10^7$ cm/s and $\sigma_{Cx} = 10^{-14}$-10^{-13} cm^2 we arrive at the conclusion that for the highest gas pressures of the present experiments charge exchange population rates are of the order of the radiative decay of the triplet 1s2l ^{3}L - levels. Therefore, the population channel via the triplet exited states is as effective as the ground state channel (1s^2 $^1S_0$ + N_2). Cascading in the doubly excited states (as depicted in Figure 7b) is therefore effective, resulting in an anomalously high population of the 1s2p[^{3}P]2p ^{4}P states and the orders of magnitude enhanced intensity of the corresponding intercombination satellite transitions. This is in agreement with the observations that only for gas pressures higher than about 10^{-2} mbar the intercombination satellite transitions are visible. For lower gas pressures, the relation (3) is highly invalid. The chain reactions 1. and 2. also permit us to understand the spatial correlation of the emission maxima "A" and "E" of the He-like Y-line and the Li-like intercombination satellites (see dotted line in Figures 6c,d). The chain reactions 1. and 2. also explain why the resonance line does not show the emission maxima "A" and "E": the relevant radiative decay rate is about $2 \cdot 10^{13}$ s^{-1}. The singlet system therefore does not effectively take part in a chain reaction because the 1s2l ^1L-states decay before charge exchange recombination takes place (note, that due to electron collisions in dense plasmas the metastable level 1s2s $^1S_0$ is not isolated to act with its low two-photon decay probability). In a similar fashion, it can be understood why radiative recombination is not of importance: the relevant rates are several orders of magnitude smaller than for the charge exchange process. The transition probability of the Y-line therefore depopulates the 1s2l ^3L states before a further population process can take place. Therefore, in a purely radiative recombination regime, the Li-like intercombination satellite transitions are invisible, in agreement with the experimental observations.

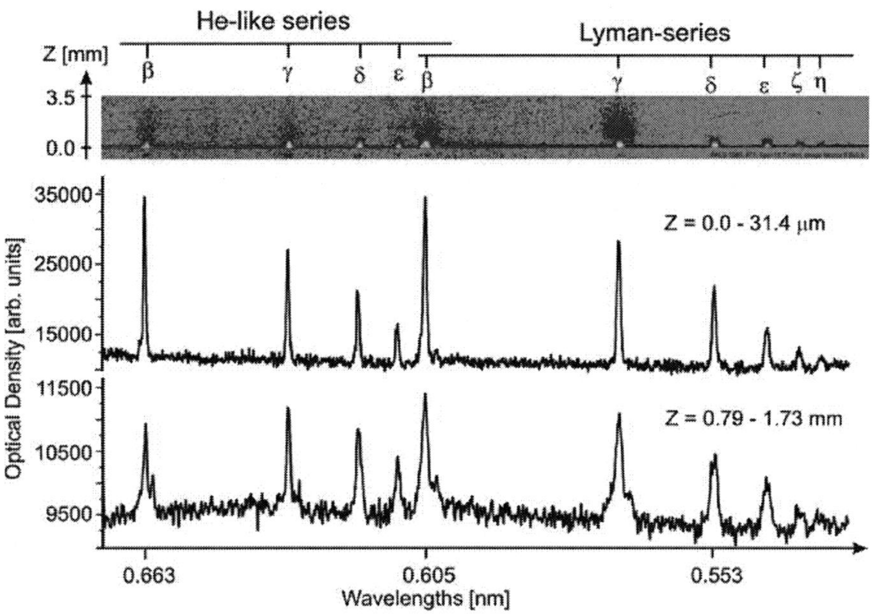

Figure 8. Spectrally resolved x-ray image (upper figure) and spatially resolved emission lines (middle and lower curves) from H- and He-like aluminum series (β-lines until the continuum) for different target distances: near the target surface, Z = 0 mm and at far target distance, Z = 0.79 – 1.73 mm. A strong emission at far target distances is visible for the Ly_γ-line (and also the He_γ-line) indicating a selective population process (charge exchange). The pressure in the target chamber was p = 0.75 mbar.

A further confirmation of the proposed charge exchange mechanism is given by the consideration of the principle quantum number n_f for the state selective capture process. According classical estimates [19]

$$n_f \approx \sqrt{\frac{Ry}{I}}\, Z \sqrt{\frac{2\sqrt{Z}+1}{Z+2\sqrt{Z}}} \tag{4}$$

For Z = 13 and $I(N_2 \rightarrow N_2^+)$ = 15.3 eV we obtain $n_f \approx 8$ whereas the spectrum in Figure 5b shows a maximum near Ly_δ, i.e. at $n_f \approx 5$. At high relative velocities, higher l-quantum numbers are populated, however, e.g., the states 8f, 8g cannot directly decay via the 1s state. They need to cascade down to a p-state (e.g., 8f→7d→6p, 8g→7f→6d→5p) until they can decay via an x-ray transition and the observable principal quantum number in the x-ray transitions n_{fx} is shifted to lower values by about δn=1-3. Therefore, the observation of an intensity maximum of $Ly_{\delta,\varepsilon}$ is in agreement with the estimations according to eq. (3). Alternatively, charge exchange may proceed from N_2+ which may be produced by collisional processes. As the corresponding ionization potential is larger, lower principal quantum numbers are populated: $I(N_2^+ \rightarrow 2N^+)$ = 23.5 eV giving $n_f \approx 6$.

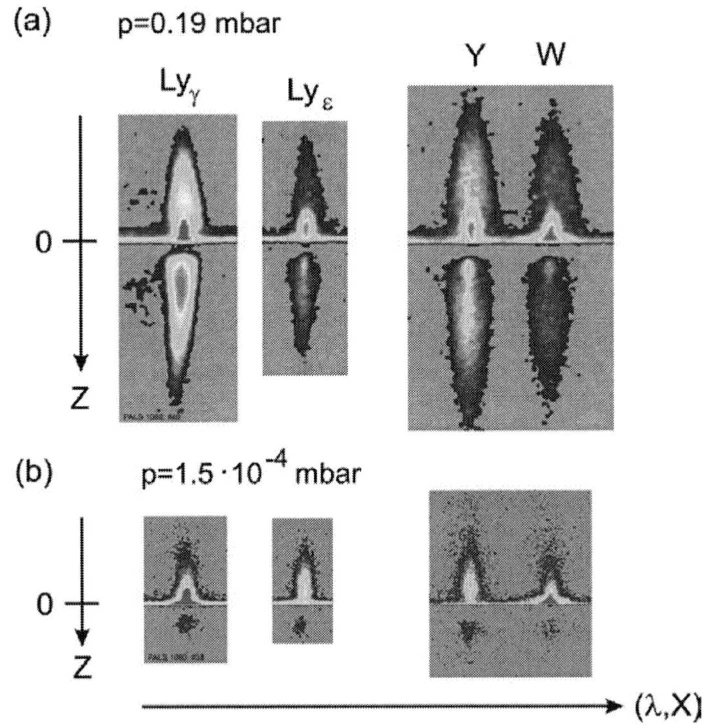

FIGURE 9. X-ray images of spectrally resolved plasma jets for different gas pressures, a) p = $1.9 \cdot 10^{-1}$ mbar, laser energy E = 25 J, b) p = $1.5 \cdot 10^{-4}$ mbar, laser energy E = 23 J. Large differences in the emission features are visible for different gas pressures, front (positive Z) and rear side (negative Z) emission and various resolved line transitions.

In order to investigate the state selective population of particular emission lines and provide further proof of the charge exchange interpretation, we have setup the spherically bent Bragg crystal spectrometer for a very large spectral range in 2nd order of Bragg crystal reflection (2D-setup according Figure 3): He_β until the series limit as well as Ly_β until the series limit, see Figure 8. The distance between target and crystal was 36 cm, the Bragg angle was 36.3° (corresponding to a central wavelengths λ_0 = 5.909 Å). The demagnification factors were M_S = 1.85 and M_d = 2.54. With a 3000 dpi scan of the x-ray film, 1 pixel corresponds to a spatial resolution of 15.7 μm (Z-axis) and 21.5 μm (X-axis). Figure 8 shows, that at far target distances the He_γ line is more intense than the He_β line and the He_δ line is of similar intensity than the He_β-line. The Lyman series shows similar effects: the Ly_γ-line is more intense than the Ly_β line (note, that He_ζ and He_η are located near Ly_β and the real Ly_β peak intensity is therefore smaller than it is visible in the spectrum).

X-ray topological structure of plasma jets

The observed phenomena relate to potentially important applications. The discovered spatial emission features hint at a novel diagnostic method employing Li-like intercombination satellite transitions: their observation is indicative of charge exchange processes while other lines (in particular the He-like intercombination line) are sensitive to several excitation mechanisms at the same time. Therefore, an unequivocal analysis of the line evolution processes can be made which in turn will serve as an important impact on fluid simulations. Moreover, as can be seen from Figure 4, detailed emission structures are visible in the spectrally and spatially resolved plasma jet.

Figure 9 shows the details of the emission structures of Ly_γ, Ly_ε, Y and W presented in Figure 4. Figure 9a shows, that for Ly_γ the rear side emission at large gas pressures is distinct different from the front side emission. For the rear side an emission maximum is observed at far target distances (feature "A" in Figure 6e). For higher Lyman series members and also for lower gas pressures (Figure 9b), this feature is absent. Near the target surface, the emission of the resonance line W and the intercombination line Y differ considerably: this differences are connected with the large radiation transport of the resonance line. At far target distances, the Y-line still shows strong emission due to the cascading processes in the triplet system (Figure 7a).

Figure 10a shows the 2-dimensional x-ray imaging of the "Ly_γ-plasma jet" of Figure 8 obtained by placing the detector out of the Rowland circle (Figure 3). Figure 10b shows the spectra of $Ly_{\gamma,\delta,\varepsilon}$ (left) and $He_{\gamma,\delta,\varepsilon}$ (right) at far target distances. Ly_γ and He_γ shows a large line width (about 250 px) which is due to the radiating halo: the jet moves into the gas and diffuses radially up to mm source size. The large halo is made visible by charge exchange driven x-ray radiation emission. Close to the target surface, the source size is small resulting in small line width, Figure 10c.

The detailed topological information might be also a valuable test for fluid simulations to understand how the hot dense plasma jet pushes and ionizes the gas, i.e., the general problem of interpenetrating plasmas. As charge exchange processes are usually not included in collisional-radiative codes for the parameter evaluation, the proposed charge exchange phenomena and their characterization by means of Li-like intercombination statellite transitions are important to reinterpret the radiative emission in terms of plasma parameters.

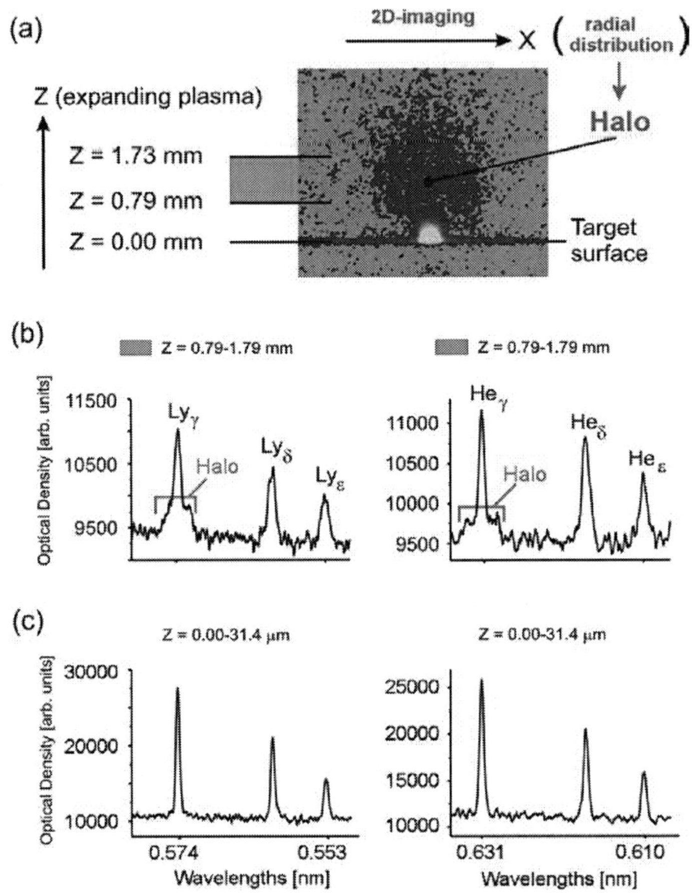

Figure 10. The broad emission structure is due to two-dimensional x-ray imaging placing the detector out of the Rowland circle (see Figure 3), see text. The x-ray film is displaced from the Rowland circle (see Figure 3) in order to obtain spatial resolution also along the x-axis (wavelengths-axis).

COLLIDING PLASMAS: 2 FOIL EXPERIMENTS

Figure 11 shows the experimental setup to collide plasmas between foils. Two different target elements have been chosen in order to distinguish (via the characteristic x-ray emission of Mg and Al) the plasma created on the left side from the plasma created on the right side.

Figure 12 shows the x-ray image of an experiment, where an 0.8 μm thick Al foil is irradiated (laser beam 1) with a laser pulse (3 ω, E = 78 J) whereas the 2 μm thick Mg foil is not irradiated (no laser beam 2). The distance between the Mg and the Al target was 300 μm.

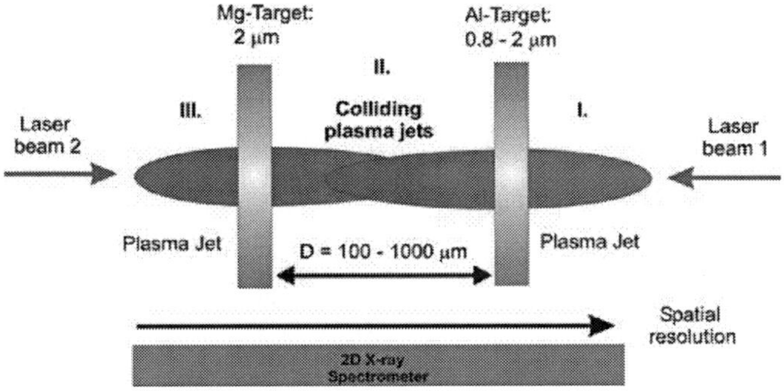

FIGURE 11. Colliding plasma experiment with indicated spatial regions I., II. and III.

The Figure 12 shows a strong plasma jet creation at the front surface of the Al target (region I in Figure 11), similar to those discussed previously (Figure 4). The two foils are visible via the x-ray emission. The plasma jet which is created at the rear side of the Al foil (spatial region II. in Figure 11) collides with the Mg foils resulting in an enhanced Al x-ray emission near the Mg target. Finally, the Mg foils burns through and Al x-ray emission is visible even at the rear surface of the Mg foil (spatial region III. in Figure 11).

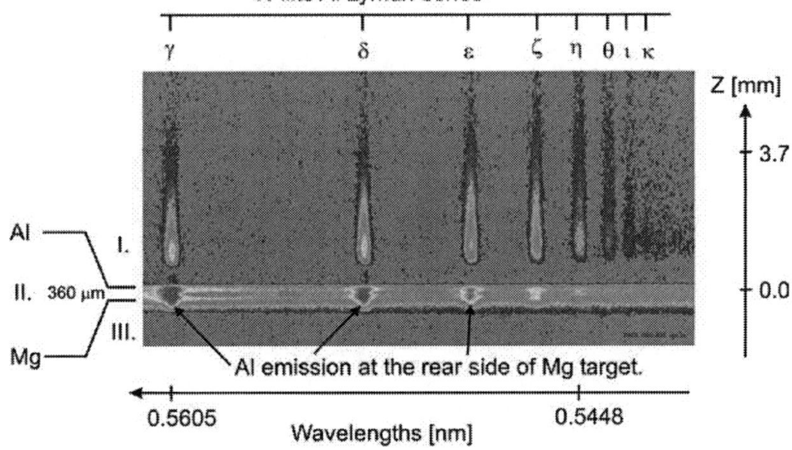

FIGURE 12. X-ray images of a two-foil experiment. The Al foil is irradiated with the laser beam 1 (I.), the foil burns through and the plasma collides (II.) with the opposite Mg target. Finally, the Mg foil is burned through and Al-emission on the rear side of the Mg target (III.) is visible. On the front side of the Al foil (I), strong plasma jets are created. Indicated spatial regions I. - III. refer to those of Figure 11.

Figure 13 shows the spatial dependence of various line transitions. Figure 13a shows the strong emission from the plasma jet at the front side of the Al foil and also a strong emission between the two foils. The dip in the emission at about Z = 4.25 mm is an artefact caused by the target holder which screens the line of sight for the front side emission.

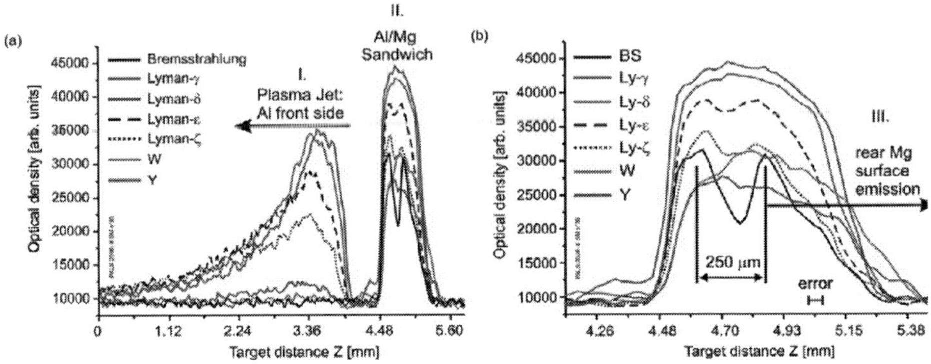

FIGURE 13. Spatial dependences of the x-ray emission in the two-foil experiments. The point Z = 0 is set arbitrary. Spatial regions I.-III. refer to those of Figure 11.

Figure 13b shows the spatial emission features of various line transitions between the two foils. Ly_γ and Ly_δ show emission maxima between the two foils whereas Ly_ε and Ly_ζ show emission maxima at the place of the foils and a minimum between the foils. The He-like intercombination line Y shows an emission maximum more close to the Al foil whereas the He-like resonance line W has a maximum between the foils. The Bremsstrahlung (BS) shows clear maxima at the place of the original foils.

CONCLUSION

In conclusion we have shown, that high resolution x-ray microscopy of plasma jets colliding with gases provides a detailed topological structure of the interpenetrating plasmas. We have identified anomalous (3 orders of magnitude) intensities of intercombination satellite transitions of autoionizing states. The proposed chain reaction of charge exchange recombination and cascading in autoionizing states is able to explain the spatial correlation with the single excited intercombination lines and the correlated anomalous intercombination satellite intensity.

The x-ray emission between two irradiated foils shows distinct emission features which depend largely on the type of the x-ray line transitions and the gas pressure.

The results have important potential to cross check corresponding fluid/MHD simulations of counter-propagating plasmas.

ACKNOWLEDGMENTS

This work has been partially supported by the ARCUS-Program (n° P003-5a). The experiments have been carried out in the framework of the European access program LASERLAB under contract number PALS n° 1080.

REFERENCES

1. D.R. Farley, K.G. Estabrook, S.G. Glendinning, S.H. Glenzer, B.A. Remington, K. Shigemori, J.M. Stone, R.J. Wallace, G.B. Zimmerman, J.A. Harte, Phys. Rev. Lett. **83**, 1982 (1999).
2. K. Shigemori, R. Kodama, D.R. Farley, T. Koase, K.G. Estabrook, B.A. Remington, D.D. Ryutov, Y. Ochi, H. Azechi, J. Stone, N. Turner, Phys. Rev. E **62**, 8838–8841 (2000).
3. L.O. Silva, R.A. Fonseca, J.W. Tonge, J.M. Dawson, W.B. Mori, M.V. Medvedev, Astrophysical Journal **596**, L121–L124 (2003).
4. C. Courtois, R.A.D. Grundy, A.D. Ash, D.M. Chambers, N.C. Woolsey, R.O. Dendy, K.G. McClements, Physics Plasmas **11**, 3386 (2004).
5. T.R. Dittrich, S.W. Haan, M.M. Marinak, S.M. Pollaine, D.E. Hinkel, D.H. Munro, C.P. Verdon, G.L. Strobel, R. McEachern, R.C. Cook, C.C. Roberts, D.C. Wilson, P.A. Bradley, L.R. Foreman, W.S. Varnum, Physics Plasmas **6**, 2164 (1999).
6. R.W. Clark, J. Davis, A.L. Velikovich, K.G. Whitney, Physics Plasmas **4**, 3718 (1997).
7. F.B. Rosmej, V.S. Lisitsa, D. Reiter, M. Bitter, O. Herzog, G. Bertschinger, H.-J. Kunze, Plasma Physics and Controlled Fusion **41**, 191 (1999).
8. F.B. Rosmej, R. Stamm, V.S. Lisitsa, Europhysics Letters **73**, 342 (2006).
9. F.B. Rosmej, D.H.H. Hoffmann, W. Süß, M. Geißel, O.N. Rosmej, A.Ya. Faenov, T.A. Pikuz, T. Auguste, P. D'Oliveira, S. Hulin, P. Monot, J.E. Hansen, G. Verbookhaven, Nuc. Instrum. Methods A **464**, 257 (2001).
10. O. Rancu, P. Renaudin, C. Chenais-Popovics, H. Kawagashi, J.-C. Gauthier, M. Dirksmöller, T. Missalla, I. Uschmann, E. Förster, O. Larroche, O. Peyrusse, O. Renner, E. Krouský, H. Pépin, T. Shepard, Phys. Rev. Lett. **75**, 3845 (1995).
11. A.S. Wan, T. Barbee, R. Cauble, P. Celliers, L.B. Da Silva, J.C. Moreno, P.W. Rambo, G.F. Stone, J.E. Trebes, F. Weber, Phys. Rev. E **55**, 6293 (1997).
12. T. Atwee, H.-J. Kunze, J. Phys. D: Appl. Phys. **35**, 524 (2002).
13. K. Jungwirth et al., Physics Plasma **8**, 2495 (2001).
14. F.B. Rosmej, "Application of interpenetrating plasma jets to charge exchange processes", LASERLAB PALS n°1080 (2006).
15. I. Yu. Skobelev, A. Ya. Faenov, B.A. Bryunetkin, V.M. Dyakin, T.A. Pikuz, S.A. Pikuz, T.A. Shelkovenko, V.M. Romanova, JETP **81**, 692 (1995).
16. F.B. Rosmej, A.Ya. Faenov, T.A. Pikuz, A. Calisti, L. Godberg-Mouret, R. Stamm, B. Talin, H.R. Griem, R.C. Elton, V.L. Jacobs, M. Geissel, D.H.H. Hoffmann, M. Schollmeier, E. Fill, K. Witte, O. Renner, E. Krousky, J. Krasa, L. Laska, B. Kralikova, J. Skala, M. Bodnar: "An overview of x-ray spectroscopy of laser produced plasmas", Proceedings of the SPIE **5196**, p. 163, San Diego, USA.
17. V.P. Shevelko, L.A. Vainshtein, *Atomic Physics for Hot Plasmas*, IOP Publishing, Bristol 1993.
18. F.B. Rosmej, V.S. Lisitsa, Physics Letters A **244**, 401 (1998).
19. V.N. Ostrovsky, J. Phys. B 28, 3901 (1995).

Aluminium He-β High-Resolution Spectroscopy and the Observation of High Field Processes

N. C. Woolsey, D. M. Chambers*, C. Courtois[+], E. Förster[1], C. D. Gregory, I. M. Hall, J. Howe, O. Renner[2], and I. Uschmann[1]

Department of Physics, University of York, York, Y010 5DD, United Kingdom
[1]Institute of Optics and Quantum Electronics, University of Jena, 07743 Jena, Germany
[2]Institute of Physics, Academy of Sciences CR, Prague, Czech Republic

Abstract. Helium-like aluminium (Z=13) plasmas created using long (nanosecond) and short (picosecond) duration laser pulses at intensities between 5×10^{13} and 5×10^{16} W/cm^2 were studied using a combination of optical and X-ray spectroscopy methods and computational modeling tools. The experiments were designed to ensure that the K-shell emission spectra from the aluminium plasmas was measured in detail and that the spectra could be reproduced through a combination of hydrodynamic, atomic kinetic, and radiation transport models. Detailed spectral line shapes, with high spectral and spatial resolution, were recorded in an attempt to positively observed single frequency electric fields effects due to the application of a strong laser field, and due to strong laser-driven plasma waves. Spectral line shape modification due to laser-driven plasma waves was observed, yet there is no evidence for direct laser field effects. Experimental procedures and the computational work are outlined, the observation of modulated intensity profiles on the He-β transition and the interpretation based on intra-Stark spectroscopy discussed.

Keywords: Laser-produced-plasma; K-shell spectroscopy; Laser-induced satellites, Line shapes
PACS: 32.30.Rj, 32.70.Jz, 52.25.Os, 52.38.-r

INTRODUCTION

An exciting new development in spectroscopy is the possible observation, and future diagnostic application, of intense and ultra-intense electric and magnetic fields in laser produced plasmas. The modification of spectra through the application of electric and magnetic fields has been studied for more than a century and a rich science has developed, it was quickly realized following the development of the laser that the fields associated with pulsed lasers could be extreme albeit transient. Since the invention of the laser, the development in laser peak power, see Fig. 1, and the focused intensity has been staggering. Current technology enables single laser beam powers of 10^{15} W, and peak intensities exceeding 10^{21} W/cm^2 in sub-100 fs or1 ps duration laser pulses. The pulse duration depends on whether the laser gain medium is either Ti:Sapphire with fundamental wavelength of 790-800 nm or Nd:Glass, wavelength 1053 nm. Recent, and future, peak power advances have been based on new technological developments in particular the application of chirped-pulse amplification (CPA).

CP874, *Spectral Line Shapes: 18th International Conference,*
edited by E. Oks and M. Pindzola

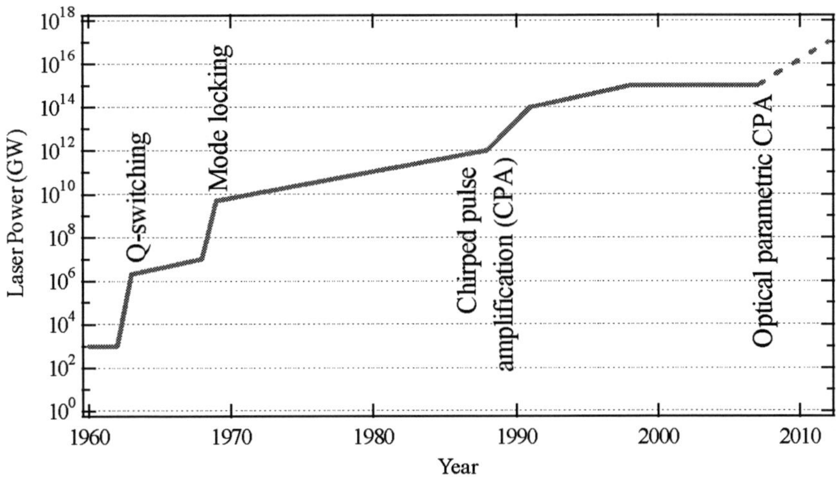

FIGURE 1. This schematic shows the evolution of laser power increase since 1960, and indicates the major technological advances. Many order-of-magnitude increases occurred following the development of Q-switching in 1962 [1], mode locking, and chirped pulse amplification (CPA) in 1988 [2]. The most recent development is optical parametric chirped pulse amplification [3].

Table 1 compares the electric, E, and the magnetic, B, fields associated with laser intensities between 10^{14} and 10^{21} W/cm^2 with the electric and magnetic fields of the first Bohr orbit in the hydrogen atom, E_0 and B_0 respectively. We notice that relatively modest laser intensities (~10^{16} W/cm^{-2}) exceed the atomic field ($E_0 = 5 \times 10^9$ V/cm), and free electron quiver motion with paths that greatly exceed the Bohr radius, a_0. This immediately suggests that spectroscopic observation of these fields requires highly ionized atomic systems, and highlights the need for high spectral resolution. At the highest intensities currently available, the instantaneous magnetic field exceeds the atomic field of $B_0 = 4 \times 10^5$ T, furthermore the quiver motion of a free electron in the laser field is relativistic.

TABLE 1. Four laser intensities relevant to high field atomic physics are shown. E/E_0 and B/B_0 are the ratio of the peak laser electric and magnetic fields to those in the first Bohr orbit of atomic hydrogen. In addition, the ratio of the free electron relativistic mass due to quiver motion in the laser field to the electron rest mass, Γ_q, and ratio of the free electron excursion from the atom to the Bohr radius x_q/a_0 are shown.

I (W/cm^2)	1.4×10^{14}	3.5×10^{16}	1.2×10^{18}	2×10^{21}
E/E_0	1/16	**1**	6	240
Γ_q	0.01	0.16	**1**	40
B/B_0	3×10^{-4}	4×10^{-3}	0.03	**1**
x_q/a_0	30	500	-	

Currently, Ti:Sapphire and Nd:Glass technologies are energy limited by the physical size of some of the optical components to approximately 10 J and 1 kJ respectively. Yet scientific interest in the fundamental aspects of the ultra-intense laser-matter interactions and the fast ignition laser fusion will extend these limits to achieve higher energies with both shorter and longer duration pulses to either drive up the focused intensity well beyond 10^{22} W/cm^{-2} or pump larger plasma volumes for longer durations.

In addition to the laser fields, the laser-plasma interaction drives intense plasma currents and plasma waves, these result in quasi-static magnetic fields [4] and stimulated Brillouin scattering (SBS) and stimulated Raman scattering (SRS) processes [5]. Plasma currents have been observed to produce quasi-static magnetic field in excess of 7×10^4 T [6] and numerical studies show the importance of plasma waves to fast ignition fusion [7].

Intense laser interactions lasting for short duration and over small focal spot diameters, conditions needed to achieve the highest intensity, poses interesting yet significant experimental difficulties. The approach taken here was to obtain high quality spectra by developing high luminosity, spatially resolving spectrometers, and working at intensities below the highest possible. To achieve this we both defocused and extended the duration of the laser pulse, and when using a high-repetition rate Ti:Sapphire laser integrated emission from 100 shots to make a single measurement. The spectrometers could be coupled to either to single photon sensitive CCDs, film or X-ray streak cameras to provide spatially resolved and time integrated emission data, or time resolved emission data at a single spatial location. CCD technology enables additional low-resolution energy discrimination. This is used to enhance background subtraction and aid the identification of low brightness spectral features.

HIGH FIELDS PROCESSES

By defocusing the laser focus and stretching the temporal duration of the laser pulse the upper laser intensity limit was approximately 10^{17} W/cm^{-2}. The K-shell of aluminium, with spectral emission between 1.5 and 2.3 keV is an ideal element to study below this intensity and the low lying excited states (principal quantum number < 4) are not field ionized. At higher intensities, it is necessary to work at higher atomic number. Baranger and Mozer [8] described the influence of single frequency electric fields, such as might be possible in a resonantly driven plasma or with a laser, on the emission spectrum from plasmas in 1961. They predicted the appearance of additional satellites to forbidden transitions. This led to a work on helium atoms by Kunze and Griem [9], Gallagher and Levine [10], Nee and Griem [11] and further developments of the original model by a number of authors [12,13 &14]. Vinogradov and Yukov [15] used a 3-level atomic model to describe anti-Stokes Raman scattering and stimulated two-photon emission that can drive laser-induced satellite structures. These satellites are centered on a dipole-forbidden transition, and separated by a laser photon energy ($\hbar\omega_L$) to higher (Raman) and lower (two-photon) energies. The small photon energy (~1 eV for a Nd:Glass laser) compared to the transition energy (>1 keV for K-shell Al) highlights the need for a minimum spectral resolution ($\lambda/\Delta\lambda$) of ~ 2000, when studying high field processes.

Elton *et al* [16] gives a useful expression, reproduced below, which shows that the satellite intensity is proportional to the square of the average laser field, $\langle E_L^2 \rangle$, and that emission process is more efficient when the forbidden transition is energetically close to a dipole-allowed transition, i.e. $\hbar\Delta \rightarrow 0$.

$$S_{\pm} = \frac{3}{8}\frac{\ell_{\max}}{(2\ell_{\max}+1)}\left(\frac{\hbar n_u}{m_e e z}\right)^2 \frac{n_u^2 - \ell_{\max}^2}{(\hbar\Delta \pm \hbar\omega_L)^2}\left\langle E_L^2 \right\rangle \tag{1}$$

where $\ell_{\max}$ is the larger of the upper orbital quantum numbers, z is the screened nuclear charge, and $\hbar$, m_e, e are the fundamental constants.

Measurements suggestive of direct laser-induced satellite emission were first reported in 1997. For example, time-integrated measurements of laser satellites to the $1s2s - 1s^2$ transitions in He-like titanium were obtained by Skobelev *et al.* [17] and modeled by Osterheld *et al.* [18]. Elton *et al.* [16] observed laser-induced satellites in the soft X-ray region of He- and Li-like magnesium plasmas. Short-pulse duration measurements reported by Pikuz *et al.* [19], indicate that laser-satellites were observed to the forbidden $1s2s - 1s^2$ and $1s5s - 1s^2$ transitions in helium-like magnesium and aluminium plasmas. Emission spectra with 5 ps temporal resolution were obtained by integrating a large number of laser shots using a synchronized, repetitive-scan X-ray streak camera. The results suggest that satellites are observed during the laser pulse. More recently Renner *et al.* [20] report experimental observation of the laser-induced satellites from hydrogenic aluminium, these satellites were observed on the wings of the $3p - 1s$ line shape and are consistent with theoretical predictions. To date, the experimental work is preliminary. Most measurements, except [16] and [20], are dominated by Stark broadening and emission from the target surface, this considerable complicates data interpretation and the atomic physics.

In addition to satellite and line modification driven by the strong electric fields discussed above, weaker fields can affect spectral lines [21]. If the magnitude of a quasi-monochromatic field, say due to a plasma wave, is of the order of the ion microfield that Stark broadens a spectral line, it is possible to modulate the line intensity [22]. The quasi-monochromatic electric field splits the upper and lower (if not a resonant transition) energy levels of a radiative transition. This produces intensity modulations in a spectral line at definite separations $\Delta\omega^{dip}$ from the line centre at $\hbar\omega_0$. The positions of the intensity minima or dips are dependent on the plasma density, and the half-width of the modulated intensity structure is related to the perturbing field amplitude. The analysis of a spectrum may be obscured if the emitting ion interacts with a strong field interacting with a number of quanta and by any variation in electron density. Leboucher-Dalimier *et al.* [23] report the observation of intensity modulations in spectral lines shapes in the laser-produced aluminium plasma on hydrogen-like Ly-γ transition. Measurement of these spectral structures offers an alternative method for observing charge-exchange processes between multi-ionised ions. In a different series of experiments, Renner *et al.* [24] observe spectral structures on the same transition which are identified as the Langmuir-wave-caused dips (so-called L-dips).

EXPERIMENTS

The approach taken in this work [25] was to focus on the dipole-allowed He-β transition, $1s^2\ ^1S_0$ -$1s3p\ ^1P_1$ with line centre at 1866.0 eV, and the dipole-forbidden transition, $1s^2\ ^1S_0$ - $1s3s\ ^1S_0$ line centre at 1862.3 eV, in aluminium. Our approach was to diagnose the plasma electron temperature and density using optical and general spectroscopic methods, then measure spectra lineshapes of interest in detail before subjecting the plasma to a high field laser pulse. The optical and spectroscopic data are used to develop hydrodynamic models of the plasma, and this hydrodynamic information is used in population kinetic calculations and spectral line synthesis.

The experiments were carried out using two lasers at the Central Laser Facility of the Rutherford Appleton Laboratory. We used Vulcan, a high energy multi-beam Nd:Glass laser, to obtain single shot measurements. These plasmas were formed by a cluster of 1 ns duration beams, focused to 10^{14} W/cm^2 in a 500 μm diameter spot. These beams ablated a 200 μm square aluminium dot mounted on a Mylar ($C_{10}H_8O_4$) plastic support. The ablating Mylar tamps the lateral expansion of the ablating aluminium dot to form a near one-dimensional expansion away from the target surface. The target geometry and laser geometry, and the position of the spectrometers are shown in Fig. 2. This experimental geometry allows an additional intense (10^{17} W/cm^2, 10 ps duration with 100 μm spot diameter) laser to propagate in to the side of the plasma plume. The timing and position of the intense beam was fixed and set to 1 ns with respect to long pulse beams. Moving the height of the target allowed different plasma densities to be probed.

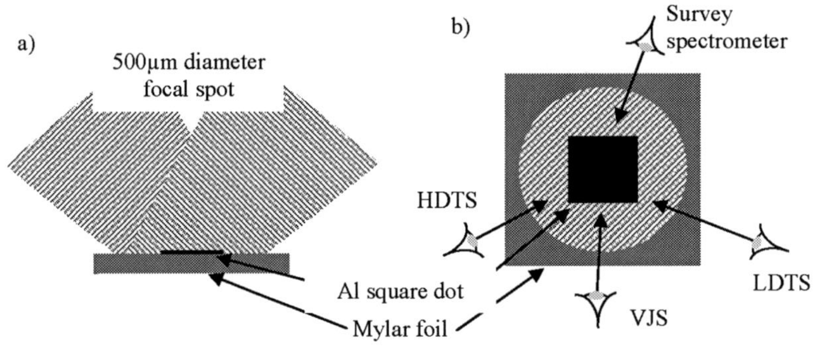

FIGURE 2. A schematic of the aluminium microdot on a Mylar plastic foil and the laser focus shows the experiment from a) the side and b) above. The Al microdot is 1 × 200 × 200 μm in size and the laser focal spot diameter is 500 μm. The X-ray emission is diagnosed using two pinhole cameras and four spectrometers including the low-dispersion toroidal spectrometer (LDTS), the high-dispersion toroidal spectrometer (HDTS), and a vertical Johann spectrometer (VJS). In addition, the electron density and temperature are diagnosed using time-resolved Raman and Thomson scattering respectively.

All instruments were able to record single-shot spectra, with the high-dispersion toroidal spectrometer (HDTS), and a vertical Johann spectrometer (VJS) used to measure detailed spectral line shapes. The HDTS operated at a magnification of approximately 4 with $\lambda/\Delta\lambda \sim 8000$ and the VJS instruments recorded spectra well above the target surface with high spatial and spectral resolution. Spatially resolved aluminium K-shell and detailed line shape data (which was in some cases were time resolved) was obtained. In combination with optical measurements we have established a robust model of the plasma based on 1½-dimensional laser-plasma hydrodynamics code, Ehybrid [26], coupled with the radiative transfer model Cretin [27]. Synthetic spectra match experimental data if line opacity, plasma flow, and two dimensional effects in both the hydrodynamic model and line transfer are accounted for at large distances above the target surface. This occurs at approximately 300μm above the initial target surface were lateral expansion of the aluminium column results in electron temperature and density gradients transverse to the plasma flow. We have some confidence that our experimental plasma is near one-dimensional below 300 μm. An example of pinhole camera data is shown in Fig. 3. This time integrated image shows strong emission at energies above 1 keV from the aluminium plasma. The X-ray transparent C-H-O plasma ablated from the Mylar target support confines the aluminium to a column approximately 200 μm wide. Evidence for lateral expansion at and above 300 μm is difficult to extract due to a drop in emission.

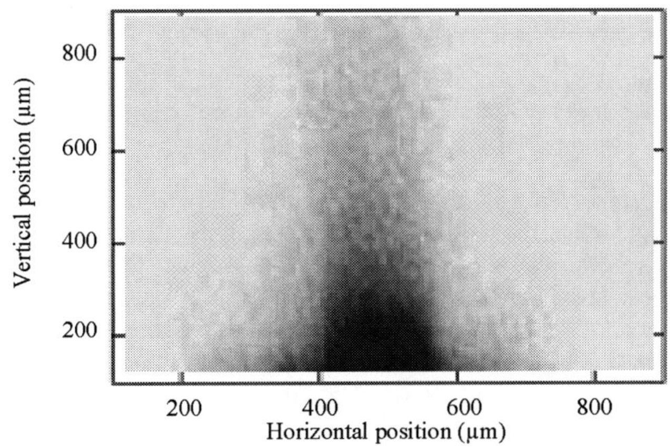

FIGURE 3. Example pinhole camera data showing a time integrated X-ray (> 1keV) image of the plasma. The image shows the ablating aluminium is restricted to a ~200μm wide column for several hundred microns above the target surface. The aluminium is tamped by the ablating Mylar.

K-shell emission spectra taken from the experiment is shown in Fig. 4. Data of this sort is used to test the modeling and is compared to synthetic spectra calculated using Ehybrid hydrodynamic data and Cretin. The simulation results in time integrated spectra, integrating emission for up to 5 ns (the duration of the hydrodynamic simulation), incorporating, in addition to standard broadening, calculations of opacity and Doppler effects as well as the instrument response. In this case, the Cretin simulation used atomic data calculated using screened hydrodynamic model, this accounts for the poor match between the experimental and synthetic spectral line positions.

The intense 10 ps duration beam was frequency doubled to 532 nm and positioned at 300 μm above the target surface. At this position, the Stark broadening is sufficiently small to enable clear identification of laser-satellite emission from

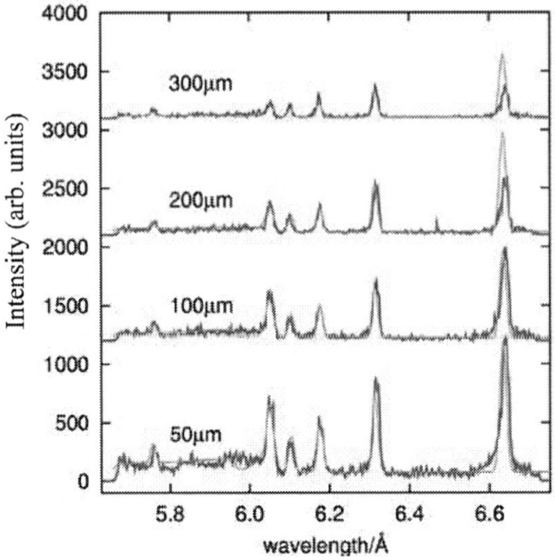

FIGURE 4. Experimental, time-integrated K-shell emission spectra compared to synthetic spectra calculated using electron temperatures and densities extracted from Ehybrid simulations and Cretin model. Spectra are shown at 50, 100, 200, and 300 μm above the initial target surface. These Cretin simulations used a screened hydrogenic model, which results in the discrepancy spectral position of some of the emission lines.

detailed, spatially resolved spectral measurements. The low plasma density limits the emission intensity, thus requiring a large laser focus, and the use of focussing geometry X-ray spectrometers with sensitive detectors. Furthermore, the plasma density was below critical density [28], however, the beam needed to propagate through an extended C-H-O plasma plume to reach the aluminium column. This may result in the formation of a plasma channel, filamentation and scattering of the beam. The key feature of the toroidally bent quartz (100) crystal spectrometer, with horizontal and a vertical bending radius of 1404.55 mm and 312.8 mm respectively, in

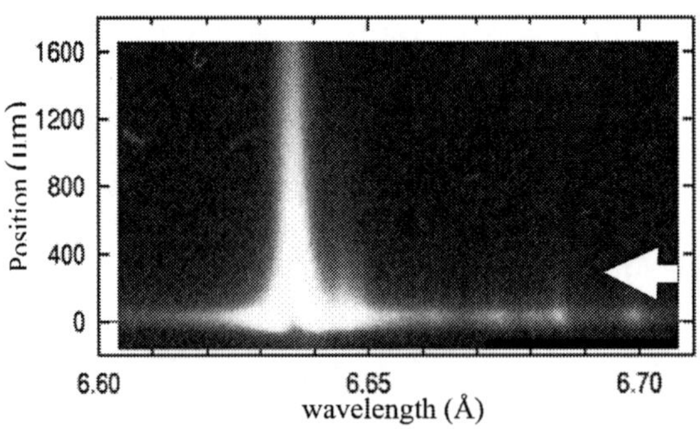

FIGURE 5. Spatially resolved, high-spectral resolution image of the He-β transition and satellites. An arrow 300 μm above the target surface indicates the position of a 10^{16} W/cm^2, 10 ps duration laser at 532 nm focussed to a 100 μm diameter spot. Laser-induced satellites were not observed near the resonance line or near the forbidden ($1s^2 - 1s3s$) transition.

addition to the resolving power and high spatial resolution, which are independent of the source size, is the high luminosity.

Measurements were made using the additional intense beam and an example of the single shot high resolution data is shown in Fig. 5. The image is over exposed to highlight weak spectral features. Data is spectrally resolved in the horizontal axis and shows emission including the He-β transition between 6.6 and 6.7 Å, which extend to more than 1 mm from the target surface. There is no clear evidence for laser-satellites around the position of the interacting beam, which was centred on 300 μm. The intercombination lines ($1s^2\ ^1S_0$ -$1s3p\ ^3P$) are observed at 6.35 Å with emission restricted to within 200 μm of the target surface. Satellite emission is observed to the long wavelength side of the He-β and occurs at the target surface.

A second series of measurements were made with the Astra laser. This is a single beam Ti:Sapphire system that delivers 200 mJ in a 50 fs pulse. To obtain high quality, spatially resolved data the pulse was stretched to 3.4 ps duration, with a 10^3:1 peak to prepulse contrast ratio. As many as 100 shots, at 2 Hz repetition rate, were integrated for a single measurement. Space limitations restricted the diagnostics to a survey spectrometer covering most of the K-shell, and the HDTS centred on individual spectral lines. Both spectrometers were mounted to image emission perpendicular to the surface normal. The target was a 500 μm diameter, 5 cm long aluminium rod. A precision linear translator was used to move the target through the laser focus to present fresh material for each shot, and achieve 3 μm spatial resolution. These targets are not inertially tamped, the expansion close to the target surface (i.e. for expansions up to a focal spot diameter) is one-dimensional, and at greater distances the emitting plasma forms a two dimensional plume. Adjusting the focus diameter from 15 μm to 100 μm allowed on target focal intensities to be varied between 7×10^{14} W/cm^2 and 3×10^{16} W/cm^2. The experimental set up is shown in Fig. 6. It was necessary to shield the spectrometer and CCD detector from hard X-rays and energetic electrons with lead and magnets.

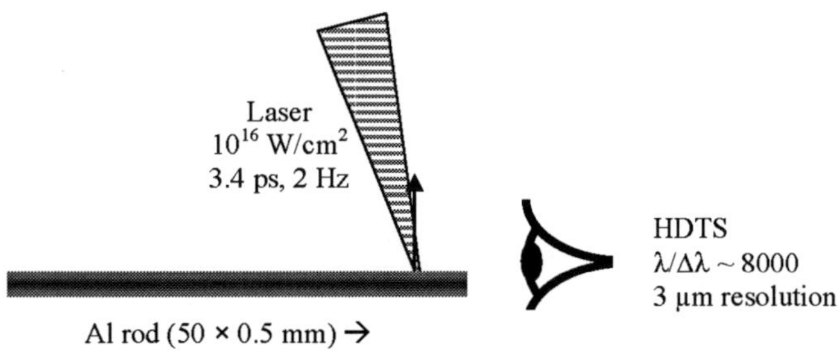

FIGURE 6. This schematic of the Astra short pulse, high repetition rate experiment illustrates the translation of the target to better than 3 μm accuracy and alignment of the high dispersion toroidal spectrometer (HTDS). The HTDS images the plasma perpendicular to the target surface.

Some of the data taken at an intensity of 3×10^{16} W/cm^2 is shown in Fig. 7. The spectral profile of the He-β transition, extracted from data over 6 μm plasma extension, is centred at 5 μm above the target. What is notable is that the spectral profile shows extensive, and regular, intensity modulation. Similar structure is observed for all laser intensities studied, and it is noted that the structure is observed with the integration of 100 shots. This is notably different from the measurements made using the Vulcan laser (see Fig. 5), which consistently show smooth line shape profiles. In Fig. 7 the vertical dashed lines show intensity minima in the experimental data. Assuming these intensity modulations observed in Fig. 7 are due to intra-Stark processes as described by Oks [21] we are able to extract estimates for the amplitude of the electric field ($E \sim 10^8$ V/cm) modulating the profile, the electron density (n_e = 2.5 $\times 10^{20}$ cm^{-3}), and the number of field quanta involved (k = 3). We find that the electric field and number of quanta is dependent on the laser intensity. However, the inferred electron density is unchanged at 3 $\times 10^{20}$ cm^{-3}. The large intensity dip at line centre is due to absorption.

Data is compared to a synthetic line shape calculated using the Ehybrid and Cretin models. This synthetic profile was post-processed to include the effects of intra-Stark processes. The comparison, when using E, n_e, and k extracted from data, is convincing and suggests that a high amplitude quasi-monochromatic wave is present in the plasma, and modifies the spectra. As the field amplitude is laser intensity dependent, it is likely that these modulations, or L-dips, are laser driven.

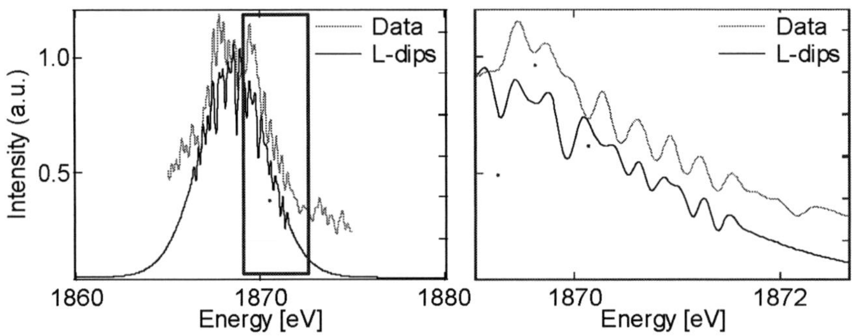

FIGURE 7. Time and shot integrated experimental cross-section through the aluminium He-β taken 5μm from the target surface. The high-spectral resolution data shows and intensity dip at line centre and intensity modulations on the low and high energy wings of the spectral line. b) A section of the high energy wing is expanded to show the modulations. Data is compared to a synthetic He-β profile (labelled L-dips) which incorporates intra-Stark effects. The calculation is displaced for clarity.

Hydrodynamic simulations suggest that with the 3.4 ps laser pulse, the plasma is characterised close to the target surface by a density length-scale that is suitable for the growth of non-linear parametric instabilities. Processes such as stimulated Raman scattering or two-plasmon decay at the quarter critical plasma density are evident, and will drive large amplitude plasma waves. This strong coupling of the laser energy into the plasma is also evident in the data with the detection of hard X-rays and of energetic electrons impact tracks on the CCD.

CONCLUSION

Experimental studies of field induced modification of spectral lines have been described. The approach was to accurately characterise and model the plasma, and then determine the intrinsic spectral line shape before attempting to study high-field modification. We believe this approach is appropriate for the definitive measurement of field induced spectral features. High spatial and spectral resolution, coupled with high luminosity is important. We have observed field-induced spectral structure on the aluminium He-β transition. Measurements show that these features occur at similar densities from laser intensities of 7×10^{14} W/cm^2 and upwards. Inferred electron densities suggest that emission originates close to quarter critical density; and is suggestive of laser-driven parametric instabilities. Similar features and field induced satellites were not observed in the long pulse experiments, and this needs further study. High-field spectroscopy will (and is becoming) an important diagnostic for interpreting ultra-intense laser-matter interactions and possibly an innovative way of studying parametric instabilities.

ACKNOWLEDGMENTS

We wish to thank the staff at the CCLRC Central laser Facility, Rutherford Appleton Laboratory for their expertise and enabling these experiments to happen. JH thanks the EU Marie-Curie Research Training Network for support. This work was supported by UK Engineering and Physical Sciences Research Council.

REFERENCES

* Current address: AWE Ltd, Aldermaston, Berkshire, RG7 4PR, United Kingdom.
[+] Current address: CEA, Bruyeres Le Chatel, BP12 91680, France.

1. F. J. McClungand, R. W. Hellwarth, J. Appl. Phys. **33**, 828 (1962).
2. D. Strickland, G. Mourou, Opt. Comm. **56**, 219 (1985).
3. I. N. Ross, P. Matousek, M. Towrie, A. J. Langley, J. L Collier, Opt. Comm. **144**, 125 (1997).
4. J. A. Stamper, K. Papadopoulos, R. N. Sudan, S. O. Dean, E. A. McLean, J. M. Dawson, Phys. Rev. Lett. **26**, 1012 (1971).
5. D. W. Forslund, J. M. Kindel, and E. L. Lindman, Phys. Fluids 18, 1002 (1975), *ibid.* Phys. Fluids **18**, 1017 (1975).
6. U. Wagner, M. Tatarakis, A. Gopal, F. N. Beg, E. L. Clark, A. E. Dangor, R. G. Evans, M. G. Haines, S. P. D. Mangles, P. A. Norreys, M.-S. Wei, M. Zepf, and K. Krushelnick, Phys. Rev. E **70**, 026401 (2004).
7. Miyakoshi et al. Phys. Plasmas **9**, 3552 (2002).
8. M. Baranger, B. Mozer, Phys Rev **123**, 25 (1961).
9. H. Kunze and H. R. Griem, Phys. Rev. Lett. **21**, 1048 (1968).
10. C. C. Gallagher, M. A. Levine, Phys. Rev. Lett. **127**, 1693 (1975).
11. T-JA Nee, H. R. Griem, Phys. Rev. A **14**, 1853 (1976).
12. D. D. Burgess, J. Phys. B **4**, L7 (1971).
13. H. R. Griem, Spectral Line Broadening in Plasmas, Academic Press, NY (1974).
14. R. W. Lee. J Phys B **12**, 1165 (1979).
15. A. V. Vinogradov, E. A. Yukov, Sov. J. Quantum Electron. 3, 163 (1973).
16. R. C. Elton, H. R. Griem, B. L. Welch, A. L. Osterheld, R. C. Mancini, J. Knauer, G. Pien, R. G. Watt, J. A. Cobble, P. A. Jaanimagi, D. K. Bradley, J. A. Delettrez, R. Epstein, J. Quant. Spectrosc. Radiat. Transfer **58**, 559 (1997).
17. Yu I. Skobelev, A. Y. Faenov, A. I. Magunov, A. Osterheld, B. Young, J. Dunn, R. E. Stewart, Physica Scripta **T73**, 104 (1997).

18. A. L. Osterheld, B. K. F. Young, J. Dunn, R. E. Stewart, I. Y. Skobelev, A. Y. Faenov, A. I. Magunov, J. Quant. Spectrosc. Radiat. Transfer **58**, 827 (1997).
19. S. A. Pikuz, A. Maksimchuk, D. Umstadter, M. Nantel, I. Y. Skobelev, A. Y. Faenov, A. Osterheld, JEPT Lett. **66**, 480 (1997).
20. O. Renner, O. Peyrusse, P. Sondhauss, E. Förster, J. Phys. B **33**, L153 (2000).
21. E. Oks, Plasma Spectroscopy: The Influence of Microwave and Laser Fields, Springer (1995).
22. E. A. Oks, St. Böddeker, H.-J. Kunze, Phys. Rev. A **44**, 8338 (1991); E. Oks, E. Leboucher-Dalimie, Phys. Rev. E **62**, R3067 (2000)
23. E. Leboucher-Dalimier, E. Oks, E. Dufour, P. Sauvan, P. Angelo, R. Schott, A. Poquerusse, Phys. Rev. E **64**, 065401 (2001).
[24] O. Renner, E. Dalimier, E. Oks, F. Krasniqi, E. Dufour, R. Schott, E. Forster, J. Quant. Spectrosc. Radiat. Transfer **99**, 439 (2006).
V. P. Gavrilenko, A. Ya. Faenov, A. I. Magunov, T. A. Pikuz, I. Yu. Skobelev, K. Y. Kim, H. M. Milchberg, Phys. Rev. A **73**, 013203 (2006).
25. N. C. Woolsey, D. M. Chambers, C. Courtois, E. Förster, C. D. Gregory, I. M. Hall, J. Howe, O. Renner, and I. Uschmann, J. Quant. Spectrosc. Radiat. Transfer **99**, 680 (2006).
26. G. J. Pert, J. Plasma Phys. 41, 263 (1989).
27. H. A. Scott, J. Quant. Spectrosc. Radiat. Transfer. 71, 689 (2001).
28. D. M. Chambers, P. A. Pinto, J. Hawreliak, I. R. Al'Miev, A. Gouveia, P. Sondhauss, E. Wolfrum E, J. S. Wark, S. H. Glenzer, R. W. Lee, P. E. Young, O. Renner, R. S. Marjoribanks, S. Topping, Phys Rev E **66**, 026410 (2002).

Observation of Modulations in Lyman-Alpha Line Profiles of Hydrogen-Like Ions of Oxygen in Clusters Irradiated by Femtosecond Laser Pulses: Effect of an Oscillatory Electric Field

V.P. Gavrilenko[1], A.Ya. Faenov[2], A.I. Magunov[2], T.A. Pikuz[2], I.Yu. Skobelev[2], K.Y. Kim[3], and H.M. Milchberg[3]

[1]*Center for Surface and Vacuum Research, Federal Agency on Technical Regulating and Metrology, Novatorov Street 40-1, Moscow 119421, Russia*
[2]*Multicharged Ions Spectra Data Center of VNIIFTRI, Mendeleevo, Moscow region, 141570, Russia*
[3]*Institute for Physical Science and Technology, University of Maryland, College Park, Maryland 20742, USA*

Abstract. Profiles of the Ly_α line of O VIII ions in N_2O clusters irradiated by femtosecond laser pulses (laser flux up to 4×10^{17} W/cm^2) were recorded. The features (peaks and dips) in the Ly_α O VIII line profiles were observed evidencing the presence of intense plasma oscillations with an amplitude of the electric field $E_0 \sim (0.5-1.0) \times 10^9$ V/cm and a frequency $\omega \sim (0.7-2.4) \times 10^{15}$ s^{-1}.

Keywords: Multicharged ions, clusters, spectral line profiles, dynamic Stark effect.
PACS: 32.70.Jz, 32.60.+i, 52.50.Jm, 52.70.La

This work reports the results of measuring the spectra of H-like ions of oxygen (O VIII ions) in plasma produced by femtosecond laser pulses. The experiments were carried out with a 10Hz, 50 mJ Ti:sapphire laser system, based on the technique of chirped pulse amplification. Laser pulses at 800 nm were focused with a single lens onto an N_2O gas jet, created by a high-pressure (usually 15 bar), liquid nitrogen cooled, pulsed (10 Hz) nozzle [1]. The outlet of the nozzle had rectangular shape with sizes 11.5 mm x 0.4 mm with the long side along the direction of laser beam propagation. At a nozzle temperature of T_n = -30°C, large clusters (radius > 30 nm) were generated. The laser pulse duration τ_{las} was 70 fs, giving a peak vacuum intensity of $\sim 4 \cdot 10^{17}$ W/cm^2. The x-ray emission from plasma was recorded using Focusing Spectrometers with Spatial Resolution (FSSR-1D) [2]. A large-aperture (15 mm ×50 mm) spherically bent (R = 150 mm) mica crystal was used in the spectrometer as both a dispersive and imaging element. In our experiments the spectrometer operated with spectral resolution $\lambda/\Delta\lambda$ ~5000 and spatial resolution of ~30μm. Figure 1 shows experimental profiles of the Ly_α line of O VIII ions recorded in three experiments performed with the same values of the pulse energy E_{las} and

CP874, *Spectral Line Shapes: 18th International Conference,*
edited by E. Oks and M. Pindzola

duration τ_{las} ($E_{las} = 50$ mJ, $\tau_{las} = 70$ fs), but different polarizations of the laser radiation (O-, P-, and S- polarizations). It can be seen from this figure that the profiles have a complex structure having the form of a series of peaks and dips.

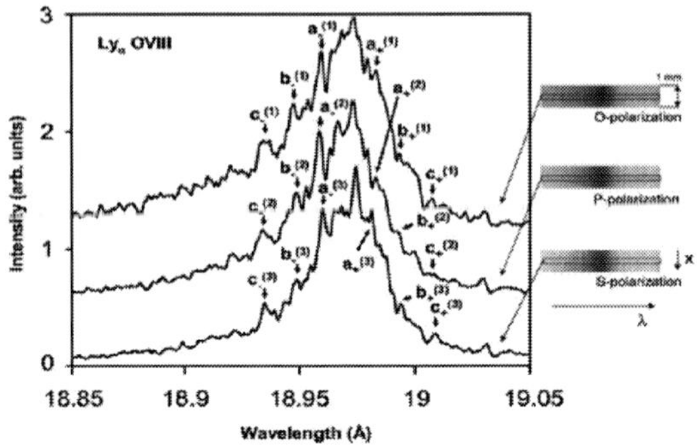

FIGURE 1. Experimental profiles of the Ly$_\alpha$ line of O VIII ions corresponding to three different polarizations of the laser radiation: O-, P-, and S-polarizations. Intensity of the laser radiation $J_{las} = 4 \times 10^{17}$ W/cm^2.

Among these peaks, there are a few pairs of peaks that are located symmetrically with respect to the line center. In particular, three pairs of such peaks are labeled by the letters a_-, a_+, b_-, b_+, and c_-, c_+ for each of the experimental profiles. It is interesting to note that the separation between the peaks b_- and b_+ is close (in the frequency scale) to the laser frequency $\omega_{las} = 2.36 \times 10^{15}$ s^{-1}. The main mechanism that can lead to the modulation of the O VIII line profiles (appearance of peaks and dips) is the action of a high-frequency oscillatory electric field (see [3,4]).

For the calculation of profiles of the Ly$_\alpha$ line of O VIII ions we used our simulation code STARKPRO. This code calculates line profiles of hydrogen-like atoms and ions being under the combined action of two electric fields in a plasma: a quasistatic field **F** produced by randomly located perturbing ions and dynamic field $\mathbf{E}_0 \cos\omega t$. The code is based on the use of the Floquet theorem for finding the solutions of the Schrödinger equation. After calculating the Ly$_\alpha$ line emission spectrum (for a fixed value of **F** and a fixed angle θ between vectors **F** and $\mathbf{E}_0 \cos\omega t$), the spectral line shape is averaged over a spherically symmetric distribution of the field **F**. Along with the Stark effect, the Doppler effect and the optical thickness of the plasma are taken into account.

In Figure 2 three theoretical profiles of the Ly$_\alpha$ line of O VIII ions are given together with the experimental one. These profiles were calculated for three different frequencies and amplitudes of the oscillatory electric field. We note that the oscillatory electric field at the frequency ω_{las} (cf. the profile (1)) can be either the laser field or the field of the plasma oscillations at critical density $N_e = N_{e,cr} = 1.74 \times 10^{21}\,\text{cm}^3$. Theoretical profiles (1) and (3) were calculated for electron density $N_e = N_{e,cr}$, whereas the profile (2) was calculated at $N_e = N_{e,cr}/4$. The frequency $\omega = \omega_{las}/2$, for which the profile (2) was calculated, can represent plasma oscillations at $N_e = N_{e,cr}/4$.

Thus, from the comparison of the experimental profile with the theoretical profiles we obtain the following estimates for the parameters of such oscillating electric fields: $E_0 \sim (0.5-1)\times 10^9$ V/cm, and $\omega \sim (0.7-2.4)\times 10^{15}$ s^{-1}.

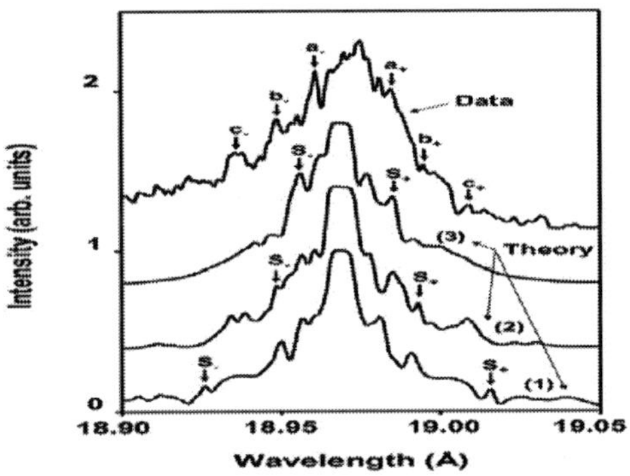

FIGURE 2. Comparison of experimental (upper curve) and theoretical profiles of the O VIII Ly_α line. The experiment was carried out at $J_{las} = 4\times 10^{17}$ W/cm^2, and calculations were performed for optically thick plasma with $T_i = 50$ eV: $N_e = 1.74\times 10^{21}$ cm^{-3}, $\omega = \omega_{las}$, $E_0 = 10^9$ V/cm (1); $N_e = 4.35\times 10^{20}$ cm^{-3}, $\omega = \omega_{las}/2$, $E_0 = 6\times 10^8$ V/cm (2); $N_e = 1.74\times 10^{21}$ cm^{-3}, $\omega = 5.78\times 10^{14}$ s^{-1}, $E_0 = 5\times 10^8$ V/cm (3). Here $\omega_{las} = 2.36\times 10^{15}$ s^{-1} is the frequency of the laser radiation. For each theoretical profile, the letters S_- and S_+ indicate the peaks at $\delta\lambda = -\lambda_0^2\omega/(2\pi c)$ and $\delta\lambda = \lambda_0^2\omega/(2\pi c)$, respectively, where λ_0 is the unperturbed wavelength of the Ly$_\alpha$ line of O VIII ions.

ACKNOWLEDGMENTS

This work was partially supported by the Russian Foundation for Basic Research (project No 06-02-16174).

REFERENCES

1. E.Parra, S.J. Mchaught, and H.M. Milchberg, *Rev. Sci. Instrum.* **73**, 468 (2002)
2. T.A. Pikuz, A.Ya. Faenov, S.A. Pikuz, V.M. Romanova, and T.A. Shelkovenko, *Journal of X-ray Science and Technology* **5**, 323 (1995).
3. E. Oks, *Plasma Spectroscopy. The Influence of Microwave and Laser Fields*, Berlin: Springer, 1995.
4. V.P. Gavrilenko, V.N. Ochkin, and S.N. Tskhai, "Progress in Plasma Spectroscopic Diagnostics Based on Stark Effect in Atoms and Molecules," in *Spectroscopy of Nonequilibrium Plasma at Elevated Pressures*, Proceedings of the International Society for Optical Engineering (SPIE) **4460**, 207 (2002).

Line Broadening Analysis of Argon X-Ray Emission from Z-Driven Implosions Cores

R.C. Mancini[1], J. Bailey[2], G. Rochau[2], G. Chandler[2], G. Dunham[2], P. Lake[2], K. Peterson[2], S. Slutz[2], D. Sinars[2], T. Mehlhorn[2], M. Gunderson[3], D.A. Haynes[3], D. Kilcrease[3]

[1] *Physics Department, University of Nevada, Reno, Nevada, USA*
[2] *Sandia National Laboratories, Albuquerque, New Mexico, USA*
[3] *Los Alamos National Laboratory, New Mexico, USA*

Abstract. We disucss the line broadening analysis of argon-doped deuterium-filled implosion cores driven by a dynamic hohlraum z-pinch. The observed line spectra include emissions from the Lyα, Lyβ, Lyγ, Heα, Heβ, Heγ and Heδ lines in H- and He-like argon ions, respectively. The data recorded using a crystal spectrometer with space-resolving slits and coupled to a framing camera yields time- and spatially-resolved argon line spectra in the implosion core. The analysis accounts for opacity and Stark line broadening to determine electron density, Ne, and areal-density, NΔR, values for the ground state populations of H- and He-like argon ions. In this connection, a set of Stark-broadened line shapes was computed considering the effect of differing ion and electron temperatures, which impacts the calculation of the ion microfield distribution function, the Doppler broadening, and the ion dynamic's effect. Analysis results will be discussed emphasizing the relevance of using these detailed Stark-broadened line shapes.

INTRODUCTION

In experiments performed at the Z facility of Sandia National Laboratories (SNL) deuterium-filled, argon-doped plastic shells were driven by the radiation from z-pinch dynamic hohlraums [1-5]. Tracer amounts of argon were added to the gas fill for diagnostic purposes. At the collapse of the implosion, time- and space-resolved argon K-shell spectra were recorded. The observed argon spectra show emissions from the Lyα (1s ^{2}S – 2p ^{2}P), Lyβ (1s ^{2}S – 3p ^{2}P) and Lyγ (1s ^{2}S – 4p ^{2}P), and Heα (1s^2 ^{1}S – 1s2p ^{1}P), Heβ (1s^2 ^{1}S – 1s3p ^{1}P), Heγ (1s^2 ^{1}S – 1s4p ^{1}P) and Heδ (1s^2 ^{1}S – 1s5p ^{1}P) line transitions in H- and He-like argon ions, respectively, and their associated He- and

CP874, *Spectral Line Shapes: 18th International Conference*,
edited by E. Oks and M. Pindzola

Li-like satellite lines. We focus on line broadening analysis of the resonance lines and apply it to time-resolved, space-integrated data. This analysis is done within the scope of a uniform model approximation and thus the extracted plasma conditions represent spatial averages inside the core. The analysis relies on the fact that at the plasma conditions of these implosion cores broadening due to Stark, predominantly Ne dependent, and opacity, predominantly NΔR dependent, dominate the line shape, so that the broadening can be modeled by several combinations of Stark and opacity broadening. The correct combination can be determined by simultaneously considering several members of the same line series (i.e. same lower state, different upper states) since only a restricted set of plasma conditions can simultaneously fit the broadening of several members of the same series [6]. In contrast to previous applications of line broadening analysis [6-8], here we work with the α, β and γ lines of both the Lyman and He-like line series of argon, and use an opacity broadening function for the case of a uniform absorbing and emitting spherical plasma source based on an analytical integration of the intensity flux through the surface of a sphere. The analysis yields Ne values characteristic of the formation of Lyman and He-like series of lines, and NΔR values for the ground state populations of H- and He-like Ar ions. From the NΔR values, the optical depths of all lines used in the analysis can be computed. Furthermore, these results combined with ionization balance calculations can be used to extract the electron temperature, Te, using the temperature sensitivity of the H-like Ar to He-like Ar ground state population ratio. For the range of densities relevant to these implosion cores the Ne dependence of this ratio is weak thus making it useful as a temperature diagnostic. This paper is organized as follows: in Sec. 2 we discuss the experiments, in Sec. 3 the line broadening analysis method is described, and in Sec. 4 we present our results and conclusions.

EXPERIMENTS

The experiments were performed at the Z facility of SNL. Double-nested tungsten wire-arrays were driven with a pulsed 20 MA current to produce an imploding z-pinch plasma. A low density foam cylinder was placed co-axially within the wire-arrays. The impact of the z-pinch plasma on the foam cylinder launched a cylindrical shock in the foam whose radiation was trapped by the optically thick walls of the tungsten plasma. This shock is the main radiation source of the dynamic hohlraum cavity formed by the imploding z-pinch plasma [1,2]. Embedded at the center of the foam cylinder is the gas-filled plastic capsule. It undergoes implosion driven by the x-rays of the dynamic hohlraum. The capsules of these experiments absorbed up to 20 kJ of x-rays. At the collapse of the implosion, argon x-ray line emission and nuclear fusion reaction neutrons produced in the core were observed [3,4].

X-ray line emission spectra from the core were recorded with a resolution power $\lambda/\Delta\lambda$ of 1000 using the TREX elliptical crystal spectrometer [5]. Figure 1 shows the time-resolved (gated) space-integrated argon line spectrum recorded at Z during the time of brightest argon emission, observed along a line-of-sight (LOS) parallel to the z-pinch axis. In this experiment the plastic capsule had an initial diameter of 2 mm, a wall thickness of 50 μm, and was filled with 20 atm of deuterium and 0.085 atm of argon. The imploded core had a diameter of about 190 μm – 290 μm estimated from the size of the x-ray emitting region, and the argon line emission lasted for approximately 1.5 ns. The spectrum of Figure 1 is characteristic of a time interval of 0.5 ns. It displays several resonance lines and satellite transitions on the low energy side of the parent resonance lins, spanning three ionization stages, i.e. Li-, He- and H-like argon. In particular, the n=2 Li-like satellites of the Heα and Heβ lines, and the n=2 He-like satellites of the Lyα line, are clearly observed. In addition, the triplet to singlet intercombination line in He-like Ar ($1s^2\ ^1S - 1s2p\ ^3P$) on the low energy side of the Heα is observed to be well resolved.

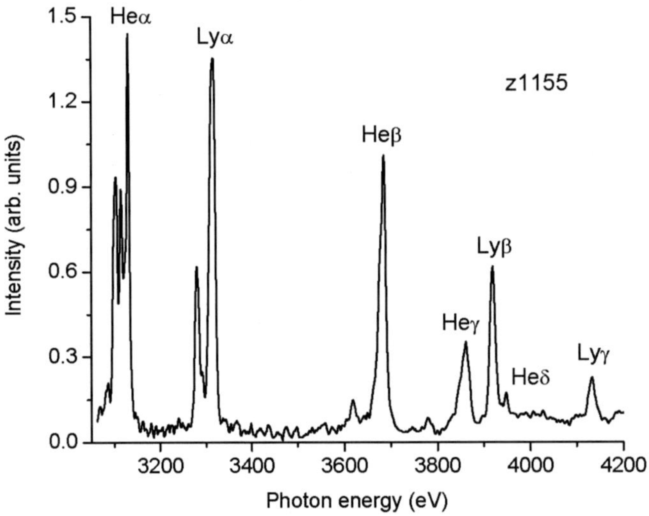

FIGURE 1. Time-resolved (gated), spaced-integrated argon line spectrum from Z shot z1155, recorded over a time interval of 0.5 ns.

LINE BROADENING ANALYSIS

Line broadening analysis of K-shell x-ray line emission has proven to be a useful spectroscopic diagnostic technique for extracting plasma conditions [6-8]. In the implosions at Z discussed here, the simultaneous observation of α, β and γ lines for two line series (i.e. He and Lyman, see Figure 1) makes the spectrum particularly attractive for the application of line broadening analysis. At the collapse of these implosions, hydrodynamic simulations of the experiments suggest that core plasma conditions are expected to reach T_e of about 1 keV and N_e of order 10^{23} cm^{-3} [2]. We note in passing that within 4% the electron and deuterium atom number densities are approximately equal in these core plasmas, since argon represents a small amount of the total gas fill and thus the pool of free electrons is dominated by the contribution from the ionization of deuterium atoms. Four line broadening mechanisms contribute to the overall line profile: natural, Doppler, and Stark which are intrinsic to the line profile and opacity broadening which is due to the transport of radiation through the plasma. The full-width-half-maximum (FWHM) of the natural broadening of resonance argon K-shell lines due to the finite lifetime of upper states are respectively 0.26 eV and 0.22 eV for the Lyα and Heα, 0.069 eV and 0.058 eV for the Lyβ and Heβ, and 0.028 eV and 0.024 eV for the Lyγ and Heγ. The FWHM of Doppler broadening due to the thermal motion of the ionic radiator, which scales as $\sqrt{T_i}$, at a temperature of 1 keV is in the range of 1.2 eV to 1.6 eV for these lines. However, for the conditions of these implosion cores, Stark and opacity broadening are in the range of a few to tens of eV and thus make the dominant contribution to the overall line broadening.

A database of detailed Stark broadened line shapes, including the effects of plasma microfields due to electrons and ions, was computed using the formalism and codes of Refs. [9-11]. The effect of dynamic electrons was taken into account via a second order quantum mechanical relaxation approximation. Since the majority of ionic perturbers are deuterium ions, both static and ion dynamic effects were considered in the line shape calculations. Natural and Doppler broadening assuming T_i=T_e=1 keV were also included. Line shifts were not considered, since they are small for the plasma conditions of these cores. The important property of these Stark line profiles is that they are mainly dependent on N_e and weakly dependent on T_e. Hence, a database of line shapes was computed for a representative T_e of 1 keV and a grid of N_e points (approximately evenly spaced on log(N_e)) covering the range from 3×10^{22} cm^{-3}

to 1×10^{24} cm^{-3}. Line profiles at conditions not included in the database were obtained by performing a cubic interpolation on the log(profile) vs. log(N_e) plot for every photon energy point in the line profile.

The optical depth τ_ν of a single line transition can be written as:

$$\tau_\nu = k_\nu \Delta R = \frac{\pi e^2}{mc} f \phi_\nu N \Delta R$$

where e and m are the electron's charge and mass, respectively, c is the speed of light in vacuum, f is the absorption oscillator strength of the transition, ϕ_ν is the normalized intrinsic line shape, and the product NΔR is the areal density based on the transition's lower state population N and the characteristic physical length ΔR. Hence, the effect of radiation transport on the line shape depends on the areal density. Taking into account the N_e dependence of the Stark broadened line shape ϕ_ν and the NΔR dependence of the opacity, the final intensity distribution used to fit the line shapes extracted from the data is dependent on both N_e and the product NΔR. This dependence leads to ambiguity in the determination of NΔR and N_e if we use only one line shape. Figure 2 illustrates this problem, showing fits of comparable quality for several (NΔR,N_e) sets for the experimental Heα, Heβ and Heγ line shapes from the spectrum of Figure 1. We characterize the quality of fits by a weighted Q^2 sum defined by:

$$Q^2 = \sum_i \omega_i (\phi_i^{\exp} - \phi_i^{theory})^2$$

where the sum is over all photon energy points in the experimental line shape, and the weight factor ω is given by $1/\sigma^2$; σ is the uncertainty of each experimental point estimated from the spread in the data according to the procedure discussed in Ref. [12]. The collection of (NΔR,N_e) sets yielding fits of comparable quality is given by the location of the local minima of the Q^2 surface. This set of (NΔR,N_e) coordinates define on the NΔR-N_e plane the lines-of-best-fits (LOBF), and the second partial derivatives of the function $Q^2=Q^2$(NΔR,N_e) can be used to set up the error matrix and from there compute the uncertainties of NΔR and N_e [12]. However, if we work with all three lines of the same series and look for the intersection of the LOBF, the intersection point will give the set of NΔR and N_e values that simultaneously fit all observed lines of the same series and thus can be interpreted as characteristic of the plasma conditions during the formation of these lines. Ideally, the intersection of the LOBF should be a point; in practice, it is a small region defined by how close the LOBF come together and the errors in the LOBF NΔR and N_e coordinates.

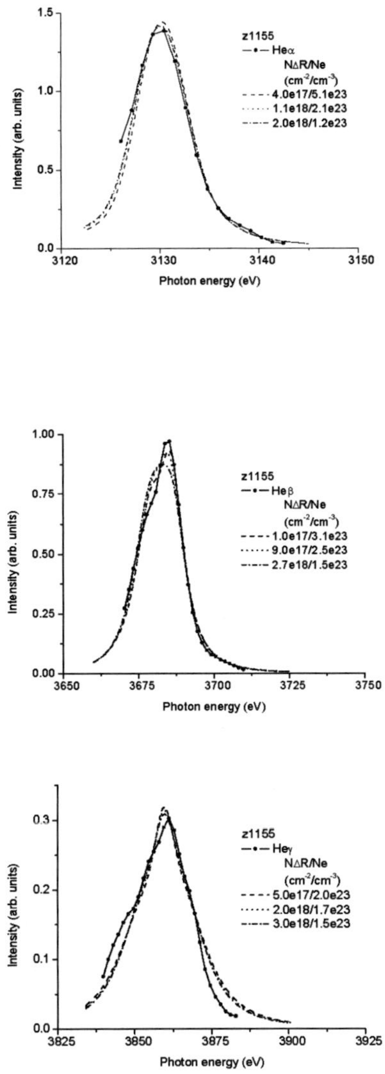

FIGURE 2. Fits of comparable quality for the Heα, Heβ and Heγ experimental line shapes extracted from the data shown in Figure 1.

RESULTS AND DISCUSSION

The analysis method discussed in the previous section was applied to the α, β and γ line shapes of the He and Lyman series extracted from the experimental time-resolved spectrum shown in Figure 1. Before comparing with the data, theoretical line shapes were convolved with a Gaussian function of FWHM=3.5 eV to account for the instrumental broadening effect. Also, a linear combination of the Lyβ and an (optically thin) Heδ was used to fit the Lyβ line to account for the partial blending on the high energy wing of the Lyβ with a weak Heδ emission, as observed in the data.

The results of the LOBF for both series of lines are displayed in Figure 3. We emphasize that the two "free" parameters on the analysis of each line profile are NΔR and N_e. The shape of the LOBF traces depends on the different opacities of the line transitions. In both series, the α line is the most sensitive to NΔR changes while β and γ are less dependent on this parameter. Also, shown in Figure 3 is a trace representing maximum possible values of NΔR for a given N_e, $N\Delta R_{max}$, so that meaningful intersections of LOBF occur to the left of this curve, i.e. the "allowed" region. To compute $N\Delta R_{max}$ we use an optically thick collisional-radiative atomic kinetics model that includes the effects of electron collisional excitation and de-excitation, electron collisional ionization and recombination, spontaneous radiative decay and photo-excitation, radiative recombination and photo-ionization, and dielectronic recombination [13]. Photo-excitation and photo-ionization are taken into account via escape factors calculated using detailed Stark broadened line shapes and photo-ionization cross-sections, respectively. In addition, a uniform distribution of absorbers and emitters inside a spherical plasma source is assumed [14]. The escape factors also depend on the radius of the plasma source. The radius R of the compressed core can be related to the initial core radius and gas filling conditions using a mass conservation argument since the electron and deuterium number densities are, to a good approximation, equal. Thus, for a given Ne, a compressed core radius R can be determined and the NΔR product for He- and H-like Ar ground states computed as a function of T_e for several N_e values using the optically thick collisional-radiative atomic kinetics model. As a function of increasing T_e and for given N_e, these functions rise and drop as a result of the shift of the ionization balance with temperature and display a single maximum point. The maxima of these functions give the values of the $N\Delta R_{max}$ plotted in Figure 3.

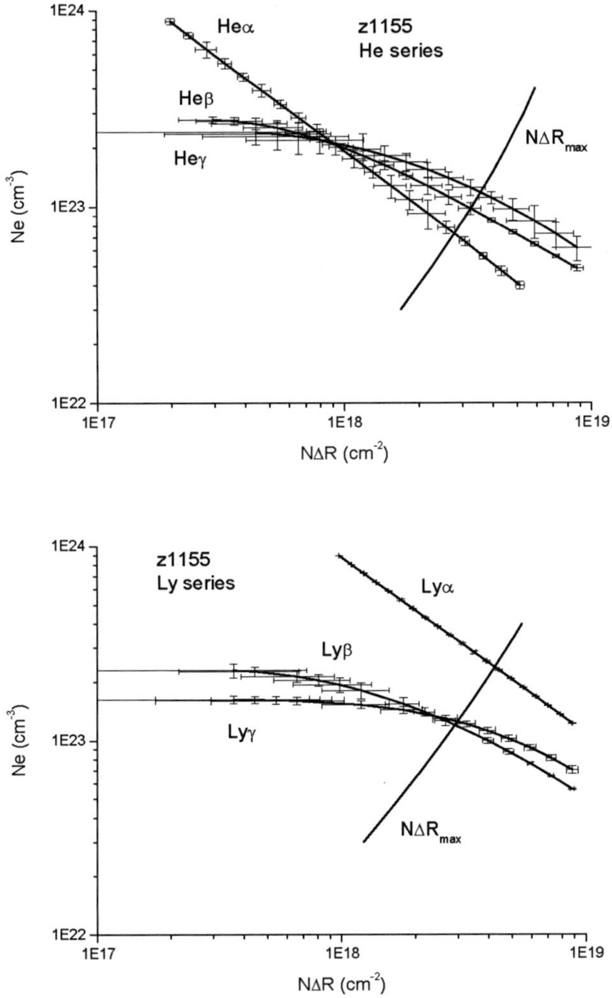

FIGURE 3. Lines-of-best-fits analysis results for the He and Lyman series α, β and γ lines of the z1155 spectrum. Upper-bounds $N\Delta R_{max}$, for each line series are also shown.

The three LOBF of the He series show an intersection for $N\Delta R = 9.1\times10^{17}$ cm^{-2} and $N_e=2.1\times10^{23}$ cm^{-3}, while the intersection of the β and γ lines of the Lyman series takes place for $N\Delta R=2.5\times10^{18}$ cm^{-2} and $N_e=1.3\times10^{23}$ cm^{-3}. These values can be used to compute the following optical depths of the α, β and γ lines: 67, 1.5 and 0.34 for the He series and 79, 2.8 and 0.61 for the Lyman series. Since the Lyα LOBF does not intersect with those of the β and γ lines, its optical depth is based on the intersection of its LOBF with the $N\Delta R_{max}$ curve. Furthermore, the electron temperature T_e can also be determined if we note that (1) the ratio of H- and He-like argon ground state populations can be obtained from the ratios of their respective NΔR's and (2) for the range of N_e values relevant for this work this ground state population ratio is a strong function of T_e and weakly dependent on N_e. For the data of Figure 1 this ratio is 2.8, which corresponds to T_e=1020 eV.

It is interesting to note that the N_e values extracted for the He and Lyman series are different. This can be interpreted as the line emissions of these two line series being formed in different spatial regions of the core and hence bringing information about different spatial averages. In this connection, we note that line emission distributions in the core obtained from post-processing hydrodynamic simulations of other ICF experiments also suggests this scenario [15]. In principle, something similar could be said of the lack of intersection of the Lyα line with the β and γ Lyman lines. However, this case is more striking since it happens within the same line series. We note that this line broadening analysis was also applied to the time-integrated Lyman series recorded in similar experiments with the FSSR instrument, and the same lack-of-intersection problem was found for the Lyα line. In both TREX and FSSR recorded data, the Lyα LOBF falls on the upper right portion of the $N\Delta R$-N_e plane indicating an amount of broadening larger than what it should have to intersect the β and γ LOBF. It is possible to interpret this lack of crossing in the LOBF in terms of two-temperature effects on the Lyα line shape, i.e. accounting for different electron and ion temperatures in the plasma core. Hydrodynamic simulations indicate that after the collapse of the main shock wave, plasma ions become hotter than electrons and then ions and electrons thermalize. In this connection, a new set of Stark-broadened line shapes was computed considering the effect of differing ion and electron temperatures, which impacts the calculation of the ion microfield distribution function, the Doppler broadening, and the ion dynamic's effect. All these effects will contribute to increase the broadening of the Lyα line profile, wich in turn can bring the croosing of the Lyα LOBF closer to the crossing of the LOBF of the β and γ Lyman lines. Figure 4 shows that this is the case for the Lyman series LOBF displayed in Figure 3. These first results are encouraging but additional work is required in order to further test this idea systematically in the analysis of several sets of data.

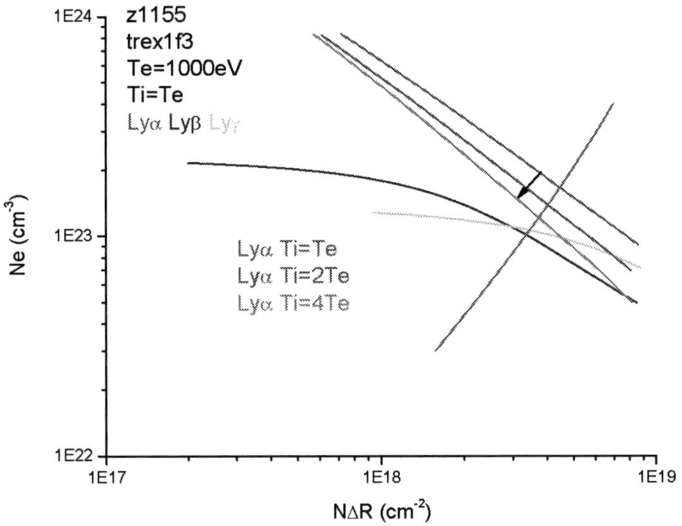

FIGURE 4. Effect of differing electron and ion temperatures on the crossing of the Lyα LOBF with the β and γ lines, for the Lyman series case displayed in Figure 3.

In summary, we have applied line broadening analysis to extract N_e, NΔR, optical depth and T_e from Ar K-shell lines observed at the collapse of ICF implosions driven by z-pinch dynamic hohlraums. The results of the analysis are consistent with conditions predicted by hydrodynamic simulations of the experiments. Further work is in progress to understand the problem of the lack of consistency for the broadening of the Lyα line compared to the broadening of the other lines in the Lyman series an its interpretation in terms of two-temperature effects on line shapes.

ACKNOWLEGDEMENTS

Sandia is a multiprogram laboratory operated by Sandia Corporation, a Lockheed Martin Company, for the U.S. Department of Energy under Contract No. DE-AC04-94AL85000. This work is also supported by SNL Contract No. 29195.

REFERENCES

1. Bailey JE, Chandler GA, Slutz SA, Bennett GR, Cooper G, Lash JS, Lazier S, Nash TJ, Nielsen DS, Moore TC, Ruiz CL, Schroen DG, Smelzer R, Torres J, Vesey RA. X-ray imaging measurements of capsule implosions driven by a z-pinch dynamic hohlraum. Physical Review Letters, 2002;89:095004.
2. Slutz SA, Bailey JE, Chandler GA, Bennett GR, Cooper G, Lash JS, Lazier S, Lake P, Lemke RW, Mehlhorn TA, Nash TJ, Nielsen DS, McGurn J, Moore TC, Ruiz CL, Schroen DG, Torres J, Varnum W, Vesey RA. Dynamic hohlraum driven inertial fusion capsules. Physics of Plasmas. Phys. Plasmas 2003;10:1875-1882.
3. Bailey JE, Chandler GA, Slutz SA, Golovkin I, Lake PW, MacFarlane JJ, Mancini RC, Burris-Mog TJ, Cooper G, Leeper RJ, Mehlhorn TA, Moore TC, Nash TJ, Nielsen DS, Ruiz CL, Schroen DG, Varnum WA. Hot dense capsule-implosion cores produced by z-pinch dynamic hohlraum radiation. Physical Review Letters 2004;92:085002.
4. Ruiz C, Cooper GW, Slutz SA, Bailey JE, Chandler GA, Nash TJ, Mehlhorn TA, Leeper RL, Fehl D, Nelson AJ, Franklin J, Ziegler L. Production of thermonuclear neutrons from deuterium-filled capsule implosions driven by Z-pinch dynamic hohlraums. Physical Review Letters 2004;93:015001.
5. Lake PW, Bailey JE, Rochau GA, Moore TC, Petmecky D, Gard P, Time- and space-resolved elliptical crystal spectrometers for high energy density physics research. Rev. Sci. Instrum. 2004;75:3690-3692.
6. Kilkenny JD, Lee RW, Key MH, Lunney JG. X-ray spectroscopic diagnosis of laser-produced plasmas, with emphasis on line broadening. Physical Review A 1980;22:2746-2760.
7. Delamater N, Hooper Jr. C F, Joyce RF, Woltz LA, Ceglio NM, Kauffman, Lee RW, Richardson MC. Opacity effects on hydrogen like x-ray lines emitted from laser-driven implosions. Physical Review A 1985;31:2460-2463.
8. Haynes DA, Hooper Jr. CF, Delamater ND, Pollack GD, Barnes C, Tubbs TL, Jaanimagi P. Chlorine K-shell spectroscopy of directly driven cylindrical implosions. Journ. Quant. Spect. Radiat. Transfer 2000;65:297-302.
9. Woltz LA,Hooper Jr. CF. Calculation of spectral line profiles of multielectron emitters in plasmas. Physical Review A 1988;38:4766-4771.
10. Mancini RC, Kilcrease DP, Woltz LA, Hooper Jr. CF. Calculation aspects of the Stark line broadening of multielectron ions in plasmas. Comp. Phys. Comm. 1991;63:314-322.
11. Haynes Jr. DA, Garber DT, Hooper Jr. CF, Mancini RC, Lee YT, Bradley DK, Delettrez JA, Epstein R, Jaanimagi P. Effects of ion dynamics and opacity on Stark-broadened argon line-profiles. Physical Review E 1996;53:1042-1050.
12. Coldwell RL, Bamford G, Spectral Analysis using ROBFIT. New York: American Institute of Physics, 1991.
13. Mancini RC, Hooper Jr. CF. Hydrodynamics and K-shell radiation emission of laser compressed Ar-filled glass microballoons. J. Phys. D 1988;21:1099-1111.
14. Mancini RC, Joyce RF, Hooper Jr. CF. Escape factors for Stark-broadened line profiles. J. Phys. B 1987;20:2975-2987.
15. Langer SH, Scott HA, Marinak MM, Landen OL. Towards a complete model of 3D line emission from ICF capsules: results from the first 3D simulations. Journ. Quant. Spectrosc. Radiat. Transfer 2000;65:353-366.

Measurements of Laser-Induced Plasma and Optical Breakdown Spectra of Aluminium

Christian G. Parigger

The University of Tennessee, The University of Tennessee Space Institute (UTSI), Center for Laser Applications, 411 B. H. Goethert Parkway, Tullahoma, TN 37388, U.S.A.

Abstract. Time-resolved spectroscopy is utilized in investigations of laser-induced optical breakdown plasma. Atomic and molecular features are analyzed for diagnostics applications in laser materials processing. The micro-plasma that is generated by nominal 1 to 10 nanosecond, focused laser radiation shows Stark broadened and shifted atomic lines early in the plasma decay. Subsequently, the emissions from the plasma reveal molecular recombination spectra. Analysis of the spectroscopic features allows one to infer species number densities and temperature. Measured spatial characteristics and post-breakdown fluid physics phenomena are numerically modeled using computational fluid dynamics. Electron number densities as high as 10^{19} cm^{-3} are found, determined from Stark widths and Stark shifts of hydrogen Balmer-alpha line. Stark broadened aluminium line spectra, following optical breakdown of solid aluminium, show electron number densities of typically 0.5 x 10^{19} cm^{-3} early in the plasma decay. The electron number densities for the aluminium experiments are inferred by comparison with hydrogen Balmer series lines.

Keywords: Atomic spectra; Molecular spectra; Plasma diagnostics; Laser spectroscopy; Laser-induced breakdown; Line shapes, widths and shifts; Materials processing.
PACS: 32.30.-r; 33.20.-t; 52.70.-d; 42.62.Fi; 78.47.+p;39.30.+w; 82.80.-d; 32.70.Jz; 81.20.-n.

INTRODUCTION

Laser-induced optical breakdown of gases and solids generates high temperature micro plasma that shows rich emission spectra. Early in the plasma decay, spectroscopic signatures include atomic species embedded in free-electron background. Later in the plasma decay, typical signatures of highly excited molecular recombination spectra can be recorded. Historically, laser-induced breakdown spectroscopy (LIBS) studies utilize nominal nano-second laser radiation for generation of the micro-plasma. In the work presented here, nominal 1 to 10 nanosecond Nd:YAG radiation is focused into gas samples or selected solid samples. Following optical breakdown, the initially "non-absorbing" medium is transformed into hot plasma, and temporal and spatial dynamics are of interest.

Time-resolved spectroscopy techniques are employed to investigate the temporal evolution of electron number density and temperature early in the plasma decay. Subsequently, molecular emissions are used to characterize post-breakdown phenomena. Of interest is the use of hydrogen Balmer series lines, in particular H_α and H_β, to infer the plasma state generated from focusing laser radiation onto solid aluminium. The application of spectroscopy methods are of interest in materials

CP874, *Spectral Line Shapes: 18th International Conference,*
edited by E. Oks and M. Pindzola

science for example in micro-machining. Nominal nanosecond Q-switched laser radiation has found several machining applications. Yet precise micro-machining is typically achieved by use of radiation from pico- or femto-second laser devices. Experimental investigations are discussed of the laser-induced plasma, including aluminium emission spectroscopy to determine of electron number density [1, 2].

EXPERIMENTAL DETAILS

Optical breakdown is generated using radiation from a Continuum Q-switched Nd:YAG laser at a repetition rate of 10 Hz. Figure 1 shows a typical experimental arrangement generation of optical breakdown. Fig. 2 shows a typical arrangement for collection of time-resolved images following optical breakdown.

FIGURE 1. Experimental arrangement for laser-induced optical breakdown studies in laboratory air. Both fundamental 1064-nm and 2nd harmonic 532-nm Nd:YAG radiation are focused above a flat-flame burner after passing through a telescope and polarizer.

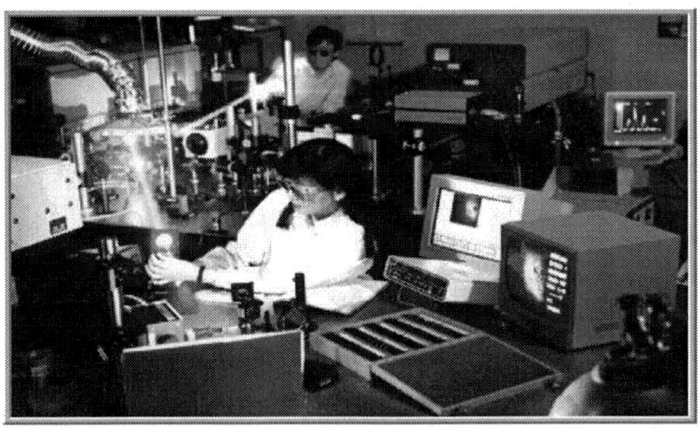

FIGURE 2. Experimental arrangement for shadowgraph studies subsequent to laser-induced optical breakdown in laboratory air. Copper vapor laser radiation is used to illuminate the focal volume of the breakdown laser, and shadowgraphs are recorded by filming the screen with a high-speed camera.

The 8 nano-second pulses for the 1064-nm fundamental mode are focused with a 100-mm focal length lens into gas samples, or later onto an aluminium block. The focal spot diameter amounts to typically 30 micron, yielding an irradiance of approximately 1 TW/cm^2. Figure 3 shows records of shadowgraphs subsequent to laser-breakdown in air.

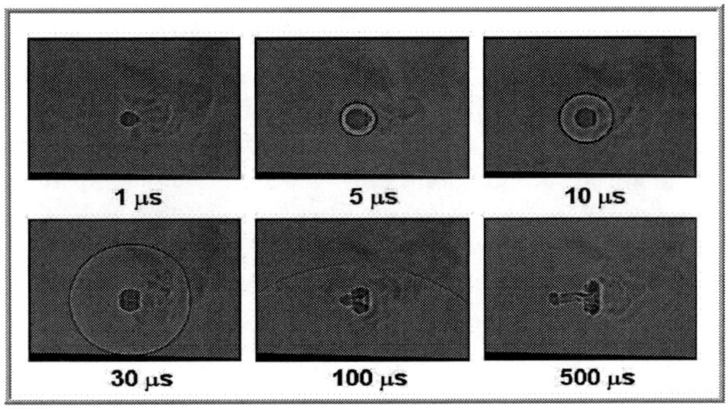

FIGURE 3. The recorded shadowgraphs for the indicated time delays show the generation of a shock wave followed by development of vortices and fluid physics phenomena.

The images (36 mm x 48 mm) show the development in time of the optical breakdown kernel as well as the development in time of the shock wave. This shock wave shows propagation speeds of approximately 1km/s a few microseconds after breakdown. For time delays of 100 micro-seconds, a reflected shock wave can be seen. The images also show computationally modeled fluid physics phenomena [3].

Optical breakdown requires high peak power, typically 100 GW/cm^2 in laboratory air when using nominal 10 nanosecond ir radiation. Expansion of the source beam is normally indicated to decrease the f-number and increase the focal irradiance. Figure 4 shows results of computed irradiance distributions. Note the typical asymmetry. The resulting asymmetric temperature distribution can be viewed as precursor for the fluid physics phenomena of vortex generation associated with expansion and strong axial flow towards the laser device (see Fig. 3 for time delays of 100 μs and 500 μs).

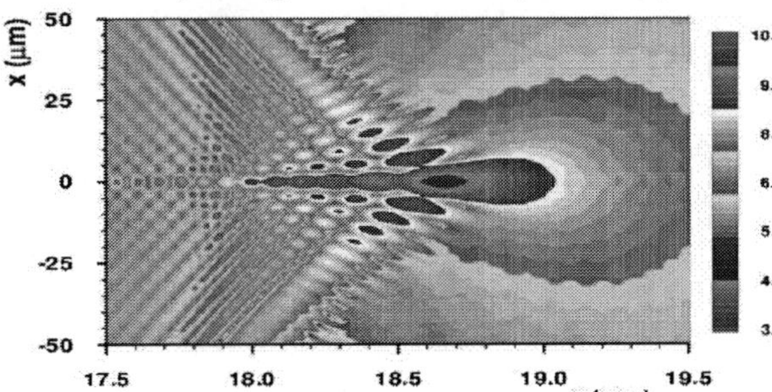

FIGURE 4. Computed irradiance distributions in the focal volume of a 19-mm focal length singlet lens show the effects of spherical aberration for f-number of 3.2 focusing.

RESULTS

The early plasma decay shows high free-electron background, neutrals and ions of atomic species [4,5,6]. Later in the plasma decay, molecular recombination spectra can be measured. A typical example of such a molecular spectrum is presented below. Subsequently, selected results of atomic spectra following optical breakdown of solid aluminium are presented, showing details of line-width and line shifts

Figure 5 shows the simulated and measured spectra following laser-induced breakdown using Nd:YAG 1064-nm radiation of 7.5 ns pulse-width focused into a cell containing CO at 3.25 psi (168 Torr) above atmospheric pressure. These spectra are recognizable within the first microsecond after optical breakdown. Typically, for delays of 30 μs the Swan spectra become prominent, without presence of atomic carbon lines. A spectral resolution of 11 cm^{-1} was used, and a temperature in excess of 6000 K was fitted.

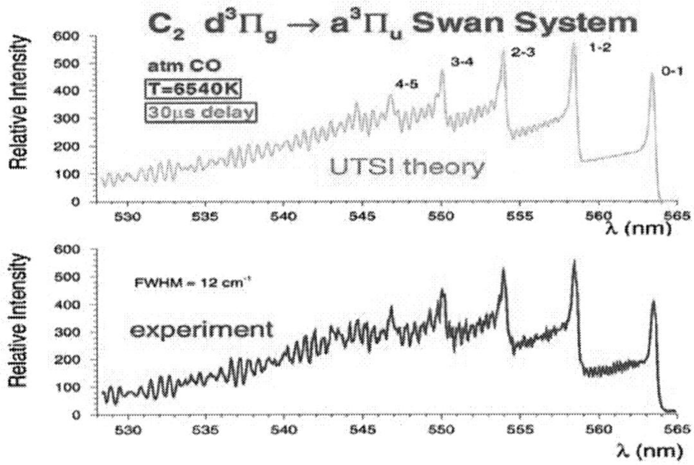

FIGURE 5. Measured and computed Δv=-1 sequence of the C_2 Swan system molecular spectrum.

The computation of the synthetic molecular spectrum utilizes the so-called Boltzmann Equilibrium Spectrum Program (BESP). This program uses line-strength files generated for selected diatomic molecular transitions. The line strength files are obtained by combining Franck-Condon and Hönl-London factors. The Frank-Condon factors are found by numerically solving the radial Schrödinger equation, and the Hönl-London factors are obtained by numerical diagonalization of the rotational and fine-structure Hamiltonian [7].

Analysis of superimposed spectra however is challenging. The computer code for non-equilibrium air radiation (NEQAIR) has been successfully employed in analysis of post-breakdown molecular spectra for NO, N_2, OH and O_2, and also AlO [8,9,10, 11]. Species number densities, including free electron number density, are calculated by use of the NASA chemical equilibrium code to serve as input for application of NEQAIR for equilibrium conditions.

Time-resolved techniques with 100-ns gate width are initially applied in spectroscopy studies of optical breakdown of solid aluminium in laboratory air [12]. Figure 6 shows the recorded, single shot spectra for several time delays following optical breakdown. The gate width of 100-ns was selected to record spectra at the indicated time delays. The resolution of these spectra amounts to approximately 0.4-nm for the selected slit-width of the ½-m monochromator fitted with a 1200 g/mm grating. The early Al lines are intense, are Stark broadened and noticeably Stark shifted.

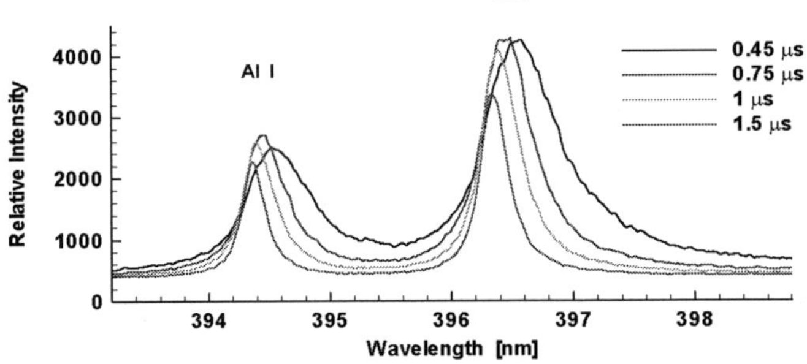

FIGURE 6. Recorded Stark profiles - aluminium. A number density of $N_e=10^{17}$ cm^{-3} is inferred 1-μs after optical breakdown in air, using here a 100-ns gate width.

In the work reported below, we use 150 mJ, 8-ns laser pulses with a peak irradiance of 1 TW/cm^2. This peak irradiance corresponds to an electric field strength of E=1/100 x 2.8GV/cm, or 1/100 of the field strength that holds the hydrogen atom together. Hydrogen Balmer series lines are used to infer the number density of aluminium expanding into a cell filled with 125 Torr of hydrogen. Figure 7 shows characteristic results recorded for the Al I 394.4-nm and 396.15-nm lines. Fig. 7 also shows the un-shifted line position, individually measured intensified optical multichannel analyzer (OMA) points and fitted results using a Savitzky-Golay filter. The spectrum was recorded using fiber-imaging and coupling to a ¼ -m spectrometer fitted with a 2400 grooves/mm grating, resulting in a spectral resolution of 0.25-nm.

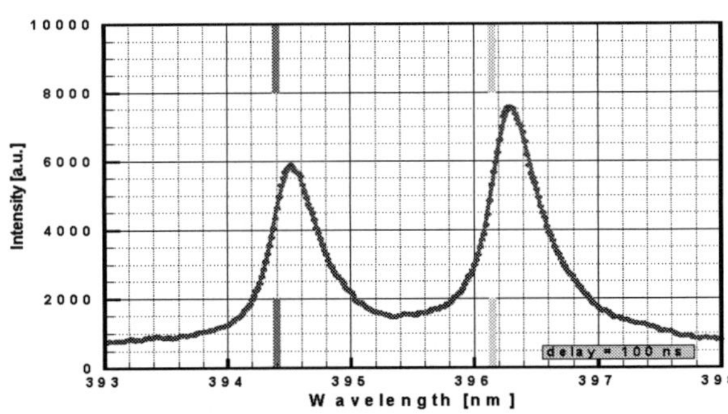

FIGURE 7. Recorded Stark widths and shifts for Al I 394.40-nm and Al I 396.15-nm lines. A number density of N_e=3 x 10^{18} cm^{-3} is inferred by comparison with H_α profiles 100-ns after optical breakdown in a cell filled with 125 Torr of hydrogen. Here, a 6-ns gate width was used.

Tables 1 and 2 show the aluminium Stark width and Stark shifts for selected time delays from optical breakdown.

TABLE 1. Line width (FWHM) and red shift for the λ=394.4 nm line of Al I .

Time delay [ns]	Width [nm]	Shift [nm]	N_e [10^{18} cm^{-3}]
30	0.708 ± 0.08	0.248 ± 0.03	
50	0.682 ± 0.08	0.235 ± 0.03	
60	0.407 ± 0.05	0.140 ± 0.02	4.0 ± 0.6
100	0.383 ± 0.05	0.124 ± 0.02	3.0 ± 0.5
200	0.187 ± 0.05	0.077 ± 0.02	0.6 ± 0.06
400	0.112 ± 0.04	0.040 ± 0.02	0.2 ± 0.04
1000	0.102 ± 0.04	0.012 ± 0.02	
1500	0.039 ± 0.03	0.004 ± 0.02	
3000	0.01± 0.03	0.001 ± 0.02	

TABLE 2. Line width (FWHM) and red shift for the λ=396.15 nm line of Al I .

Time delay [ns]	Width [nm]	Shift [nm]	N_e [10^{18} cm^{-3}]
30	0.739 ± 0.08	0.248 ± 0.03	
50	0.559 ± 0.08	0.235 ± 0.03	
60	0.428 ± 0.05	0.140 ± 0.02	4.0 ± 0.6
100	0.381 ± 0.05	0.124 ± 0.02	3.0 ± 0.5
200	0.263 ± 0.05	0.077 ± 0.02	0.6 ± 0.06
400	0.176 ± 0.04	0.040 ± 0.02	0.2 ± 0.04
1000	0.139 ± 0.03	0.012 ± 0.02	
1500	0.078 ± 0.03	0.004 ± 0.02	
3000	0.039 ± 0.03	0.0 ± 0.02	

The listed values are de-convolved from the spectrometer resolution, and the error bars are typical estimates. Note the relative large errors for longer time delays, due to the experimental design to measure larger Stark widths and shifts. The large absolute errors for short time delays are caused by difficulties in extracting accurate widths at early time delays. Electron number densities, extracted by comparison with hydrogen Balmer series lines, are indicated.

Figure 8 illustrates results for selected time-resolved spectra, for time delays of 60-n (bottom right) to 400-ns (top left). For these time delays the inferred electron number densities are included in Tables 1 and 2. As time-resolved spectra are recorded earlier in the plasma decay, notice the increase in free electron background radiation. The recorded spectra also show ionized lines of aluminium, clearly discernable for delays of 100-ns and 60-ns. Selected identified lines are the singly ionized Al II 390.0675-nm, and a bunch of lines near 399-nm: Al II 399.6182-nm, Al II 399.6143-nm, Al II 399.6143-nm, Al II 399.6075-nm, and Al 399.5838-nm. Fig. 8 further shows occurrence of extra spectroscopic features near Al I 394.4-nm and Al I 396.15-nm lines for a delay of 60-ns. The extra features cause difficulties in using these lines early in the plasma decay (see also top row of Fig. 11 below).

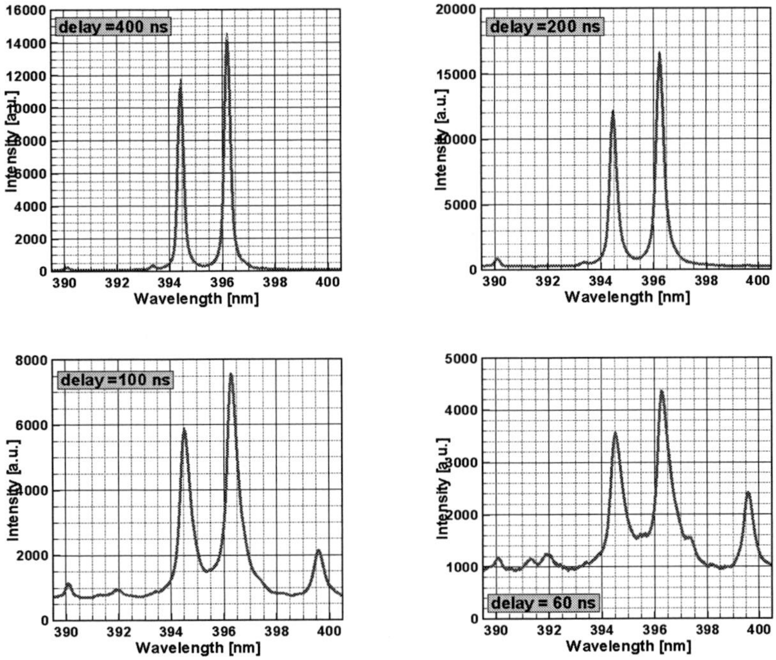

FIGURE 8. Recorded Stark widths and shifts for Al I 394.40-nm and Al I 396.15-nm lines.

Table3 shows the hydrogen alpha Balmer-series Stark width and Stark shifts for selected time delays from optical breakdown. Also indicated are the number densities that are inferred from width and red shift, using Parigger *et al.* tables 3 and 4 for H_α [5]. The number densities for hydrogen-alpha are extracted by use of results from applying Oks' convergent theory. The indicated error bars for early time delays reflect difficulties in extracting widths of atomic lines embedded in large free-electron background radiation. Figure 9 illustrates selected time-resolved spectra for delay times of 400-ns, 200-ns, 175-ns, and 100-ns. The hydrogen-alpha results for the 50-ns and 30-ns time delays are grouped with early aluminium results (see Fig. 11 below).

TABLE 3. Line width (FWHM) and red shift for λ=656.28 nm H_α Balmer series line.

Time delay [ns]	Width [nm]	N_e [10^{18} cm^{-3}]	Shift [nm]	N_e [10^{18} cm^{-3}]
30	15.3 ± 0.24	5.0 ± 0.8	4.12 ± 0.24	6.0 ± 0.8
50	13.2 ± 0.24	4.0 ± 0.6	3.60 ± 0.24	5.0 ± 0.6
75	11.7 ± 0.18	3.0 ± 0.5	2.32 ± 0.18	3.5 ± 0.5
100	10.2 ± 0.18	2.0 ± 0.4	2.02 ± 0.18	3.0 ± 0.4
175	6.48 ± 0.18	0.9 ± 0.06	1.12 ± 0.18	1.0 ± 0.06
200	5.76 ± 0.12	0.7 ± 0.04	0.62 ± 0.12	0.8 ± 0.04
400	2.40 ± 0.12	0. 3 ± 0.04	0.37 ± 0.12	0.4 ± 0.04

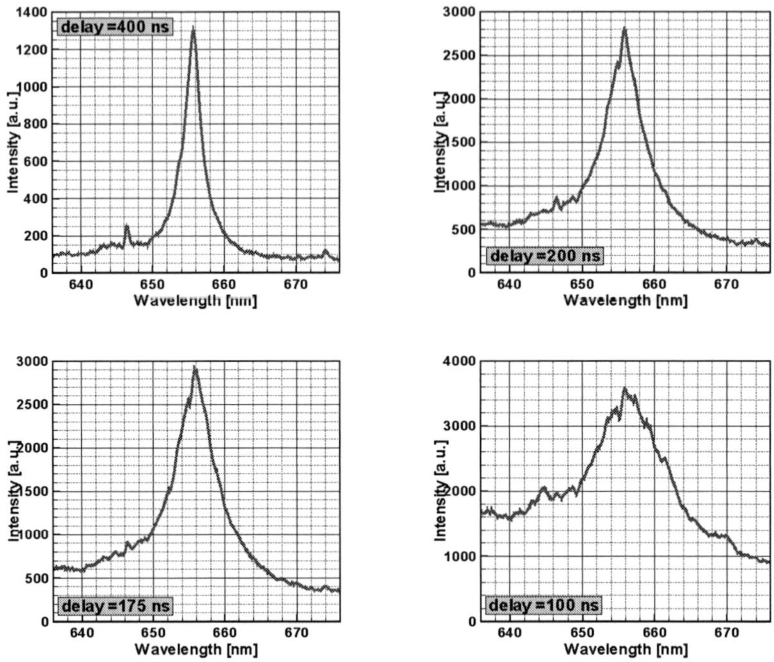

FIGURE 9. Recorded Stark widths and shifts for hydrogen-alpha Balmer series line.

Table 4 shows the hydrogen-beta Balmer-series Stark width and Stark shifts for selected time delays from optical breakdown. Also indicated are the number densities that are inferred from Stark widths, using Parigger *et al.* table 2 for H_β [5].

Table 4 also shows the extracted separation of the two hydrogen-beta peaks. Previous experimental studies by Parigger *et al.* showed that this separation amounts to approximately ¼ of the FWHM Stark-width early in the plasma decay, see Parigger *et al.* table 1 for H_β [5] for this correlation with Stark width. The red and blue peak separation is: $\lambda_{red} - \lambda_{blue} = \boldsymbol{a}\ \Delta\lambda_{1/2}$, with a value for the constant $\boldsymbol{a}$ of approx. $\boldsymbol{a} = 0.25$.

Figure 10 illustrates selected time-resolved spectra for time delays of 400-ns, 200-ns, 175-ns, and 150-ns. Note that the displayed result for a time delay of 150-ns may reflect the highest ever recorded electron number density using H_β. Obviously, FWHM-determination may be difficult, considering presence of other, here not identified lines. Yet the red/blue separation of peaks may be used as additional diagnostics for determination of high electron number density using H_β.

TABLE 4. Line width (FWHM) and peak separation for λ=486.14 nm H_β Balmer series line.

Time delay [ns]	Width [nm]	N_e [10^{18} cm^{-3}]	Separation [nm]	N_e [10^{18} cm^{-3}]
150	20.02 ± 0.24	0.9 ± 0.09	7.39 ± 0.08	0.95 ± 0.08
175	17.48 ± 0.24	0.7 ± 0.07	6.82 ± 0.07	0.75 ± 0.06
200	12.97 ± 0.18	0.5 ± 0.05	4.12 ± 0.04	0.5 ± 0.05
400	4.79 ± 0.18	0.15 ± 0.02	1.45 ± 0.02	0.16 ± 0.02

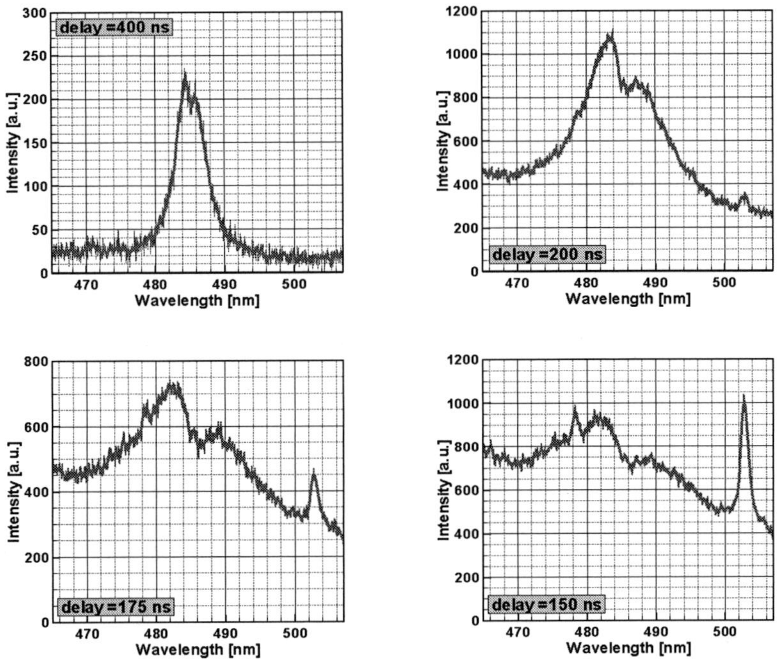

FIGURE 10. Recorded Stark widths and shifts for hydrogen-beta Balmer series line.

For even earlier time delays, determination of Stark width and shift becomes increasingly difficult, due to presence of large free-electron background and occurrence of ions' spectroscopic signatures. Previous experimental studies [5] reported by Parigger *et al.* utilized gas samples of hydrogen, leading to number densities of almost 10^{19} cm^{-3} a few nanoseconds after generation of optical breakdown. Here, breakdown of solid aluminium causes generation of typically 10^{22} cm^{-3} free electron density near the surface. Measurements of densities in the order of 10^{19} cm^{-3} are expected away from the target [2], and for time delays of several tens of nanoseconds. Figure 11 illustrates recorded spectra for aluminium and hydrogen-alpha emissions, for time delays of 50-ns and 30-ns.

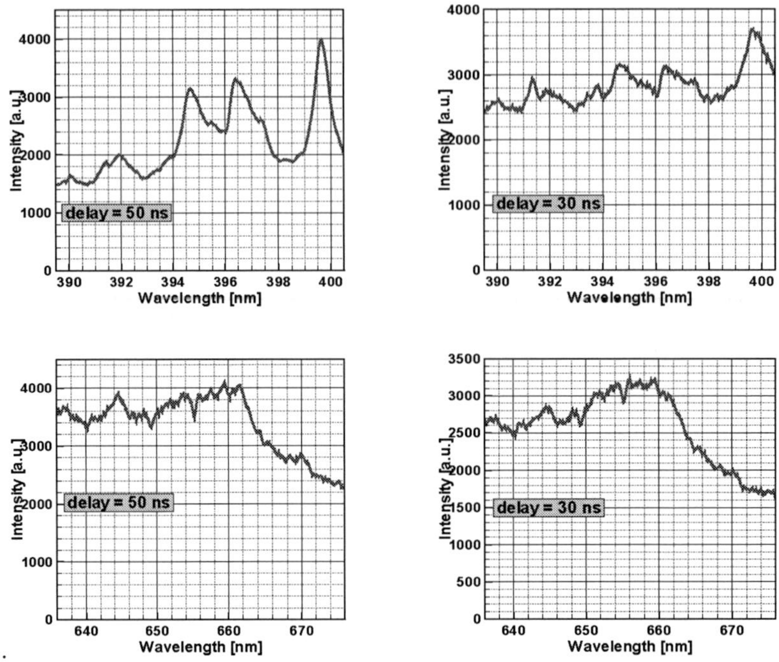

FIGURE 11. Recorded Stark widths and shifts for aluminuim (top row) and hydrogen-alpha Balmer series line (bottom row) for time delays of 50-ns and 30-ns. Inferred number densities from H_α width and shift are listed in table 3.

CONCLUSIONS

Aluminium Stark profiles were measured previously by C.Fleurier *et al.* [1]. A welding type arc at atmospheric pressure was used, combined with 95% argon, 5% hydrogen, and $AlCl_3$ vapor flow. Using 15kW electrical power, the Al I 394.4-nm and 396.1-nm lines showed a width of 0.054-nm and a shift of 0.031-nm. This resulted in an electron number density of N_e=1.28 x 10^{17}cm^{-3} and a temperature of T=1.32 x 10^4 K. The results presented here are consistent with our nominal nanosecond laser-plasma studies. However, spatial resolution over and above the temporal resolution employed in this work is desirable in the measurements of aluminium plasma in presence of hydrogen.

Laser-plasma studies by Renner *et al.* [2], using 170 J, 0.4 ns pulses, and a peak irradiance of 900 TW/cm^2, showed N_e=2 x 10^{22} cm^{-3} and a temperature of T=1 keV. Spatially resolved Lyman series of aluminium were recorded. Clearly, densities above N_e=0.8 x 10^{18} cm^{-3} are difficult to measure using only H_β line-widths, yet from the peak separation in the H_β line profile densities of approx. N_e=1 x 10^{18} cm^{-3} may be inferred for delays of 150-ns after generating the aluminium plasma (see Figure 10). A number density of N_e=1 x 10^{18} cm^{-3} may very well be the highest ever recorded from hydrogen-beta Balmer series line measurements.

Similarly, limits are indicated in determining electron number density by use of the neutral aluminium Al I 394.4-nm and Al I 396.15 lines in determining electron number density (see Figure 11). Yet within the limit of application as laser-solid interaction diagnostics of the neutral aluminium lines, the H_α and H_β width, H_α shifts, and H_β peak-separations can be used to deduce consistent values for electron number density.

ACKNOWLEDGMENTS

The author thanks Mr. J.O. Hornkohl over and above for all other contributions for his dedication in his part for recording the aluminium experimental series. Further thanks are expressed to Dr.'s Y.-L. Chen, I. G. Dors, G. Guan, D. R. Keefer, J. W. L. Lewis, E. Oks, L. Nemes, D. H. Plemmons, W. Qin, Y. Tang, and to Mr.'s F. A. Schwartz, and N. W. Wright, for their interest and contributions to this work. Also, thanks to NASA Ames Research Center for making available the NEQAIR96 program, along with detailed comments by Dr. E. E. Whiting regarding the applications and extensions of the code.

This material is based upon work that is supported in part by the National Science Foundation under Grant No. CTS-9512489, in part by a grant of High Performance Computer (HPC) time from the Department of Defense HPC Center, AEDC at Arnold AFB, on the Origin 2000 computer, in part by UTSI and UTSI's Center for Laser Applications, and in part by UTSI's Space Grant.

REFERENCES

1. C. Fleurier, S.Sahal-Bréchot and J. Chapelle, J.Phys.B: Atom. Molec.Phys. **10**, 3435-3441 (1977).
2. O. Renner, D.Salzmann, P. Sondhauß, A.djaoui, E. Krouský and E. Förster, J.Phys.B: At.Mol.Opt. Phys **31**, 1379-1390 (1998).
3. I.G. Dors and C.G. Parigger, Appl.Opt. **42**, 5978-5985 (2003).
4. C.G. Parigger, "Experiments on Stark widths and shifts of hydrogen Balmer lines in laser-induced plasma," in *Proceedings of the 17th International Conference on Spectral Line Shapes-200,* edited by E Dalimier, Physique Atomique dans les Plasmas Denses, Laboratoire pour l'Utilisation des Lasers Intenses, Université Pierre et Marie Curie, 4, place Jussieu, 75252 Paris Cedex 05, 2004, pp. 97-104.
5. C.Parigger, D.H.Plemmons, and E. Oks, Appl.Opt. **42**, 5992-6000 (2003).
6. C.Parigger, J.O.Hornkohl, A.M. Keszler and L. Nemes, Appl. Opt. **42**, 6192-6198 (2003).
7. J.O.Hornkohl, C.G.Parigger, and L. Nemes, Appl.Opt. **44**, 3686-3695 (2005).
8. I.G. Dors, C. Parigger, and J.W.L. Lewis, Opt. Lett. **23**, 1778-1780 (1998).
9. J.O.Hornkohl and C.G.Parigger, *Boltzmann Equilibrium Spectrum Program (BESP),* The University of Tennessee Space Institute, 2002, http://view.utsi.edu/besp.
10. C.G. Parigger, G. Guan, and J.O. Hornkohl, Appl. Opt. **42**, 5986-5991 (2003).
11. C.G. Parigger, "Laser-induced breakdown in gases: experiments and simulation" in *Laser-induced Breakdown Spectroscopy,* edited by A. Miziolek, V. Palleschi, and I. Schechter (Cambridge: Cambridge University Press) 2006, pp. 171-193, and see references therein.
12. J.O. Hornkohl, C.G.Parigger, L.Nemes and A. Keszler, "Measurements of Aluminium Laser-Induced Breakdown Spectra," in *Laser Applications to Chemical, Security and Environmental Analysis 2006 Technical Digest* (The Optical Society of America, Washington, DC) 2006, TuE4.

Density Matrix Approach to Description of Doubly Excited States in Dense Plasmas

Alexander V. Demura*, Frank B. Rosmej†, Roland Stamm†

*HEPTI, RRC "Kurchatov institute", Kurchatov Square 1, Moscow 123182, Russia,
†Universite de Provence et CNRS, UMR 6633, Centre de Saint Jerome, case 232, PIIM, Avenue Escadrille Normandie-Niemen, F-13397 Marseille cedex 20, France

Abstract. The density matrix approach allows to consider self-consistently the atomic kinetics and radiation dynamics in one closed system of equations. This approach is typical for laser physics, but for Stark broadening by plasmas it was introduced in the beginning of ninetieth. The present work is aimed on the specifics of such description for the located above ionization threshold doubly excited states (DES) of multiply charged ions. The Stark profiles of dielectronic satellites, originating from DES, play an important role in the diagnostic implementations. The existence of nonlinear interference effects (NIEF) in the Stark profiles of dielectronic satellites of multiply charged ions is demonstrated in the frames of the three-level model.

Keywords: density matrix, dielectronic satellites, Stark broadening by plasmas, diagnostics
PACS: 32.70.Jz, 52.20.Hv, 52.25.Gj, 52.25.Os, 52.55.Fa, 52.70.-m

"When you start to study any subject you would suddenly discover existence of permanent coup d'etat to present the same subject dogmatically and one sided..."

Bernard Shaw

INTRODUCTION

The density matrix (DM) approach (DMA) to quantum mechanical description of radiative processes is considered conventionally as more complicated and complete in comparison with amplitude one [1-5]. In fact from the very beginning it includes elements with a simultaneous dynamical and statistical description [1-5]. There are various realizations and applications of DMA but the study here will be devoted to so called atom-photon density matrix, assuming that intensity of radiation fluxes are sufficiently small [1-2, 4-5]. In this formalism the density matrix of a system should be a solution of a set of kinetic equations that connect populations with coherences, describing dynamics of radiative transitions [1-5]. In general this formalism represents an infinite chain of coupled sets of equations, corresponding to one-, two-...and "n"-photon processes. It is similar to a coupling of one-, two-..."n"-particle distribution functions of BBGKY chain or Green-functions [1, 4]. As a rule the system under consideration represents itself several interacting subsystems or statistical ensembles [1-4]. The derivation of DM equations for the particular subsystem is achieved by the formalism of projection operators either by average procedures over "bath" variables and states [1-3] of particular commutators corresponding to the interaction with the bath. That is why in zero approximation of interaction with radiation field the system of equations for diagonal matrix elements of DM has the characteristic appearance of balance equations with terms that could be attributed to sources and sinks [1]. The

CP874, *Spectral Line Shapes: 18th International Conference,*
edited by E. Oks and M. Pindzola

derivations of those "relaxation' terms as functionals of DM itself are very hard tasks [1-4]. Very often they are put down phenomenologically preventing to follow correctly the physics of various interference effects characteristic to radiative processes [1]. However up to now the general form of relaxation terms in the kinetic equations for DM is not completely defined and still is a subject of various special separate studies, constructions, different assumptions and approximations that allow to simplify the problem basing on its characteristic physical distinctive features, that predetermines the extent of confidence to the probable solutions and its range of validity [1-5]. In what is followed the atom-photon DMA for one photon transitions will be used in the form elaborated in [1-2].

The study of doubly excited states behavior in plasmas provides advantageous diagnostic possibilities [6]. The distinctive feature in the description of doubly excited states (DES), located above ionization threshold, is an introduction of continuum states or "continuum" that hits the same energy interval as DES [2-3, 6-13]. The configuration interaction with the continuum leads to the decay of DES due to autoionization [7] that is to transitions to continuum [2-3, 6-13]. Thus DES are quasistationary states on the background of two continua – the spontaneous radiation field and the continuum of states of the compound system, composed from ionized electron and an ion in the next ionization stage [2-3, 6-13]. These interactions both lead to the DES decay.

If the interaction is strong then it is necessary to seek for some exact solutions, only possible in the simplest cases [1-5]. The solutions in general form could be obtained only within the perturbation theory that would be assumed possible to implement further on [1-5]. Then it is necessary to establish a hierarchy of interactions that will control the perturbation solutions of the DM equations [1-5]. This is especially important under DES consideration, when the sources in the equations for DM include competing processes of dielectronic and radiative recombination, autoionization, charge exchange, influence of collisions etc, while the equations themselves already in zero approximation contain interaction with plasma ion microfields and coupling with continua of different origin [3, 6-13], that makes physics under the line shape problem quite interesting and complicated.

This work has a review character and is aimed to present relatively transparent recipes of DM method implementation for a description of DES spectral features in the conditions, when the interaction with plasma electrons could be described by the collision integral operator, while the interaction with plasma ion microfields could be considered with the help of the quasistatic approximation or by simulations of temporal electric field evolution by molecular dynamics method or by method of rectilinear trajectories [14-17]. The description of DES with plasma ions within the notion of plasma microfields assumes at the same time the use of the dipole approximation for a description of this interaction [17].

DENSITY MATRIX QUANTUM KINETIC EQUATIONS

Let us represent density matrix ρ as a sum of terms corresponding to the successive orders of the perturbation theory [1-2, 4-5]

$$\rho = \rho^{(0)} + \rho^{(1)} + \rho^{(2)} + ... \tag{1}$$

with respect to additive interactions, listed in the general form as [1-2, 4-5]

$$V(t) = V^{(0)}(t) + V^{(1)}(t) + V^{(2)}(t) + ... \tag{2}$$

For simplicity only the uniform case will be considered further on [1-2, 4-5], i.e. when the bra and ket vectors of quantum states entering in ρ refer to the same space coordinates. Then in the interaction representation with respect to unperturbed Hamiltonian $\boldsymbol{H_0}$ the following equation is valid [1-2, 4-5]

$$\frac{\partial \rho}{\partial t} - [\rho, V] + St(\rho) + R(\rho) + W(\rho) + Q(\rho), \tag{3}$$

$$[\rho, V] = \rho V - V\rho,$$

where the square brackets correspond to commutator, ***St*** –designates the collision operator of the radiator with plasma electrons, ***R*** – is the operator of radiation relaxation, ***W*** –is the operator of autoionization relaxation, ***Q*** – is the operator of incoherent pumping (sources and sinks).

The interaction with the field of spontaneous emission will be taken into account using perturbation theory and is described by the operator [1-2, 4-5]

$$V^{(1)}(t) = C_{jk} \beta_j^{+} \beta_k \exp(-\Omega_{jk} t), \tag{4}$$

where β_j^{+}, β_k are annihilation and creation operators of the emitting ions in some specific quantum states j and k, $\Omega_{jk} = \omega - \omega_{jk}$, ω - is the current radiator frequency, ω_{jk} – is the frequency of the transition between states $\boldsymbol{j}$ and $\boldsymbol{k}$. Under interaction with the field C_{jk} only radiation is possible since corresponding operator is defined with Hermitian conjugate [1-2]

$$[\rho, V] = V\rho - \rho V^{+}, \tag{5}$$

while a constant C_{jk} is assumed to be small.

In accordance with what was said above the interaction with the ion microfield has the following form in the interaction representation [4-5]

$$V^{(0)}(t) = \exp(-iH_0 t)\vec{d}\vec{F}(t)\exp(iH_0 t)/\hbar \tag{6}$$

One of the ways for simplification the MDM system of equations consists in the decreasing its rank by introduction of the effective operators ***W*** and ***Q*** that do not depend on ρ and are reduced essentially to some constants, characterizing the intensity of level pumping. This corresponds to the separation of subsystem for which the DM system of equations is put down exactly while the interaction with other subsystems and corresponding systems of equations are reduced due to the introduction of pointed out effective operators [1-2, 4-5].

In the general form the operator of incoherent pumping [1-2, 3-5] for DES includes the collisional pumping from the other levels $Q_j^{(in)} = \sum_{n \le j} Q_{jn}^{(exc)} \rho_{nn} + \sum_{n \ge j} Q_{jn}^{(dexc)} \rho_{nn}$, describing the processes of collisional excitations $Q_{jn}^{(exc)}$ from the lower lying levels and the deexcitation processes $Q_{jn}^{(dexc)}$ from higher lying levels, as well as the specific for DES item $Q_j^{(RC)}$, stipulated by the process of resonant capture – the first stage of dielectronic recombination [2-3, 6-13], that is represented as effective operator, depending only on density and temperature of plasma electrons [2-3, 6-13], and the devastation of the given level $Q_j^{(out)} = -\rho_{jj}(\sum_{n\leq j} Q_{jn}^{(dexc)} + \sum_{n\geq j} Q_{Jn}^{(exc)} + Q_j^{(ion)})$ due to the deexcitation processes $Q_{jn}^{(dexc)}$, excitation $Q_{jn}^{(exc)}$ and ionization $Q_j^{(ion)}$ etc. It is very important to take into account charge exchange processes that could populate or depopulate DES very effectively as well [11]. The net effect of charge exchange will be designated as $Q_j^{(CH)}$. Thus the incoherent pumping terms could be represented as

$$Q_j = Q_j^{(in)} + Q_j^{(RC)} + Q_j^{(CH)} + Q_j^{(out)} = \sum_{n\leq j} Q_{jn}^{(exc)}\rho_{nn} + \sum_{n\geq j} Q_{jn}^{(dexc)}\rho_{nn} + Q_j^{(RC)} - \\ - \rho_{jj}(\sum_{n\leq j} Q_{jn}^{(dexc)} + \sum_{n\geq j} Q_{jn}^{(exc)} + Q_j^{(ion)}) \tag{7}$$

As it is evident from the given description the collisional part of the incoherent pumping operator to some extent correlates with the collision integral operator ***St***. And by per se the introduction of the collision integral operator presumes the description of more subtle effects in the collisional interactions [4-5]. In particular, this operator includes a part of dynamical description of the system in the interaction representation [1]. It was tacitly presumed up to now that the collision integral operator describes collisions of the radiator with plasma electrons. However, there are also important processes due to collisions with heavy particles. First of all, the collisions of ions of the next ionization stage with respect to the radiator ion with the other heavy species (atoms or ions) in plasma could populate DES of the radiator ion in the process of charge exchange [11]. And on the contrary the collisions of the radiator ion with other heavy particles could depopulate DES due to capture of one or two electrons by the projectile heavy particles [11]. In fact the inclusion of the charge exchange processes would mean that the system for DM equations becomes unclosed. Moreover, the DES ionization due to the collisions with heavy particles also exists and depopulates DES levels as well. If the first process could not be represented as a part of the collision integral operator due to heavy particles $\boldsymbol{St_h}$ in the frames of the closed DM system, the second one could constitute the part of it. The other important part that enter into the collision integral operator due to heavy particles $\boldsymbol{St_h}$ is determined by the processes of redistribution of DES populations over orbital angular momentum values [12-13]. The inclusion of those effects although possible would complicate more the consideration, and seems to have no principle significance for the subject of DM method.

The introduction of the different operators for collisional interactions itself already on this stage realizes to some extent the mentioned above reduction of the total system of DM equations to the DES DM subsystem, described with the help of the ***St*** operator, and the other subsystems for the rest of levels, described with the help of the operator ***Q***.

As it was shown in [4] within the general settings the electron collision integral operator ***St*** from [1] could be expressed in terms of the electron impact broadening operator $\boldsymbol{\Phi}$. In the case of isolated non-degenerate DES levels using for simplicity of expressions but without loss of generality the semi-classical approximation with the hyperbolic trajectories of electrons scattered by the ion radiator, the ***St (ρ)*** takes the form (compare with [4-5, 8-10, 14])

$$St_{ab}(\rho) = 2\sum_{kg} \exp[i(\omega_{ak} - \omega_{bg})t]\langle ag^{+}|\Phi|bk^{+}\rangle \rho_{kg}$$

$$< ag^{+} | \Phi | bk^{+} > = \frac{4\pi}{3}\left(N_e a_0^3\right)\frac{e^2}{a_0\hbar}[\delta_{bg}\sum_{s} < a | \frac{\vec{r}}{a_0} | s >< s | \frac{\vec{r}}{a_0} | k > \left\langle \frac{e^2}{\hbar v} a_c(\xi_{as}, \xi_{sk}; \varepsilon_{\min}) \right\rangle + \quad (8)$$

$$+ \delta_{ak}\sum_{s'} < b | \frac{\vec{r}}{a_0} | s' >< s' | \frac{\vec{r}}{a_0} | g > \left\langle \frac{e^2}{\hbar v} a_c(\xi_{bs'}, \xi_{s'g}; \varepsilon_{\min}) \right\rangle - 2 < a | \frac{\vec{r}}{a_0} | k >< b | \frac{\vec{r}}{a_0} | g > \bullet$$

$$\bullet \left\langle \frac{e^2}{\hbar v} a_c^{(\mathrm{int})}(\xi_{ak}, \xi_{bg}; \varepsilon_{\min}) \right\rangle], \ \xi_{ik} = \left(\frac{e^2}{\hbar v}\right)\left(\frac{\hbar\omega_{ik}}{mv^2}\right), \ \varepsilon_{\min} = \sqrt{\left(1 + \left[\frac{mpv^2}{(Z_r - 1)e^2}\right]^2\right)},$$

where: ***e*** – is the charge and ***m*** –mass of electron; a_0 – is the Bohr radius; N_e- is the electron density; ***v***- is the thermal electron velocity; ***p***-is the impact parameter of collisions; ε_{min} –the minimal eccentricity of hyperbolic trajectory; $\vec{r} = \vec{r}_1 + \vec{r}_2$ - is the summary electron radius-vector for the radiating ion, that is helium-like, for example; δ_{ik} – is the Kronecker delta symbol; Z_r is the charge of the radiating ion; the symbol <...> designates the average over the velocity of incident plasma electrons. The discussion of the ε_{min} choice and the appearance of functions a_c and $a_c^{(int)}$ is performed, for example, in [14]. It is clear that the structure of this expression is universal with respect to its extension on multi-electron radiators [6-13], and this is important for its applicability for the DES description.

While writing the expressions for all collision operators the velocity of the radiator is assumed to be unchanged in collisional processes [1], since taking account of its change would lead to by far more laborious and cumbersome expressions, but seems insignificant for the questions under discussion. However, it should be underlined that the obtained results have general significance and the electron broadening operator could be taken in the fully quantum mechanical form as well and expressed in terms of the transition matrix ***T*** [14], for example.

For simplification of the system of equations and analysis of its solutions it is presumed that photo pumping of DES is unimportant [2-3].

In the context of mathematical operations used in the construction of kinetic equation for DM the collision integral operator, the operator of radiation relaxation, the operator of autoionization relaxation and the operator of incoherent pumping correspond to the statistical description of the DM, while the commutators with interaction operators give the dynamical description of the system [1-2].

It is possible to separate two items in the operator of radiation relaxation, one of which describes the level devastation due to the radiation decay to the lower lying states $R^{(1)} = -\rho_{jj}\sum_{n\le j} A_{nj}$ [1], and the other represents itself the source describing the increasing of population of the given level due to the radiative transitions from the levels located higher in energy $R^{(2)} = \sum_{n\ge j} A_{jn}\rho_{nn}$ [1], where A_{nj} designates the Einstein coefficient for the transition ***n-j*** :

$$R = R^{(1)} + R^{(2)} = -\rho_{jj} \sum_{n \le j} A_{nj} + \sum_{n \ge j} A_{nj} \rho_{nn} \tag{9}$$

Here in the expressions for the radiation relaxation operator the recoil effects are neglected [1].

In total in zero approximation with the spontaneous radiation field the described above system is reduced to the system of balance equations for DES and other states not counting some nuances, connected with the ***St*** operator action. The DES energy levels between which the dipole matrix elements are not equal to zero would be mixed due to the Stark effect in the ion microfield and due to the action of collisions with plasma electrons [4-5]. With respect to the Stark mixing the system of equations for DES will be supposed untangled with the levels lying significantly lower in energy, corresponding to the one-electron excitations [15, 18].

In the DM formalism in order to obtain the contour of emission line due to the transition ***i-k*** it is necessary to obtain the power of spontaneous emission of the medium, that is expressed through the solution of DM system of equations, calculated for this transition in the first order of perturbation theory on the parameter G_{ik}, giving the measure of the effective interaction strength, in the following way [1-2, 4-5]

$$P(\omega_{ik}) = -\hbar\omega_{ik}\, \Re e \left\langle iG^{*}_{ik} \bar{\rho}^{(1)}_{ik} \right\rangle . \tag{10}$$

At the same time $\rho^{(1)}_{ik}(t) = \bar{\rho}^{(1)}_{ik} \exp[-i(\omega - \omega_{ik})t]$ presents the "spontaneous polarization of the medium" [1-2].

The line contour is determined from the relation

$$I(\omega) = \frac{1}{\pi \left\langle \rho^{(0)}_{ii} \right\rangle} \frac{\left\langle \Re e \left[iG^{*}_{ik} \bar{\rho}^{(1)}_{ik} \right] \right\rangle}{\left| G_{ik} \right|^{2}}, \tag{11}$$

where $\rho^{(0)}_{ii}$ is the population of radiating level "***i***", that is obtained from the solution of stationary system of balance equations in zero approximation on the interaction with spontaneous radiation field, the symbol <...> designates average over the ensemble of emitting plasma ions in DES states, i.e. over the total set of stochastic variables entering the equations for DM, in the first turn over the distribution of microfields, the Maxwell function distribution of radiating ions velocities etc.

The sought polarization (dielectronic satellites) corresponds to transitions from DES subsystem to the lower lying singly excited states outside the DES subsystem. For resolving the system of equations in zero approximation it is necessary to change variables for the non-diagonal matrix elements of DM between levels pertaining to DES in order to get rid of the oscillating factors in the equations for DM in zero approximation

$$\rho^{(0)}_{ik}(t) = \bar{\rho}^{(0)}_{ik} \exp(i\omega_{ik} t) . \tag{12}$$

After that only non-oscillating terms are retained in accordance with the resonance or rotational wave approximation [1-2].

The performed analysis leads to the following detailed form of the zero order DM equations for DES with introduction of the effective population sources and devastation of levels for transitions to or from the levels outside the DES system in order to reduce the system rank

$$i\hbar\dot{\overline{\rho}}^{(0)}_{\alpha\beta} = \hbar[\omega_{\alpha\beta} - i(\Gamma_{\alpha\beta} + \Gamma_{\alpha\alpha} + \Gamma_{\beta\beta} + W_{\alpha} + W_{\beta})]\overline{\rho}^{(0)}_{\alpha\beta} + \sum_{s}(V_{\alpha s}(t)\overline{\rho}^{(0)}_{s\beta} - \overline{\rho}^{(0)}_{\alpha s}V_{s\beta}(t)) +$$

$$+ i\hbar \cdot 2\sum_{\alpha'\beta'} < \alpha\beta \,|\, \Phi \,|\, \alpha'\beta' > \overline{\rho}^{(0)}_{\alpha'\beta'} + i[Q^{(exc)}_{\alpha} - Q^{(dexc)}_{\alpha} - Q^{(ion)}_{\alpha} + Q^{(RC)}_{\alpha} - W_{\alpha}]\delta_{\alpha\beta} \quad (13)$$

In this set of equations it is supposed that in zero order approximation the non-diagonal matrix elements correspond to transitions within the DES subsystem of states. Namely, the presence of DM non-diagonal elements is the important distinction of the zero order DM system from the system of balance equations. Moreover, the non-diagonal DM elements between DES states in the case under study are connected with the mixing of population in the ion plasma microfield in the settings chosen in this paper.

In the case of singly excited states in the pumping terms of the previous system of equations the resonance capture $Q^{(RC)}_{\alpha}$ should be changed on the radiative recombination of the radiator ion $Q^{(RR)}_{\alpha}$ and the autoionisation constants W_{α}, W_{β} should be omitted in all terms of the right hand side. This set of equations for brevity will not be written out since its form is very similar to the one considered above.

The system of equations for coherences connecting DES with singly excited states has the form

$$i\hbar\dot{\overline{\rho}}^{(\mu)}_{\alpha\beta} = \hbar\left(\omega_{\alpha\beta} - i\frac{(A_{\alpha\beta} + \Gamma_{\alpha\alpha} + \Gamma_{\beta\beta} + W_{\alpha} + W_{\beta})}{2}\right)\overline{\rho}^{(\mu)}_{\alpha\beta} + \sum_{s}(V_{\alpha s}(t)\overline{\rho}^{(\mu)}_{s\beta} - \overline{\rho}^{(\mu)}_{\alpha s}V_{s\beta}(t)) +$$

$$+ i\hbar \cdot 2\sum_{\alpha'\beta'} < \alpha\beta \,|\, \Phi \,|\, \alpha'\beta' > \overline{\rho}^{(\mu)}_{\alpha'\beta'} + \sum_{k}(G_{\alpha k}\overline{\rho}^{(0)}_{k\beta} - \overline{\rho}^{(0)}_{\alpha k}G^{+}_{k\beta}), \quad (14)$$

where G_{ik} are functions of polarization, frequency and orientation angles of the radiation wave vector. The form of this system has sufficiently general character and is not limited by the model assumptions in the validity range of the perturbation theory. The important moment is that there are decay constants not only between named levels, but also between unnamed, expressed through constants Γ_{ii}, W_i, that describe decay into the other subsystems not belonging to DES and not connected with polarization under consideration in the radiative transition $\alpha \rightarrow \beta$.

CHOICE OF ZERO ORDER BASIS WAVE FUNCTIONS

The next important question in this problem is the choice of the zero basis wave functions for the DES description [6, 15, 18]. Indeed, it is known, that for example, for helium like ions with nuclear charge of the order of 10 it is necessary already to take into account the configuration mixing and influence of the spin-orbit interaction, so that theses states have to be classified in the frames of intermediate coupling scheme [15, 18]. One of the most currently used radiators of such type in diagnostics is the helium-like ion Ar XVI, for which the so called terms coupling coefficients (TCC) or mixing coefficients due to configuration interaction and intermediate coupling scheme could be calculated with help of the total row of current universal programs for calculations of atomic characteristics of atoms and ions, based on various versions of the Hartree-Fock self-consistent field method [15, 18]. Presently there is a free access to many of those programs on the websites of some universities and research centers.

In particular, Cowan code had got rather wide circulation [18]. Thus TCC allow express the matrix elements over the wave functions of populated doubly excited states through the ***LS*** coupling wave functions or ***jj*** coupling wave functions as well [18]. It is important that already the zero order wave functions have to correspond to the energy levels of the real physical system since only in this case the diagonal matrix elements of the density matrix would have sense of the populations of those levels [1]. From this point it would be more appropriate if the system of equations for DM was written down for this particular system of wave functions. These wave functions are divided into blocks with fixed value of the total angular momentum quantum number ***J*** [18], composed from the total orbital angular momentum of atomic electrons ***L*** and their total spin ***S*** [18] : $\vec{J}=\vec{L}+\vec{S}$. Hence it is clear that the DES wave functions should be labeled by the number ***N*** of corresponding DE state and its ***J*** and M_J quantum numbers [18]. The number ***N*** is thought to designate the set of all other quantum numbers that are necessary to distinguish states originating from the different mixtures of configurations with the same JM_J. In the basis of these wave functions the dipole matrix elements that would differ from zero would determine the Stark mixing of the wave functions in the ions electric field in the dynamical part of the DM equations and in the electric field of electrons in the statistical part of the DM equations, determined by the collision integral operator of electrons with the radiator [1, 4-5]. As it is clear from the described above construction of the DES wave functions this mixing would happen as with the wave functions belonging to the different configurations as with the wave functions, corresponding to the different values of the total spin in the ***LS*** coupling scheme [15, 18-19]. And this mixing would have the consequences in the formation of the Stark profiles of dielectronic satellites - the spectral lines due to the radiative transitions from DES to the lower lying singly excited states [4-5, 8-10]. Besides the effects of Stark broadening of the quasistationary states, considered in the frames of the conventional broadening theory in [8-10], here the appearance of non-linear interference effects (NIEF) [1-2, 4-5] should be expected, caused by coupling of the populations (diagonal matrix elements of DM) with the "coherences" [1-2, 4-5], determining the dynamics of interaction with spontaneous radiation field [1]. The possibility of the display of those phenomena was pointed out in [4-5] on the basis of analogy under consideration of NIEF in the Stark broadening of the model three-level system and under calculations of Ly-alpha emission line of the hydrogen like argon taking account of the fine structure splitting, Lamb shift and NIEF. In this work basing on results [1-2, 4-5, 8-10] the procedure of construction of the DM system of equations for DES description and calculation of the dielectronic satellites Stark profiles taking account of NIEF in dense plasmas is demonstrated explicitly.

As one of the most important in the context of diagnostic applications the configurations ***2l2l'*** of helium like ions $2s^2$, $2s2p$, $2p^2$, that during the radiative decay are transformed into configurations $1s^2$, $1s2s$ and $1s2p$, are considered. Moreover, there are two possible values of multiplicity- the singlets and triplets, or total spin equal to 0 or 1. As in the case of NIEF, the appearance of which in the Stark broadening is more pronounced for values of the multiply charged ions nuclear charge ***Z*** of the order of 30 [4-5], the dielectronic satellites are also more intense for large $Z >> 1$ [6, 18]. For present DES the allowed ***J*** values make up ***0, 1, 2***, and the dipole mixing exists only for states with ***ΔJ*** equal ***0*** and ***± 1*** [15, 18]. As the zero order wave functions are marked by the number of state (ie by configuration) and by the angular momentum quantum numbers JM, then the formal number of zero order DM equations for DES subsystem would be 224, corresponding to the squared number of the DES states if to neglect DES mixing with singly excited configurations. For the singly excited subsystem, corresponding to the lower levels of dielectronic satellite transitions, the formal number of the states is 13, that gives 169 zero order DM equations. So, one has to start from the total number of 393 zero order DM equations. The first order set of DM equations formally would contain 22x13=286 equations, corresponding to the number of components of possible radiative dielectronic satellite transitions, especially in the case when the mixing of configurations and spin-orbit interaction are taken into account and thus transitions violating conventional selection rules are allowed. Hence it would give in total 679 zero and first order DM equations. Thus one can see the enhancement of computational efforts, connected with the using DM approach instead of amplitude one on the rather simple example of ***2l2l'*** DES for helium like ions. All above numbers of equations, of course, would be more or less reduced after taking account of symmetry relations and details of the selection rules for dipole matrix elements (see [5]).

This model setting may be extended for real numerical calculations in the case of ***2l2l'*** and lower lying configurations. Then the DM system of equations will include in detail the interlevel kinetics completely. In principle it would be possible to take into account the mixing of configurations of doubly and singly excited states. In this case it would be necessary to join the DM systems for DES and singly excited states in one common system of equations. Evidently it would increase more the formal rank of the system under study.

NIEF DEMONSTRATION IN THREE-LEVEL MODEL SYSTEM

It is evident that the DM set of equations even for considered rather simple system is quite complicated (compare with [4-5]). That is why in order to demonstrate the essentials of appearance of NIEF it is plausible again as in [4-5] to consider the model three-level system, where two upper levels with labels "*2*" and "*3*" would represent effective DES states interacting with each other through dipole Stark mixing in the ion plasma microfield and due to impacts with electrons and only one lower lying effective singly excited state level with label "*1*". Then the DM zero order system of equations for this three-level model could be represented in the form after substitution removing time dependence of the non-diagonal matrix element along with (12)

$$\begin{aligned}
&2V\Im m\,\overline{\rho}_{32}^{(0)} + 2\Phi(\rho_{22}^{(0)} - \rho_{33}^{(0)}) + (A_{21} + \Gamma_{22} + W_2)\rho_{22}^{(0)} - A_{32}\rho_{32}^{(0)} = Q_2 \\
&-2V\Im m\,\overline{\rho}_{32}^{(0)} - 2\Phi(\rho_{22}^{(0)} - \rho_{33}^{(0)}) + (A_{31} + A_{32} + \Gamma_{33} + W_3)\rho_{33}^{(0)} = Q_3 \\
&\overline{\rho}_{32}^{(0)}(2i\omega_{32} + A) + i8\Phi\,\Im m\,\overline{\rho}_{32}^{(0)} + i2V(\rho_{22}^{(0)} - \rho_{33}^{(0)}) = 0 \\
&-\Gamma_{11}\rho_{11}^{(0)} + A_{21}\rho_{22}^{(0)} = Q_1
\end{aligned} \tag{15}$$

$$A = A_{31} + A_{32} + A_{21} + \Gamma_{22} + \Gamma_{33} + W_2 + W_3$$

Here ω_{32} is the unperturbed frequency splitting between the effective DES levels "*2*" and "*3*". In distinction from [4-5] here the additional decay constants corresponding to autoionisation $W_{2,3}$ are appeared. Excluding ρ_{32} and ρ_{23} from (14) extracts the field mixing frequency of populations $\rho_{22}^{(0)}$ and $\rho_{33}^{(0)}$ that could be put down as follows:

$$\nu_F = \frac{V^2}{\omega_{32}^2 + 2\Phi A + (A^2/4)} \tag{16}$$

This mixing is proportional to the squared matrix element of the dipole interaction with the ion microfield that is presumed here for simplicity real $V = d_{23}F$, where d_{23} is the effective dipole moment matrix element between states "*2*" and "*3*". The effective matrix element of electron impact broadening operator between the mentioned states Φ describes the collisional mixing of the above states as well in (15). The effective decay constant A includes radiative decay, autoionisation and decay to other systems or continua through constants Γ_{22} or Γ_{33} of uniform broadening. The same kind of constant Γ_{11} governs the balance of the population of the level "*1*" $-\rho_{11}^{(0)}$. The incoherent pumping terms could be split into several items corresponding to particular physical processes as well, while the level "*1*" is considered as excited one

$$\begin{aligned} Q_2 &= Q_2^{(RC)} + Q_2^{(RC)} - Q_2^{(ion)} + Q_2^{(exc)} - Q_2^{(Dexc)} \\ Q_3 &= Q_3^{(RC)} + Q_2^{(CH)} - Q_3^{(ion)} + Q_3^{(exc)} - Q_3^{(Dexc)} \\ Q_1 &= Q_1^{(RR)} + Q_1^{(CH)} - Q_1^{(ion)} + Q_1^{(exc)} - Q_1^{(Dexc)} \end{aligned} \tag{17}$$

The first order DM system of equations for coherences $\overline{\rho}_{21}^{(1)}$, $\overline{\rho}_{31}^{(1)}$ after reduction of time dependences according with (12) within the frames of three-level model could be represented in the form

$$\begin{aligned} -i\Omega\overline{\rho}_{21}^{(1)} + iV\overline{\rho}_{31}^{(1)} + \frac{\Gamma_2}{2}\overline{\rho}_{21}^{(1)} &= -G\rho_{22}^{(0)} \\ -i(\Omega - \omega_{32})\overline{\rho}_{31}^{(1)} + iV\overline{\rho}_{21}^{(1)} + \frac{\Gamma_3}{2}\overline{\rho}_{31}^{(1)} &= -G\rho_{32}^{(0)}, \end{aligned} \tag{18}$$

where the new constants $\Gamma_{2,3}$ and Ω are defined for brevity of expressions as follows

$$\Gamma_2 = A_{21} + \Gamma_{22} + \Gamma_{11} + 2\Phi + W_1, \quad \Gamma_3 = A_{31} + A_{32} + \Gamma_{22} + \Gamma_{11} + 2\Phi + W_3, \quad \Omega = \omega - \omega_{21}.$$

The solution of this first order DM model system then could be written in the way

$$\overline{\rho}_{21}^{(1)} = -iG\left[\rho_{22}^{(0)} + \frac{V\overline{\rho}_{32}^{(0)}}{(\omega - \omega_{31}) + i\dfrac{\Gamma_3}{2}}\right]\left[\Omega + \frac{i\Gamma_2}{2} - \frac{V^2}{(\omega - \omega_{31}) + i\dfrac{\Gamma_3}{2}}\right]^{-1} \tag{19}$$

This allows to represent the expression for the line profile (11) for the transition ***1→2*** in the particular case as the sum of two items. The first item is normalized over frequency to unity and has the conventional meaning of the normalized to unity averaged elementary line profile φ(ω) while the second item is normalized over frequency to zero and represents the redistribution of intensity in the profile due to NIEF [4]

$$I(\omega)=\frac{\left\langle\rho_{22}^{(0)}\phi(\omega)\right\rangle}{\left\langle\rho_{22}^{(0)}\right\rangle}+I_{NIEF}(\omega),\quad \phi(\omega)=\frac{1}{\pi}\Im m\frac{(\omega-\omega_{31})+i\dfrac{\Gamma_3}{2}}{\Delta(\omega)} \tag{20}$$

The equation (21) determines the zero order solution for the population of the radiating level "*2*" before average over the ion microfield distribution [4]

$$\rho_{22}^{(0)}=\frac{Q_2+Q_3}{A_{21}+\Gamma_{22}}-\frac{A_{31}+\Gamma_{33}}{A_{21}+\Gamma_{22}}\frac{Q_3+(\tilde{V}+2\Phi)\dfrac{Q_2+Q_3}{A_{21}+\Gamma_{22}}}{A_{32}+A_{31}+\Gamma_{33}+(\tilde{V}+2\Phi)\left(1+\dfrac{\Gamma_{33}}{A_{21}+\Gamma_{22}}\right)}, \tag{21}$$

and the population of the level ***"3"*** could be derived from the expression

$$\rho_{33}^{(0)}=\frac{Q_3+(\tilde{V}+2\Phi)\dfrac{Q_2+Q_3}{A_{21}+\Gamma_{22}}}{A_{32}+A_{31}+\Gamma_{33}+(\tilde{V}+2\Phi)\left(1+\dfrac{A_{31}+\Gamma_{33}}{A_{21}+\Gamma_{22}}\right)}. \tag{22}$$

The expression for the NIEF term in its turn could be written in the following way [4]

$$I_{NIEF}(\omega)=-\frac{V}{\pi}\frac{1}{\left\langle\rho_{22}^{(0)}\right\rangle}\Im m\left\langle\frac{\bar{\rho}_{32}^{(0)}}{\Delta(\omega)}\right\rangle=$$

$$\frac{1}{\pi\left\langle\rho_{22}^{(0)}\right\rangle}\bullet\left\langle\frac{V^2}{\omega_{32}\Delta_{\text{det}}}(\rho_{22}^{(0)}-\rho_{33}^{(0)})\left(\frac{A}{2\omega_{32}}\Re e\frac{1}{\Delta(\omega)}+\Im m\frac{1}{\Delta(\omega)}\right)\right\rangle \tag{23}$$

These two types of representations (22) show the significant proportionality of the NIEF to the non-diagonal density matrix element of the states ***"2"*** and ***"3"*** or to the difference of populations of the mentioned levels and the squared matrix element of the dipole interaction with the ion microfield [4]. The real and imaginary parts of the effective resolvent $\boldsymbol{\Delta(\omega)}$, determinant $\boldsymbol{\Delta_{det}}$, and the effective interaction constant $\tilde{V}$ that enter the set of equations (20)-(22) are defined as follows [4]

$$\Re e\,\Delta(\omega)=(\omega-\omega_{21})(\omega-\omega_{31})-\frac{\Gamma_3\Gamma_2}{4}-V^2,\quad \Im m\,\Delta(\omega)=\left[\Gamma_3(\omega-\omega_{21})+\Gamma_2(\omega-\omega_{31})\right]/2$$

$$\Delta_{\text{det}}=1+\frac{A^2}{4\omega_{32}^2}+\frac{2A\Phi}{\omega_{32}^2},\quad \tilde{V}=\frac{AV^2}{\omega_{32}^2\,\Delta_{\text{det}}} \tag{24}$$

So, these equations demonstrate the following significant features of NIEF [4-5]:

$$I_{NIEF} \propto \rho_{32}^{(0)} \propto (\rho_{22}^{(0)} - \rho_{33}^{(0)}), \quad I_{NIEF} \to 0, \; V \to 0 \; \forall \; V \to \infty, \quad \int I_{NIEF}(\omega)\, d\omega = 0 \qquad (25)$$

So, the performed in this section analysis showed that the DES specifics appear in the DM equations and NIEF characteristics mainly through the redefinition of relaxation constants with account of the physical processes inherent to DES.

INFLUENCE OF INTERFERENCE OF DIELECTRONIC AND RADIATIVE RECOMBINATION ON NIEF

The interference effects between the dielectronic (DR) and radiative recombination (RR) was predicted and studied in numerous publications (see [3, 20]). The essential of this is that the DES radiative decay amplitude to the lower singly excited state includes the admixture of continuum radiative decay amplitude to the same state, that describes the radiative recombination of the ion in the previous ionization stage with respect to the parent one [3, 7, 20]. Hence the squared module of this amplitude that determines the intensity of dielectronic satellite contains the interference term of this kind. The magnitude of this interference effects was studied for several isoelectronic sequences and versus temperature (see [20]). The cause of these interference effects from one side is an inherent many-body nature of doubly excited states of multiply charged ions

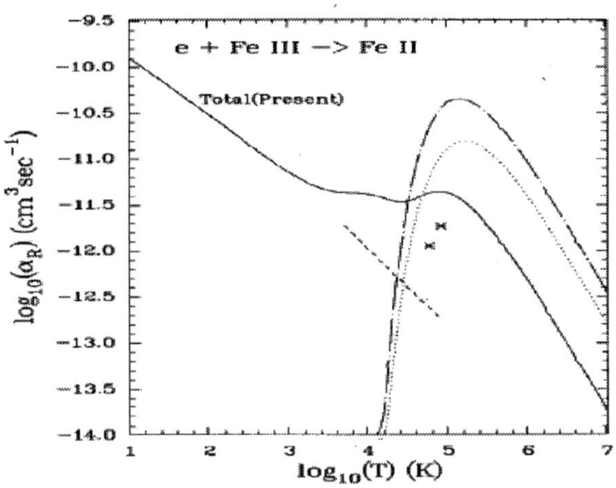

Fig 1. The graph reprinted from [20] of recombination coefficients versus temperature illustrating the trough around 10 000K due to DR and RR interference effects: total DR+RR - solid line [20], RR contribution - dashed line and DR contribution - chain-dotted lines due to [21], asterisks – DR rate due to [22], dotted line – DR rate due to [23].

with Z >> 1, and from the other side is "pathways interference due to the coinciding of the final quantum state. It was shown that these DR-RR interference effects are important in the intermediate region of parameters, where both DR and RR processes probabilities are decreasing [20]. In the figure 1 from [20] this is illustrated for the process **e + Fe III → Fe II**, for example. The question of probable entanglement of DR-RR interference effects with nonlinear interference effects thus becomes reasonable. However, it is evident that origins of these effects are quite different, although the interference of pathways due to the coincidence of the final quantum states also is typical for NIEF as well [1-2]. In the case of absence of the spectral dependence in DR-RR interference it will not affect the profile of satellites at all but do affect to some extent their intensities. Thus in the first approximation the existence of DR-RR interference effects (IEF) could change only the constants like G in (4), (11) that conventionally canceled out during the profile calculation in the resonance approximation in (11). However, more detailed approach would lead to renormalization of relaxation terms in the quantum kinetics equation for DM and thus to entanglement of DR-RR IEF with NIEF through redefined relaxation constants. Nevertheless, as it is clearly seen in the Figure 1 the significance of this contribution depends on the temperature. Indeed, for the sufficiently large temperatures the magnitude of DR-RR interference is quite small and could be neglected, so that NIEF stay "alone" as the representatives of the interference effects for DES in this region of parameters. On the other hand the large-scale calculations performed in [20] are very complicated dealing with $\boldsymbol{R}$-matrix evaluation in close-coupling approximation, that prevents to exploit them as an instrument of regular use to study atomic kinetics and NIEF in the frames of DM approach. Moreover, besides temperature the magnitude of DR-RR IEF depends on the given isoelectronic sequence and the concrete ion $\boldsymbol{Z}$ and usually is quite small and neglected in the large-scale computations of atomic kinetics.

CONCLUSIONS

In the corollary it could be resumed that in this work the general recipes for the construction of density matrix equations for description of the dielectronic satellites Stark profiles in dense plasmas with account of the nonlinear interference effects (NIEF) [1-2, 4-5], the autoionization, the radiative relaxation, the dielectronic recombination, the microfield and collisional mixing of quantum states, the mixing of configurations and intermediate coupling scheme for the large values of helium like ions nuclear charge are given.

The explicit general procedure is pointed out, and various aspects of this problem are discussed in detail. This approach could be further used in numerical calculations and for the aims of plasma diagnostics. Thus DM approach is a powerful tool for the consistent line shape description.

It was verified that the NIEF influence dielectronic satellite spectra of highly charged ions with $Z \gg 1$. However, NIEF have different origin than the interference effect between dielectronic and radiative recombination due to the many-body nature of DES located above ionization threshold.

As in a laser physics the NIEF couple the population kinetics with dynamics of radiative transitions. The NIEF make impossible the construction of line shape characteristics without simultaneous consideration of the population kinetics.

It should be noted that practically the DM approach presently is limited by the impact approximation applied along with derivation of the collision integral operator [1-2]. At the same time in the frames of DM equations the interaction with the microfields could be treated using various approaches to description of microfield statistics [17, 24]: the quasistatic approximation [14-15]; the model microfield method (MMM) [25]; the computer simulation method (CSM) for the electric fields with straight path trajectories [26]; the molecular dynamics (MD) [16-17]; the frequency fluctuation model (FFM) [27]. In fact in the different settings the FFM was applied to simulation of DES spectra previously repeatedly, and the very recent results in the course of development of so called "the virtual contour shape kinetic theory (VCSKT)" approach are presented in [28]. The current advances in MD, FFM [29] and CSM [30] are reflected in contributions to this volume as well.

ACKNOWLEDGEMENTS

The speaker's (AVD) participation in ICSLS-2006 was partially supported by Organizing Committee and by RFBR through the grant 06-02-26733. The work of F. B. Rosmej was partially supported by Progamm ARCUS, n° P003-5a. AVD acknowledges support by CNRS during stay in *Université de Provence,* where this work was initiated, and wishes to thank the team of PIIM laboratory for hospitality. The authors thank D. Nikolic and V.S. Lisitsa for useful discussions.

REFERENCES

1. S.G. Rautian, A.M. Shalagin, *Kinetic Problems of Non-Linear Spectroscopy (*North Holland, N.Y.) 1991, 439 p.
2. Yu.I. Geller, A.K. Popov, *Laser Induction of Non-Linear Resonances in Continuous Spectra,* (Nauka, Novosibirsk) 1981, 160 p. (in Russian).
3. V.L. Jacobs, J. Cooper, S.L. Haan, *Physical Review* A **36** (1987).
4. A.V. Anufrienko, A.L. Godunov, A.V. Demura et al, *Sov. Phys. JETP* **71** (1990) 728.
5. A.V. Anufrienko, A.E. Bulyshev, A.L. Godunov, A.V. Demura et al, *Sov. Phys. JETP* **76** (1993) 219.
6. N. Yamamoto, T. Kato, F.B. Rosmej, *J.Q.S.R.T.* **96** (2005) 343.
7. U. Fano, *Physical Review* A **124** (1961) 1866.
8. J.M. Banon, M. Koenig, H. Nguyen, *J. Phys* B **18** (1985) 4195.
9. J.M. Banon, H. Nguyen, *J. Phys* B **20** (1987) 2989.
10. J.M. Banon, L.L. Sanchez-Soto, E. Bernabeu, *J. Phys* B **22** (1989) 199.
11. F.B. Rosmej, H.R. Griem, R.C. Elton et al, Physical Review E **66** (2002) 0564022002.
12. V.L. Jacobs, M. Blaha, *Physical Review* A **21** (1980) 525.
13. V.L. Jacobs, J. Davis, *Physical Review* A **18** (1978) 697.
14. H.R. Griem, *Spectral Line Broadening by Plasmas,* Academic Press, N.Y., 1974.
15. I.I. Sobelman, *Introduction to Theory of Atomic Spectra,* Pergamon Press, 1972.
16. R. Stamm, E.L. Pollock, B. Talin, C.A. Iglesias, *Physical Review* A **34** (1986) 4144.
17. A.V. Demura, *Statistical and thermodynamical aspects of plasma microfield conception,* Chapter V., in: Encyclopedia of low temperature plasmas, Series B, Reference applications, bases and data banks, Vol. III-1, Editors I.L. Iosilevsky, A.N. Starostin, Fizmatlit, M., 2004, pp. 163-202.
18. R.D. Cowan, *The Theory of Atomic Structure and Spectra,* University of California Press,

Los Angeles, 1981, 731 p.
19. H. Elabidi, N. Ben Nessib, S. Sahal-Brechot, *J. Phys* B **37** (2004) 63.
20. S. Nahar, *Physical Review* A **55** (1997) 1980.
21. D.T. Woods, J.M. Shull, C.L. Sarazin, *Astrophys. J.* **249** (1981) 399.
22. Y. Hahn, *J.Q.S.R.T.* **41**, (1989) 315.
23. M. Arnaud, J. Raymond, *Astrophys. J.* **398** (1992) 395.
24. A.V. Demura, D. Gilles, Spectral Line Shapes 2004, ed. E. Dalimier, Frontier Group, Paris, 2004, pp. 121-127.
25. C. Stehle, *Astron. & Astrophys. Suppl.* **140** (1999) 93.
26. M.A. Gigosos, M.A Gonzalez, V. Cardenoso, *Spectrochimica Acta Part* B **58** (2003)1489.
27. B. Talin, A. Calisti, L. Godbert, R. Stamm, R.W. Lee, L. Klein, *Physical Review* A **51** (1995) 1918.
28. F. B. Rosmej, A. Calisti, *J.Q.S.R.T.* **99** (2006) 548.
29. S. Ferri, A. Calisti, B. Talin, M.A. Gigosos, M. A. Gonzalez , this volume.
30. M. A. Gonzalez, M. A. Gigosos, N. Konjevic, this volume.

Old and New Problems in Spectroscopic Diagnostics of Plasmas

V.S. Lisitsa

NFI RRC "Kurchatov Institute", Kurchatov Square 1, Moscow 123182, Russia

Abstract. A number of spectroscopic problems for laser, astrophysical, thermonuclear and low temperature plasmas are under consideration.The general question for hydrogen spectra at high densities related to a transition from impact to quasistatic broadening for electrons is analyzed. The problem of spectroscopic measurements of superstrong magnetic fields (>10^3 T) generated in subpicosecond laser produced plasma is under discussion. The effect of the oscillations on multicharged ion's hydrogen-like spectra results in a generation of plasma satellites at harmonics of electron cyclotron frequency. Discussion of tokamak plasma diagnostics is also presented. The effect of impurity transport on total and spectral radiative losses is discussed. The effect of plasma turbulence on Doppler hydrogen spectra results in a wrong estimations of plasma temperature in the edge tokamak plasmas. Broadening problems for highly excited (Rydberg) hydrogen atomic states are considered. The problem of a universal quasiclassical representation of intensities of radiative transitions make it possible to look for universal results for arbitrary hydrogen line shapes at arbitrary plasma temperature and densities with the help of FFM method. Additional problems of warm dense matter (WDM) diagnostics are discussed as well. The relationship of WDM plasma conditions with pellet plasma conditions in tokamaks is discussed.

Keywords: plasma spectroscopy, laser plasma, tokamak plasma, astrophysical plasma
PACS: 52.25.-b

INTRODUCTION

Conditional principle of spectroscopic plasma diagnostics [1] are needed in adaptation to modern experimental plasma objects. The following objects are under consideration below: 1) relatively low temperature and high density plasmas (with electron densities N_e>10^{18} cm^{-3}); 2) laser plasma of pico- and femtosecond laser pulses; 3) modern tokamak's plasmas and 4) astrophysical plasmas.

The problem of hydrogen spectral line shapes in plasmas is a subject for many years investigations [1]. The standard approach bases on the static approximation for ions and impact approximation for electrons [1]. The effect of ion's thermal motion was taken into account near the line centre. It was recognized also that electrons make a transition from impact to static broadening in the far wings of spectral lines. Modern state of the arts in the broadening problems is connected with experimental investigations of dense plasmas with electron densities more than 10^{18} cm^{-3} [2]. The situation is related to the situation in Warm Dense Matter (WDM) plasmas generated by multicharged ion's beams interacting with solid state matter where the electron

CP874, *Spectral Line Shapes: 18th International Conference,*
edited by E. Oks and M. Pindzola

densities may succeed $10^{19}cm^{-3}$ at relatively small values of electron temperature (several eV) [3].

Under such plasma conditions the general question related to a transition from impact to quasistatic broadening for electrons not only in the far wings but also at the main body of spectral lines. The effect results in a modification of the old problem of a transition from discrete to continuum spectra both in astrophysical and laboratory plasmas [4,5].

The general question for hydrogen spectra at high densities related to a transition from impact to quasistatic broadening for electrons is analyzed. It is demonstrated with the help of frequency fluctuation model (FFM) [6,7] that the electron broadening makes a transition to quasistatic limit for electron densities larger than 10^{18} cm^{-3} and temperatures of the order of 1 eV not only in the far wings but also at the main body of spectral lines.

The problem of spectroscopic measurements of superstrong magnetic fields (>10^3 T) in plasma generated by pico- and femtosecond laser pulses becomes actual. The presence of magnetic field results in a strong Zeeman splitting of atomic energy levels as well as in a generation of strong plasma oscillations related to Bernstein plasma waves. The effect of the oscillations on multicharged ion's hydrogen-like spectra results in a generation of plasma satellites at harmonics of electron cyclotron frequency.

Problems of modern tokamak plasma spectroscopic diagnostics connect with a determination of ion's transport coefficients. The ion's transport produces important effects both on total radiative losses of impurities and intensitites of specific spectral lines observed. The transport is closely connected with turbulent plasma properties. So the spectroscopic measurements of the turbulence effects are of great importance for tokamak's spectroscopy.

Problems of spectral line intensities and broadening for highly excited (Rydberg) hydrogen atomic states are of permanent interest both for laboratory and astrophysical plasmas. Quasiclassical or pure classical approaches seem to be effective for a description of such spectra. The problem of a universal quasiclassical representation of intensities of radiative transitions make it possible to look for universal results for arbitrary hydrogen line shapes at arbitrary plasma temperature and densities. The problem of intensities and shapes of recombination radio-lines stays still unclear for total theoretical explanations [8,9].

It is a goal of the present paper to give a general review of the plasma diagnostic problems pointed above in the connection of general principles of broadening theory.

SPECTRA AT HIGH DENSITIES

One of the important problems in diagnostics of dense plasmas is connected with a transition from impact to quasistatic broadening for electrons. It was well known that such a transition takes place in the far wings of spectral lines [1]. But the same transition can be observed on the main body of the lines at high electron densities $N_e>10^{18}cm^{-3}$.

It is known that the dimensional parameter g responsible for the transition is determined by the number of particles inside the Weskopf sphere :

$$g = N_e(C/v_e)^3 \quad (1)$$

The parameter $g^{1/3}$ for hydrogen lines can be expressed also as a ratio of typical static width $CN_e^{2/3}$ to the electron electric field fluctuation frequency $\nu = N^{1/3}v_e$ or at the same time to the ratio of an electron impact width NC^2/v_e to static one (C being a typical value of Stark constant, v_e is electron thermal velocity).

It follows from these remarks that the transition from static to impact theory can be considered in a close relation with known Dikke narrowing effect, namely: when the jumping electron fluctuation frequency is small (g>>1) the broadening mechanism is close to static one and, opposite, when the frequency is large (g<<1) the dominant broadening mechanism is static one. In the last case one can speak about narrowing of the spectral line of Dikke type (impact width becomes smaller than static one).

Simple estimations of the broadening effect can be obtained from a comparison of static and impact contributions to the total line widths W following the relationship:

$$W = W^{static} + W^{impact} = CNe^{2/3} + NeC^2L/v_e \quad (2)$$

Where L is Coulomb logarithm. The correct width corresponds to the minimum value of the impact and static widths.

The transition from quasistatic to impact broadening as a function of parameter g can be obtained in the frame of frequency fluctuation model- FFM [6,7]. The FFM just follows the Dikke idea of transitions between different spectral domains of static line profile due to the thermal particle motion described by electric field jumping frequency ν.

Another effect of Stark broadening in dense plasma is connected with atomic energy level ionization if a static electric field. The value of critical electric field F_c strength responcipnle for an ionization of a level with principle quantum number n is estimated as $F_c(au)=1/16n^4$, see for example [10]. One can estimate at fast that the critical value of electric field in plasma with density N becomes essential for hydrogen Balmer series when the plasma density is larger than 10^{18} cm^{-3}.

The starting points of FFM calculations are static profiles for electron and ion broadening given by the equation:

$$J(\omega) = \sum_{i,f} \int dFW(F) J_{i,f}(\omega, F) j_i(F) \quad (3)$$

where $J_{i,f}(\omega, F)$ is the line shape at the fixed value of electric field F, W(F) is static electric field distribution function, j(F) is a dilution factor due to the field ionization effect of atomic states given by the equation:

$$j_i(F) = A_{i,f} / [A_{i,f} + \Gamma_i(F)] \quad (4)$$

where $A_{i,f}$ are radiative decay rates between initial (i) and finite (f) atomic energy states, Γ_i(F) are ionization decay rates in plasma electric field F.

The FFM shape is obtained as it was pointed above by an introduction of electric field jumping frequency $\nu = N^{1/3}$ v (N ,v being the electric charges density and particle thermal velocity) resulting in intensities exchange between different spectral domains of static line shape [6,7].

The FFM calculations of initial members of hydrogen Balmer series are presented on Figs. 1-4 for plasma parameters: densities $N=10^{18}$ cm^{-3} and temperatures T=1 eV.

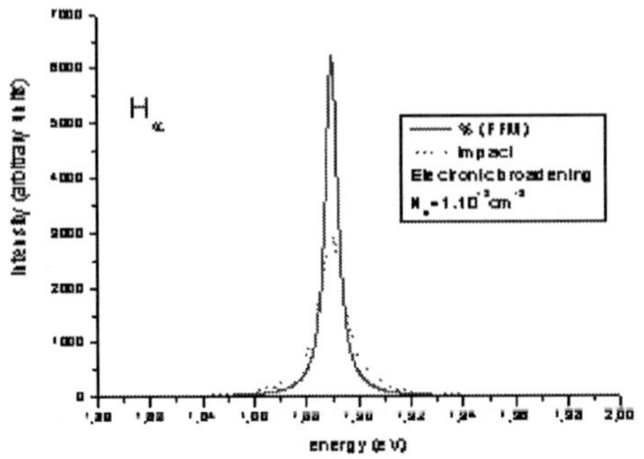

FIGURE 1. H_α line shape calculated in the frame of standard impact approximation (dotted line) and FFM (solid line) for electron density 10^{18} cm^{-3} and temperature 1 eV.

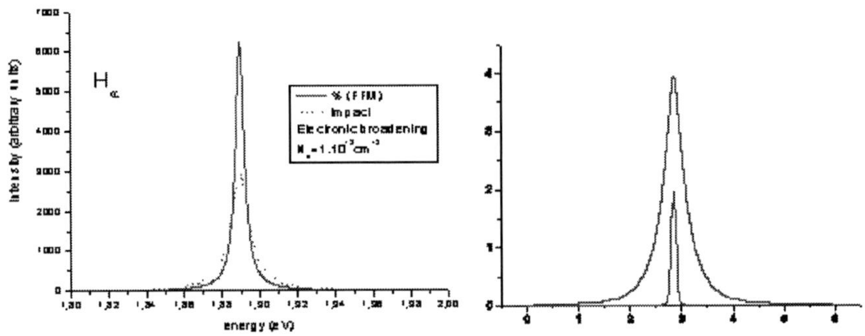

FIGURE 2. The effect of dilution factor on H_α (left) and on H_γ (right) line profile at electron density 10^{19} cm^{-3} and temperature 1 eV.

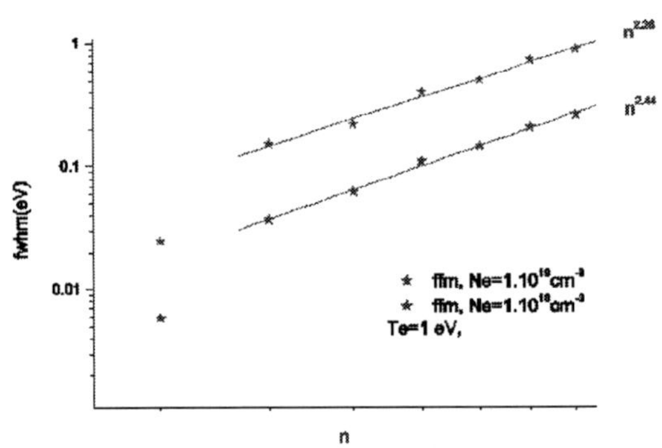

FIGURE 3. Dependence of electron's half widths of hydrogen line profiles on principle quantum numbers n at electron densities 10^{18} and 10^{19} cm^{-3} and temperature 1 eV in the frame of FFM.

Fig.1 represents the H_α line shape calculated in the frame of standard impact approximation and with non impact electron broadening calculated by FFM. An essential modification of the spectral line shape is observed due to the change of the mechanism of electron broadening. Fig.2 represents the effect of intensity dilution due to the ionization effect for H_α, H_γ .line at high electron density 10^{19} cm^{-3}. One can see the decrease of both total intensity and line widths. Fig.3 demonstrates the dependence of line electron widths on principle quantum number for Balmer series at electron density 10^{18} and 10^{19} cm^{-3}. It is seen that the widths follow the power low closer to static scaling (n^2) than to impact one (n^4).

The general conclusion from the consideration above is that the transition of electron broadening from impact to static limit can change the spectral line shapes dramatically for dense plasma conditions.

SPECTROSCOPY OF SUBPICOSECOND LASER PRODUCED PLASMA

Spectroscopic problems related to pico- and femto- second plasmas are connected with high time and space resolution of miniature plasma volume. The transition from nano- to subpicosecond laser pulses results in a number of new effects, at first to generation of fast charged particles and superstrong magnetic fields. The energy of fast ions reaches dozen MeV and quasistatic magnetic fields are close up to 10^9 G at laser intensities 10^{19}-10^{20} Wcm^{-2} [11,12].

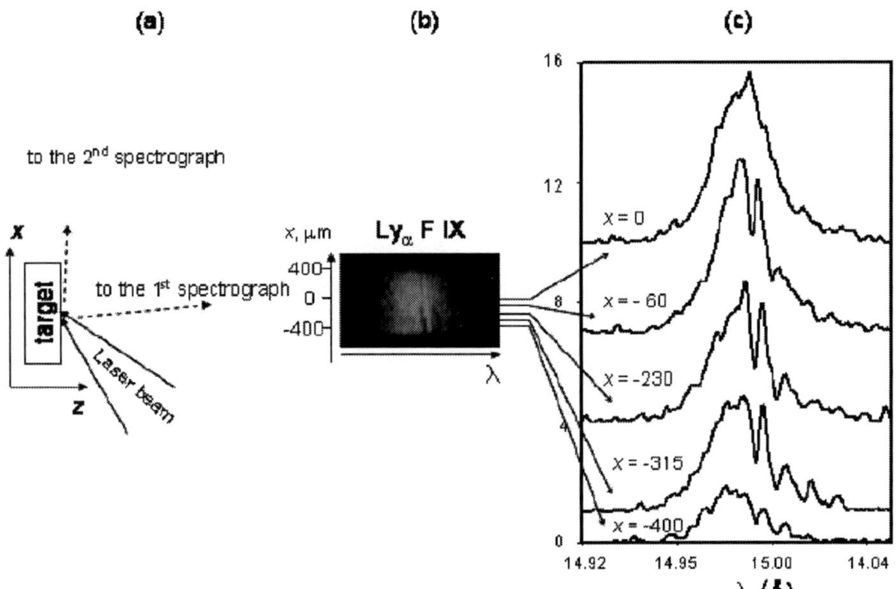

FIGURE 4. The scheme of experiment and FIX line profiles on NEODIM laser installation [13,14] .

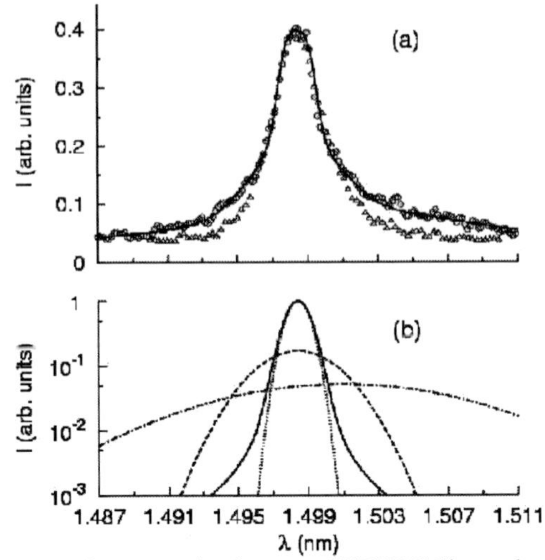

FIGURE 5. Doppler spectra in picosecond NEODIM laser plasma [13,14].

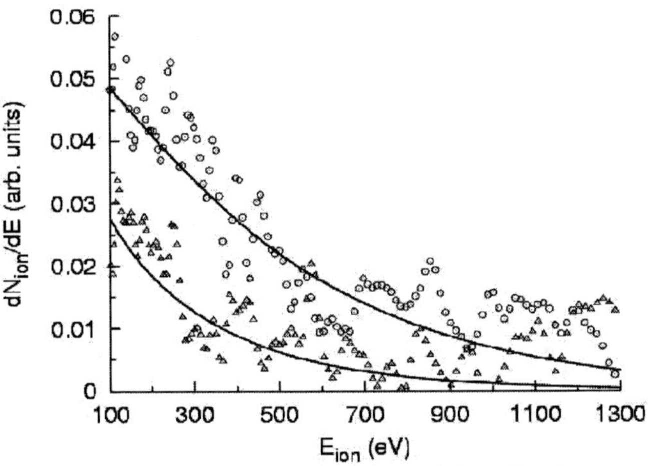

FIGURE 6. Fast ions energy distribution function extracted from Doppler spectra on NEODIM installation [13,14].

The problem of ions velocity distribution function measurements is of great importance because it is responsible for large number of nuclear reaction in laser plasmas. Direct measurements of ions velocity distribution function are possible with the help of Doppler spectra of radiative transitions in highly charged ions. The scheme of experiments and spectra obtained on laser facility NEODIM [13,14] (3 10^{17} Wcm^{-3}, 1.5 ps) are presented on fig.4,5. One can see a presence of different ion's group with different effective temperatures modeled by Gaussian profiles with corresponding Doppler widths. The ion's velocity (energy) distribution function extracted from Doppler spectra is presented on Fig.6. It is clear the strong deflection of the distribution function from Maxwellian one. Moreover a larger number of ions (more than 10%) goes to the non-Maxwellian "tail" of the distribution. The main problem for such measurements is connected with small intensities of ion's radiation in the far wings of Doppler profiles.

The problem of superstrong magnetic field measurements are of principle interest because it is connected closely with plasma dynamics responsible for generation of so strong fields. The magnetic field strengths B can be estimated at fast with the help of comparison of light and magnetic pressures that results in the simple relationship:

$$B(G)=0.1(I,\ Wcm^{-2})^{1/2} \tag{5}$$

Where I is the laser intensity flux.

Direct measurements of Zeeman effect in such magnetic fields is not effective because of the presence of strong Stark broadening in dense laser plasmas (N_e reaches values up to 10^{21} cm^{-3}). So the methods applied were based on measurements of laser harmonic radiation. At the same time it is clear that there are a lot plasma instabilities in laser plasmas generated by beams of fast charged particles. The instabilities are responsible for generation of strong oscillating electric fields resulting in an appearance of a number of plasma satellited n the bodies of spectral line observed [15].

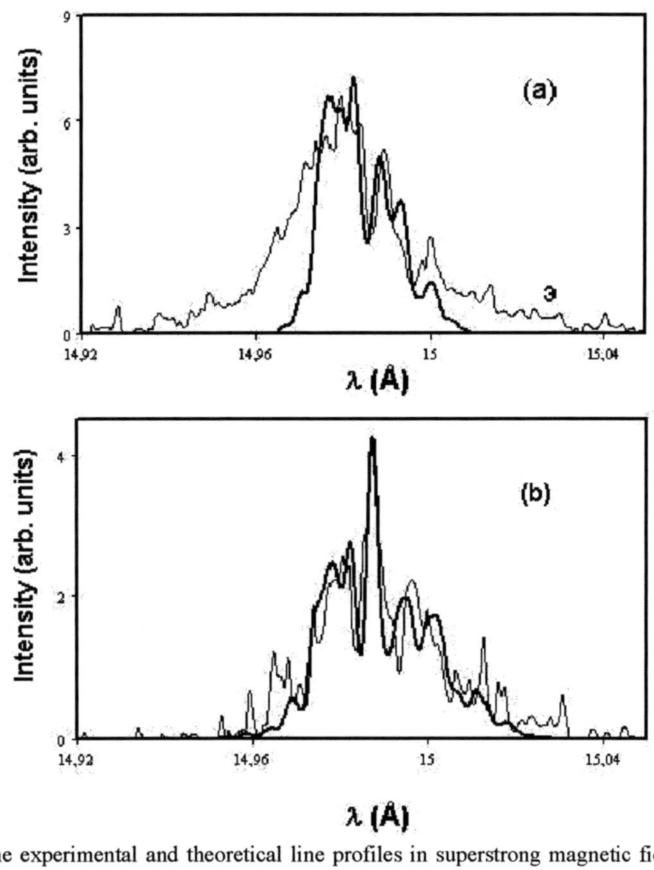

FIGURE 7. The experimental and theoretical line profiles in superstrong magnetic field generated in picosecond laser plasma of NEODIM installation [13,14].

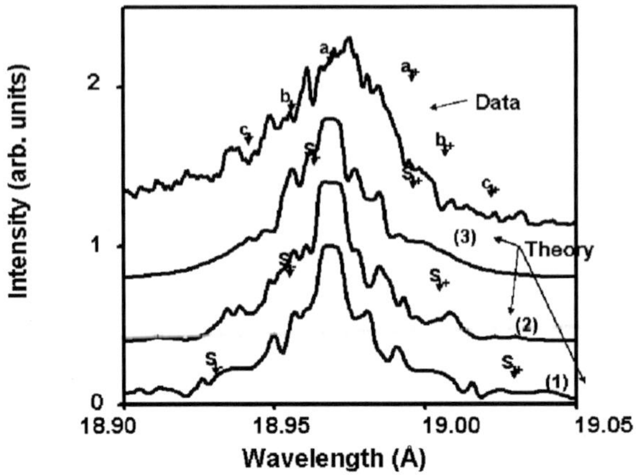

FIGURE 8. The experimental and theoretical line profiles in superstrong magnetic fields generated in femtosecond laser plasma [16].

The presence of magnetic field results in a generation of strong plasma oscillations related to Bernstein plasma waves that is longitual waves propagated across the magnetic field with typical frequency close to electron cyclotron frequency. The effect of the oscillations on multicharged ion's hydrogen-like spectra results in a generation of plasma satellites at harmonics of electron cyclotron frequency. It is clear that a magnetic field strength can be extracted from observations of plasma satellites structure. Such experimental measurements were realized both in interaction of picosecond pulses (NEODIM installation) with solid targets [14] and femtosecond laser pulses with cluster targets [16]. The results of experiments and comparison of experimental data with theoretical modeling are presented on figs. 7,8. It is seen a strong effect of plasma oscillations on the spectra observed, corresponding field strength of oscillating field is as large as 10^8 V/cm and the frequencies are close to the cyclotron frequencies in magnetic field with field strengths of the order of 10^7 G in correspondence with ones following from eq. (5).

TOKAMAK PLASMA DIAGNOSTICS

Modern trends in spectroscopic investigations of tokamak plasmas are connected with transport phenomena. The transport of impurity ions as well as transport of main plasma components are in essence anomalous one. In order to extract transport coefficients (anomalous diffusion and convection coefficients) there were performed measurements of total radiative losses, intensities of specific spectral lines as well as spectral line shapes. The measurements deal with both bulk and edge plasmas.

Transport coefficients demonstrate a significant increase under external plasma heating. These effects result in a noticeable redistribution of total radiative losses of impurity ions on a minor tokamak radius. The effect is demonstrated on fig.9 for redistribution over minor radius of radiative losses on carbon, argon and iron impurity ions in T-10 tokamak. The experimental results can be fitted on the basis of theoretical models including anomalous transport coefficients [17]. The distribution of transport coefficients over minor radius is presented on fig.10. The transport coefficients extracted from the modeling succeed essentially their values followed from neoclassical theory.

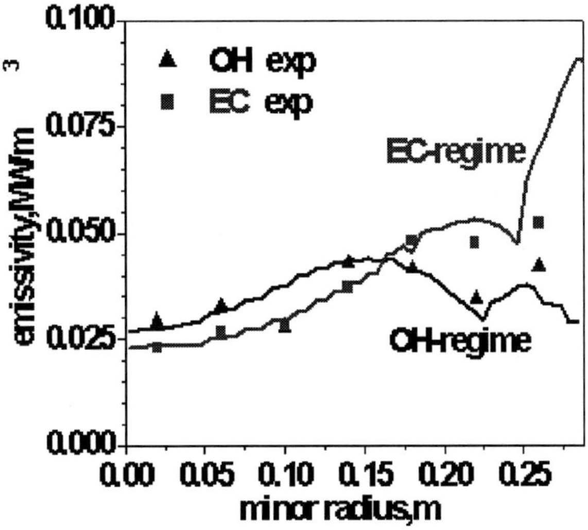

FIGURE 9. Distribution of radiative energy losses over minor radius in tokamak T-10 in Ohmic and ECR-heating regimes [17].

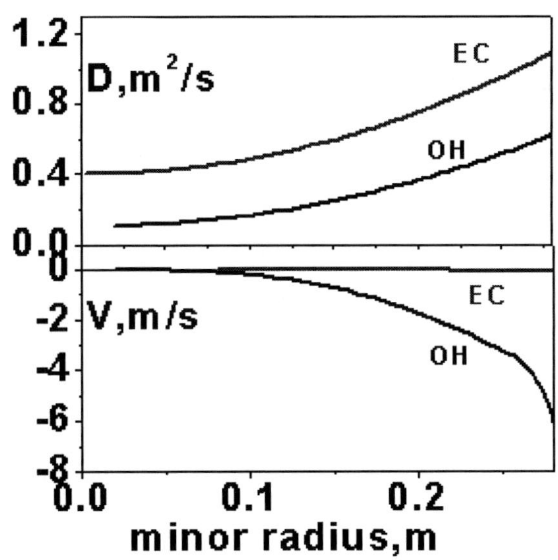

FIGURE 10. Anomalous transport diffusion and convection coefficients extracted from experimental data on radiative losses in T-10 tokamak [17].

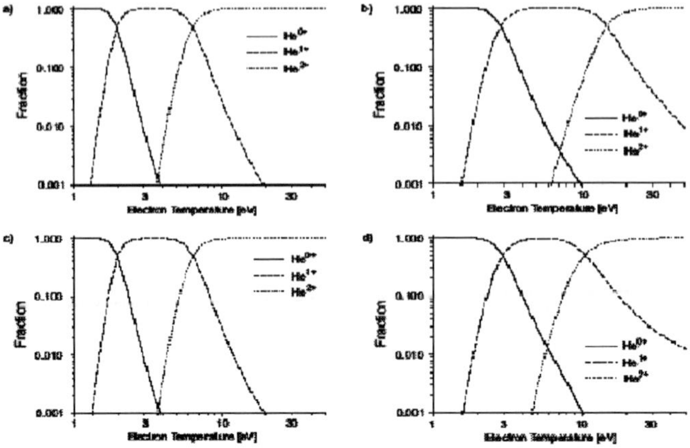

Fig. 1 – Relative fractions of He^{0+}, He^{+} and He^{++} in dependence of the electron temperature, $n_e = 10^{13}\,cm^{-3}$. a) No charge exchange, $\tau(He) = 0$; b) ground-state charge exchange coupling employing the effective charge exchange rates of table I (group I), $\tau(He) = 0$; c) excited-states charge exchange coupling, $\tau(He) = 0$; d) no charge exchange, $\tau(He) = 5 \cdot 10^{-3}$ s.

FIGURE 11. Effect of charge exchange and transport on ionization balance of He ions in edge plasma [18].

The effect of finite escape time on ionization balance of He ions in edge plasma is presented on fig.11 [18]. The effect is in a concurrence with charge exchange recombination of helium ions on neutral hydrogen presented in the edge plasma. A correct interpretation of the effects is of large importance for a determination of neutral helium fraction in future tokamak-reactor. The account for finite diffusion time produces also essential effect on intensities of singlet and triplet spectral lines of neutral helium in edge plasma [19].

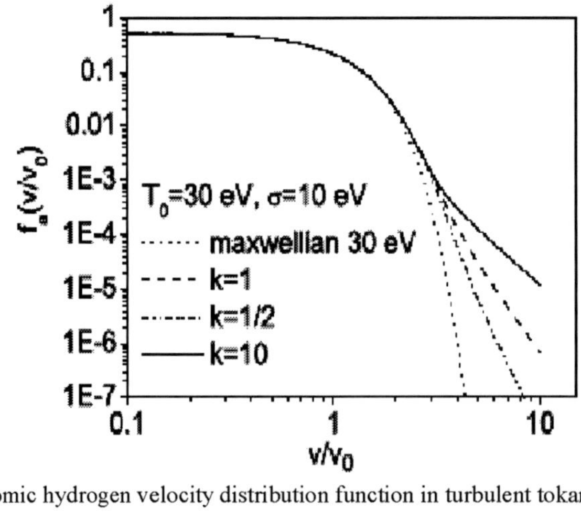

FIGURE 12. Atomic hydrogen velocity distribution function in turbulent tokamak plasma [20].

The anomalous transport is closely connected with turbulent processes in plasmas. There are a lot experimental measurements of turbulent noise in tokamak plasmas. But a new understanding proposed in [20] is connected with possible effect of turbulent fluctuation on atomic Doppler spectra. Really the turbulent fluctuations of electric field result in a modification of ion's velocity distribution function. The distribution can be transferred to the atomic one by charge exchange processes resulting in a modification of Doppler atomic spectra in far wings of spectral lines. The corresponding modification of atomic velocity distribution function in edge plasma is presented on fig.12 [20]. The results open a new possibility of plasma turbulence measurements based on atomic spectroscopy.

Generally the spectroscopic investigations of transport phenomena together with spectroscopic turbulence measurements open new possibilities for investigations of transport in tokamaks in self consistence with plasma turbulence.

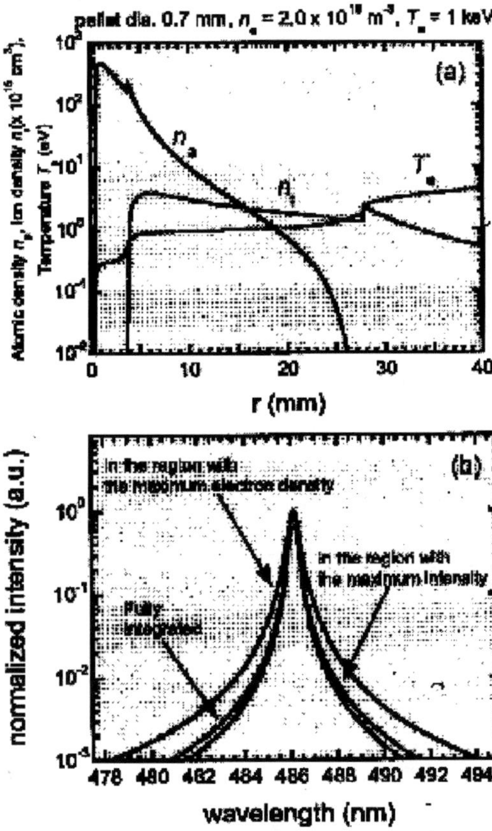

FIGURE 13. Plasma parameters (a) and Balmer-beta line profile in pellet cloud plasma of LHD installation [21].

At last the interesting object for spectroscopic investigations in thermonuclear plasmas is connected with line shape measurements of neutral hydrogen in pellet plasmas. The pellet plasmas is a cloud of low temperature (near 1 eV) and high density ($>10^{16}$ – 10^{17} cm^{-3}) plasma generated in plasma of pellet injected into high temperature thermonuclear plasma. The size of pellet cloud is of the order of 1 cm and it is characterized by strong nonuniform distribution of plasma parameters inside the cloud. The problem of plasma parameter measurements in such object is connected with the fact that the region with maximum intensity doesn't correspond to the region of maximum electron densities. The situation is illustrated on fig.13 obtained on Japan stellarator LHD [21]. The correct interpretation of the spectra observed has to be based on correct modeling of density and temperature distributions in the pellet cloud selfconsistently with line shape calculations.

BROADENING PROBLEMS FOR HIGHLY EXCITED (RYDBERG) HYDROGEN ATOMIC STATES

Highly excited (Rydberg) atomic states are observed both in laboratory and astrophysical plasmas. We'll discuss briefly the situation with investigations of extremely highly excited hydrogen levels observed in astrophysical plasma [8,9]. The first observations of the radiative transitions in radiorecombination lines discovered an anomalous narrow lines for radiative transitions between levels with principle quantum numbers n=101-100. The effect has been explained on the basis of interference effect between broadening of nearby energy levels resulting in the decrease of elastic widths due to ion's elastic impact broadening [22,23]. The main broadening in the case comes from inelastic electron collisions resulting in the scale n^4 for the line widths.

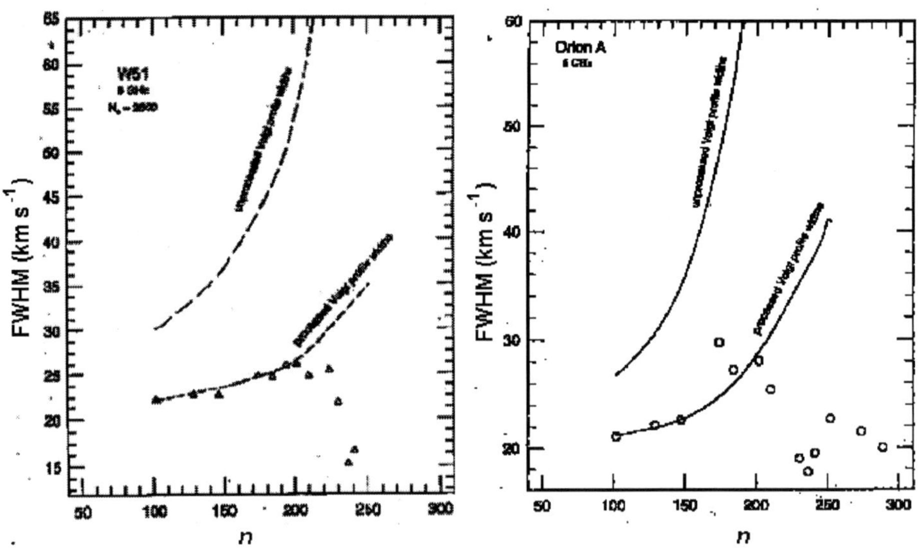

FIGURE 14. Line widths measured at the telescope for W51 (left) and for Orion A at 6 GHz (right) versus principle quantum number n [9].

Recent measurements of the widths of radiorecombination lines [9] in the same spectral range but for essentially large values of n>180 and the principle quantum number difference $\Delta n>6$ discovered a discrepancy between existent theory and data of measurements. The discrepancy is presented on figs.14 [8]. Further discussions [24-26] of the discrepancy didn't reveal any reasons for the effect observed.

Possible approaches for a description of the highly excited Rydberg states can be based on quasiclassical or pure classical description of atomic dynamics and kinetics. We point here one of the approaches for atomic kinetics based on classical kinetic equation suggested in [27]. The model [27] bases on the classical description of both radiative and collisional processes. The radiative ones are in correspondence with a classical picture of atomic electron fall on the nuclear due to its radiation. The collisions transitions are taken into account by classical Landau collision integral of atomic electron with plasma charged particles. The population source is provided by radiative recombination. The radiative recombination and radiative cascade are strongly selective in spherical quantum numbers whereas the effect of collisions results in mixing over orbital momentum quantum nubers leading to the statistical atomic level populations. The results of such classical model for atomic electron distribution function in two dimensional space "energy-orbital momentum' (that is, in principle and orbital momentum quantum numbers) are presented on fig.15. One can see a transition from nonequilibrium distribution of lower energy levels to statistical ones for highly excited levels. It is not clear however will the model be succesful for the description of the effects pointed above.

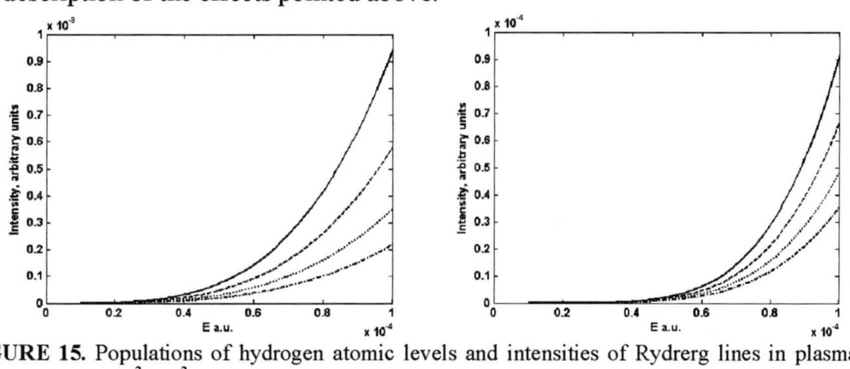

FIGURE 15. Populations of hydrogen atomic levels and intensities of Rydrerg lines in plasma with electron density 10^3 cm^{-3} and temperature 1eV following from classical kinetic model without (left) and with (right) account of collisions with plasma electron.

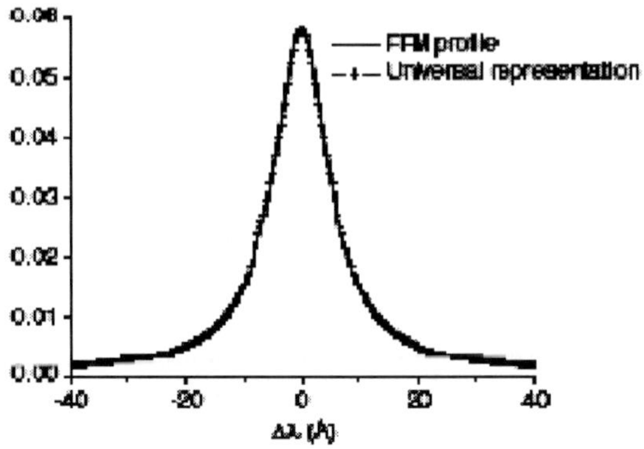

FIGURE 16. Universal and precise shapes of hydrogen H-beta line in the frame of FFM [29].

The last point concerned a possibility of a universal description of Rydberg spectra. Such universal representation bases on analytical formulas for intensities of radiative transition probabilities between highly excited hydrogen-like atomic states in parabolic coordinates (see for details [10]) being the initial point for static line shapes. The next step is an application of procedures for account of non-static ion's (or electron's) dynamics in the frame of FFM or model microfield models (MMM, see for example [28].). The comparison of universal approach with precise numerical calculations in the frame of FFM is presented on fig.16 for H-beta line [29]. One can see a good correspondence between universal analytical approach and the precise calculations.

CONCLUSION

The consideration above demonstrates a broad field of problems of atomic spectroscopy in modern plasma investigations. The problem of principle interest for dense plasmas is connected with possibilities of the transition from impact to static mechanism for electron broadening. In pico- and femto-second laser plasmas two problems seem to be of especial interest: the problem of spectroscopic measurements of fast ions energy distribution function in far wings of Doppler spectra and the problem of spectroscopic measurements of superstrong magnetic fields generated in such plasmas. Modern tokamak spectroscopy faces with a determination of impurity transport coefficients from total or spectral intensities. An interrelation between these transport coefficients and plasma turbulence is of great interest. The last effects are closely connected with possibilities of measurements of Doppler spectra in far wings. At last the problem of interpretation of hydrogen line widths for Rydberg-Rydberg transitions at extremely high values of principle quantum numbers is still actual.

ACKNOWLEDGMENTS

This work was supported by the INTAS grant N 03-54 6348 and grant RFBR 06-02-16614.

REFERENCES

1. H.R. Griem, *Spectral line broadening by plasmas*, NY: Academic Press, 1974.
2. S. Flich et al., *J.Phys.B* **36**, 283 (2003).
3. F. Rosmej, GSI site, 2004,LOI
4. L. D'yachov, G. Kobsev, P. Pankratov, *J.Phys.B* **21**, 1939 (1988).
5. A. Pigarov, J. Terry, B. Lipschultz, *Plasma Phys.Control.Fusion* **40**, 2055 (1998).
6. B.Talin, A.Calisti, et al., *Phys. Rev.* **A51**, 1918 (1995).
7. C Mossй, A Calisti, et al, *J. Phys. B* **37**, 1343 (2004)
8. R.L. Sorochenko, M.A. Gordon, *Radiorecombination Lines*, Springer, 2004.
9. M.B. Bell, L.W. Avery, E.R. Seaquist, J.P. Valiee, *Public Astron.Soc. Pacific* **112**, 1236 (2000).
10. L.A.Bureyeva, V.S. Lisitsa, *A Perturbed Atom,* Gordon and Breach, 2002.
11. M. Borhgesi, A.J. MacKinnon et al, *Phys.rev.Lett.* **81**, 112 (1998).
12. M. Tatarakis, A. Goral, I. Watt et al, *Phys. Plasmas* **9**, 2244 (2002).
13. V.S. Belyaev, V.I. Vinogradov, A.S. Kurilov et al, *JETP Lett.* **78,** 703 (2003).
14. V.S. Belyaev, V.I. Vinogradov, A.S. Kurilov et al, *JETP* **99,** 708 (2004).
15. E. Oks, *Plasma Spectroscopy. The influence of Microwave and Laser Fields*, Berlin: Springer, 1995.
16. V.P. Gavrilenko, A. Ya. Faenov, A.I. Magunov et al, *Phys. Rev.* **A73**, 1050-2947 (2006).
17. N. Timchenko, A. Dnestrovski et al, 32 EPS Conf. Plasma Phys. Proceedings v.29C, ECA, 2005, P-4.048.
18. F.B. Rosmej, R. Stamm, V.S. Lisitsa, Europys. Lett. **73**, 342 (206).
19. F. Rosmej ((private communication) presented to J. Phys.B (2006).
20. Y. Marandet et al, *Commun. Nonlinear Sci Numer. Simul.* **8**, 469 (2003).
21. N. Tamura, V. Sergeev, D. Kalinina, S. Sudo et al. *Proc. 30-th Conf. Contr. Fusion Plasma Phys,* ECA, 2003, v.27A, P-1.59.
22. H.R. Griem, *ApJ.* **148**, 547 (1967).
23. L.A.Minaeva (Bureyeva),I.I.Sobelman. J.Q.S.R.T.,**8**, 783(1968).
24. E. Oks, *ApJ.* **609**, L.25 (2004).
25. H. Griem, *ApJ.* **620**, L.133 (2005).
26. J. Watson, *J.Phys.B* **39**, 188927 (2006).
27. S.T. Belyaev, G.I. Budker, in *Plasma Physics and Nuclear Fusion Research vol.3,* edited by M.A. Leontovich, Moscow: Academy of Sciences of the USSR publisher, 1958, p.46.
28. C. Stehle, *Astron. Astroph.* **292,** 699 (1994).
29. C. Mosse, A. Calisti, R. Stamm et al, *J. Phys. B* **37**, 1343 (2004).

MD and FFM Electron Broadening for Warm and Dense Hydrogen Plasmas

S. Ferri, A. Calisti, C. Mossé and B. Talin
M. A. González and M. A. Gigosos

Laboratoire de Physique des Interactions Ioniques et Moléculaires, UMR 6633, case 232 St Jérôme, Université de Provence, 13397 Marseille cedex 20, France.
Departamento de Óptica y Física Aplicada, Facultad de Ciencias, Universidad de Valladolid, 47071 Valladolid, Spain

Abstract. Direct integration of the semi-classical evolution equation based on Molecular Dynamics simulations (MD) and the Frequency Fluctuation Model (FFM) have long been used to synthesize spectra accounting for ion dynamics. Cross comparisons of these approaches generally show results in good agreement. Recently, interest in low temperature ($T_e \sim 1$eV) and high density ($N_e \sim 10^{18}$ cm^{-3}) hydrogen plasma spectroscopy has motivated extended applications of FFM. Arising discrepancies were found to originate in electron collision operators suggesting an improper use of impact approximations for warm and dense plasma conditions. In order to clarify this point, new useful cross comparisons between MD and FFM have been carried out for electron broadening.

Keywords: Spectral line broadening, simulations, Molecular Dynamics.
PACS: 52.25Ya, 32.30.Jc, 52.65.Yy, 32.60.+I, 52.65.Yy

INTRODUCTION

A study of electron broadening for hydrogen Balmer lines calculated using two different techniques, namely direct integration of the semi-classical evolution equation based on Molecular Dynamics simulations (MD), [1], and the Frequency Fluctuation Model (FFM), [2], is reported. The MD simulations are known to give accurate results accounting for correlations and interactions between particles in a plasma. Nevertheless, the long calculation time necessary to obtain relevant results does not allow to consider complex emitters and makes simulation definitely inappropriate for spectroscopy diagnostics. Several models have been developed which provide an approximate solution, [3,4,5,6,7,8,9,10,11]. The FFM has shown its capability to give accurate results for slow varying plasma microfields (ion dynamics).The description of the effect of electrons on line profile is, in principle, the same as the problem of ion dynamics so that any solution of the many body ion dynamics problem is also applicable to electrons. In this work, the MD simulations are used to benchmark the behavior of the FFM results in the transition region between quasi static and impact broadening. The study is focused on hydrogen Balmer series lines, especially on the

CP874, *Spectral Line Shapes: 18th International Conference,*
edited by E. Oks and M. Pindzola

H-alpha line, for which a few experiments in the warm and dense plasma region are available.

MODELS

In the simulation technique used here, first, microfield sequences are generated using a standard molecular dynamics technique. Second, the evolution differential equations are solved using the well known Euler-Rodrigues parameters technique, [12]. In parallel, the Frequency Fluctuation Model is applied both for ions and electrons, using the electric field distribution obtained with APEX, [13]. The use of APEX is justified with a two-components-plasma MD simulation where both electrons and ions are considered, [14].

RESULTS

In order to analyze the transition between impact to static regime, Balmer series lines have been calculated for a fixed density (N_e=2.5 10^{18} cm^{-3}) and temperatures varying from 8000K to 200000K. This temperature-density domain includes recent experimental measurements [15,16,17].

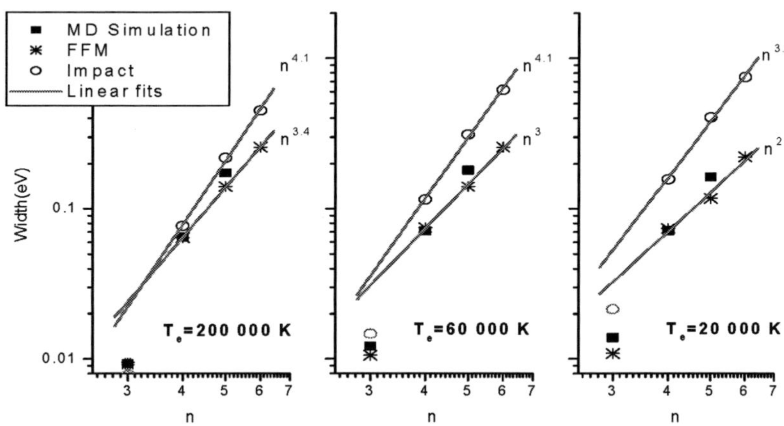

FIGURE 1. Balmer series line widths as a function of the principal quantum number. Comparison between MD simulations, FFM and impact models. (N_e=2.5 10^{18} cm^{-3})

The widths of the four first Balmer series lines show the right dependence with the principal quantum number. At high temperature FFM and MD simulations are close to the n^4 impact behavior while as the temperature decreases their dependences tend to the n^2 static behavior. The H_α line does not respect the trend because of the important role of the interference terms.

The line width versus temperature is shown in figure 2. For high temperatures both MD simulations and FFM converge toward the $T^{-1/2}$ trend which is typical of the impact behavior. As expected, the convergence between calculations and the impact behavior is shifted to higher temperature as the value of the upper principal quantum number increases. Within the whole range of temperatures, the FFM and the MD show the same tendency.

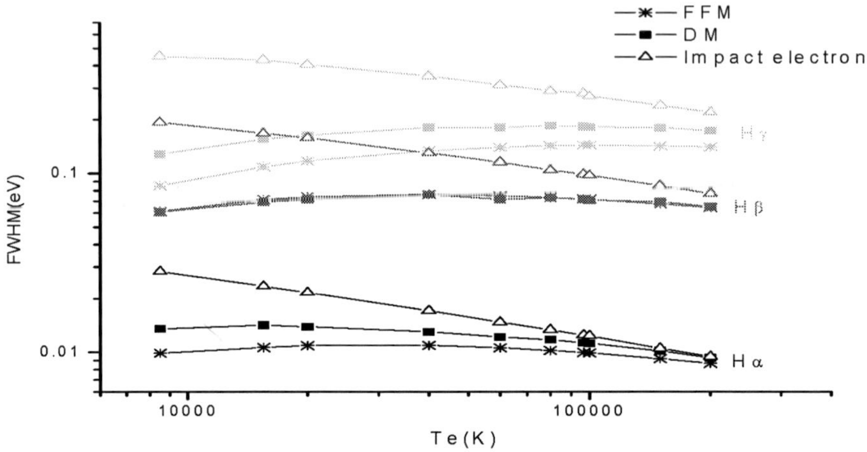

FIGURE 2. Balmer series linewidths vs electron temperature. Comparison between MD simulations, FFM and impact models. (N_e=2.5 10^{18} cm^{-3}).

Calculations (not presented here) show that the FFM possesses the right temperature, density and principal quantum number dependences. Nevertheless, improvement has to be done to reduce the remaining discrepancies appearing for lines with central components.

REFERENCES

1 B. Talin, A. Calisti and J. Dufty, *Phys. Rev. E* **65**, 056406 (2002).

2 B. Talin, A. Calisti, L. Godbert, R.Stamm, R.W.Lee et L. Klein, *Phys. Rev. A* **51**, 1918 (1995).

3 C. Stehlé, *Astron. Astrophys.* ***292***, 699 (1994).

4 J. Seidel, *Spectral Line Shape* **4**, 57 (1987).

5 S. Günter, L. Hitzschke and G. Röpke, *Phys. Rev. A* **44**, 6834 (1991).

6 Y. Ispalatov and E. Oks, *J. Quant. Spectrosc. Radiat. Transf.* **51**, 129 (1994).

7 J. Halenka, W. Olchawa, B. Grabowski and F. Gajda, *J. Quant. Spectrosc. Radiat. Transf.* **74**, 539 (2002).

8 W. Olchawa, R. Olchawa, B. Grabowski, *Eur. Phys. J. D* **28**, 119–124 (2004).

9 M. A. Gigosos, M. A. Gonzalez, V. Cardenoso, *Spectrochimica Acta Part B: Atomic Spectroscopy*, 2003.

10 S. Alexiou, *Phys. Rev. E* **71**, 066403 (2005).

11 H. Griem, J. Halenka and W. Olchawa, *J. Phys. B: At. Mol. Opt. Phys.* **38**, 975 (2005).

12 M.A. Gigosos, J. Fraile and F. Torres, *Phys. Rev. A* **31**, 3509 (1985); M.A. Gigosos and V. Cardenoso, *J. Phys. B: At. Mol. Opt. Phys.* **20**, 6005 (1987); *J. Phys. B: At. Mol. Opt. Phys.* **29**, 4795 (1996).

13 C.A. Iglesias and J.Lebowitz, , *Phys. Rev. A* **27**, 2705 (1983).

14 B. Talin, E. Dufour, A. Calisti, M.A. Gigosos, J.W. Dufty, *J. Phys. A: Math. Gen.* **36**, 6049 (2003).

15 S.A. Flih, E. Oks and Y. Vitel, *J. Phys. B: At. Mol. Opt. Phys.* **36**, 283 (2003).

16 A. Escarguel B. Fehrat and J. Richou, *J. Quant. Spectrosc. Radiat. Transf.* **64**, 353 (2000).

17 S. Büscher, Th. Wrubel, S. Ferri and H-J. Kunze, *J. Phys. B: At. Mol. Opt. Phys.* **35**, 2889 (2002).

Classical Radiative-Collisional Kinetics of Rydberg Atomic States

M.G. Levashova, V.S. Lisitsa

NFI RRC "Kurchatov Institute", Kurchatov sq. 1, Moscow, 123182, Russia

Abstract. The results of numerical modeling of the steady-state classical radiative-collisional kinetics for photorecombination are presented. The calculations are performed for edge tokamak ($N_e \sim 10^{11} - 10^{14}$ cm^{-3}, $T_e \sim 1$–100 eV) and astrophysical plasmas ($N_e \sim 10^4$ cm^{-3}, $T_e \sim 1$ eV).

Keywords: Atomic Kinetics, Plasma, Energy Level Populations.
PACS: 03.65.Sq, 02.60.Cb

INTRODUCTION: CLASSICAL KINETIC EQUATION

Substantial growth of complexity of quantum mechanical treatment of radiative-collisional kinetics at highly excited atomic states with increased quantum numbers and, at the same time, an increasing accuracy of classical (or quasi-classical) approaches makes the classical approaches more preferable.

Classical kinetic equation obtained in [1] takes account of radiative cascade (energy and angular momentum loss due to radiation emission by highly excited, bound electron in the field of ion in the approximation of classical electrodynamics) and pair distant Coulomb collisions of free electrons in plasmas (Landau kinetic equation).

$$-\frac{\eta}{\xi^4}\cdot q = \frac{\partial}{\partial\xi}\left[\frac{\eta}{\xi^4}\cdot\frac{\partial\psi}{\partial\xi}\right] + \frac{\partial}{\partial\eta}\left[\frac{5\eta(1-\eta^2)}{2\xi^6}\cdot\frac{\partial\psi}{\partial\eta}\right] - \frac{\partial}{\partial\xi}\left[\frac{\xi^3}{\eta^4}\cdot(1-\frac{\eta^2}{3})\psi\right] - \frac{\partial}{\partial\eta}\left[\frac{\xi^2}{\eta^3}\cdot(1-\eta^2)\psi\right] \quad (1)$$

where ψ - is the electron distribution function in an atom or ion in units of Maxwellian distribution, q - source of population (e.g. photorecombination).

$\xi = (-E/E_0)^{1/2}$; $\eta = \left(\frac{-2EM^2}{mz_0^{\,2}e^4}\right)^{1/2}$ are dimensionless variables vary within the range $0 < \xi < \infty$, $0 < \eta < 1$, m, e - electron mass and charge respectively, z_0 - nuclear charge, E_0 – energy such that the role of collisions and radiative transitions become equal.

We consider here only the steady-state case.

Purely radiative cascade was analytically described in [2] (both in classical and quasi-classical approaches) for arbitrary source of population.

We present in this paper the numerical modeling of the steady-state classical radiative-collisional kinetics for photorecombination source of population.

CP874, *Spectral Line Shapes: 18th International Conference,*
edited by E. Oks and M. Pindzola

NUMERICAL MODELING RESULTS

We performed calculations of electron distribution function in atom/ion and intensity of radiative transitions for edge tokamak ($N_e \sim 10^{11} - 10^{14}$ cm^{-3}, $T_e \sim 1$–100 eV) and astrophysical plasmas ($N_e \sim 10^4$ cm^{-3}, $T_e \sim 1$ eV). On figures 1-4 the results for pure radiative cascade are on the left, with account of Coulomb collisions of electrons are on the right.

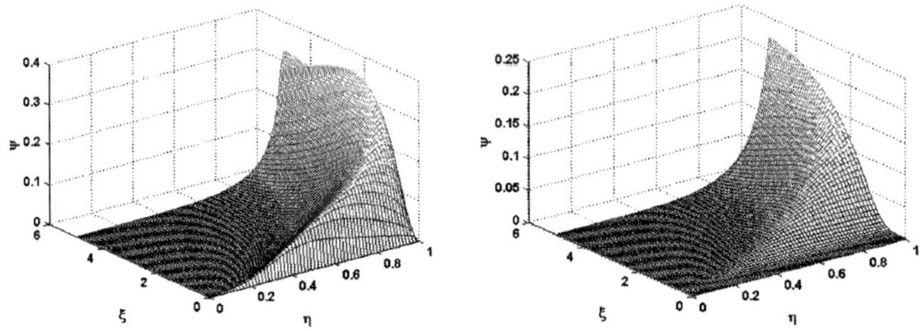

FIGURE 1. Electron distribution function in atom/ion in units of Maxwellian distribution for purely radiative cascade (on the left), with account of Coulomb collisions of electrons (on the right) with photorecombination source of atomic levels population. $N_e \sim 10^{14}$ cm^{-3}, $T_e \sim 100$ eV.

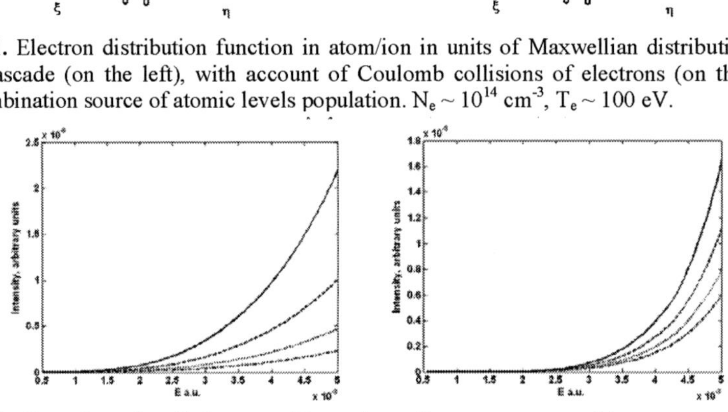

FIGURE 2. Intensity of radiative transitions depending on energy without account of electron collisions (purely radiative cascade) (left), with account of electron collisions (right) , Ne = 10^{14} cm^{-3}, Te = 100 eV. (solid – $\omega n^3 = 1.8$, dashed - $\omega n^3 = 3.0$, dotted - $\omega n^3 = 4.1$, dashdot - $\omega n^3 = 5.2$)

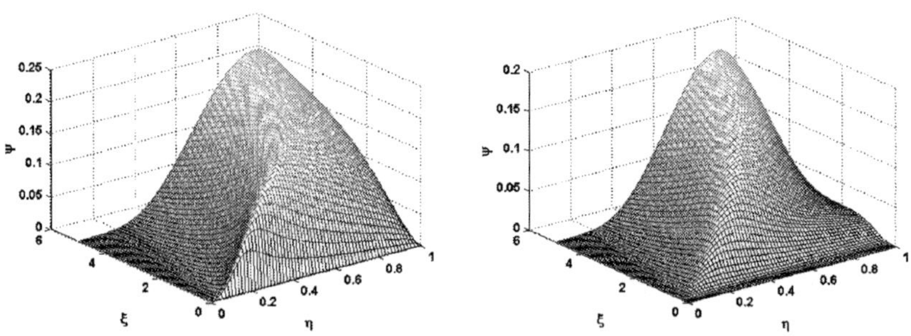

FIGURE 3. Electron distribution function in atom/ion in units of Maxwellian distribution for purely radiative cascade (on the left), with account of Coulomb collisions of electrons (on the right) with photorecombination source of atomic levels population, $N_e = 2500$ cm¯3, Te = 1 eV.

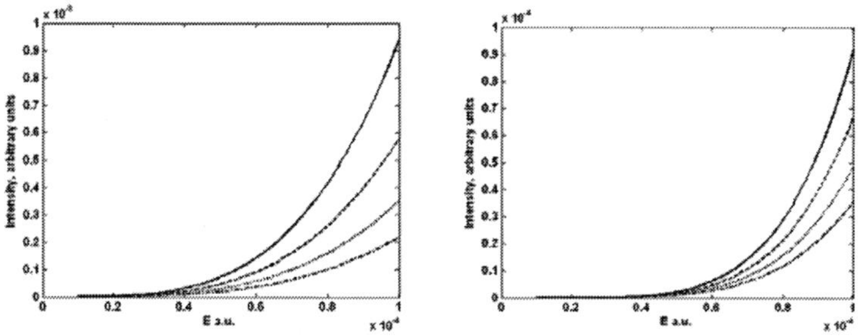

FIGURE 4. Intensity of radiative transitions depending on energy without account of electron collisions (purely radiative cascade) (left), with account of electron collisions (right), $N_e = 2500\ cm^{-3}$, Te = 1 eV. (solid – $\omega n^3 = 1.8$, dashed - $\omega n^3 = 3.0$, dotted - $\omega n^3 = 4.1$, dashdot - $\omega n^3 = 5.2$)

As one can see from the results obtained, taking account of Coulomb collisions makes distribution function smoother and its maximum lower than for pure radiative cascade. It results in lower intensity of radiative transitions.

CONCLUSION

In the present work the classical model of steady-state radiative-collisional kinetics of Rydberg atomic states is developed. Interrelation between collisional and radiative cascades is investigated. Universal results for Rydberg line intensities are presented. The classical kinetic model is a first step to the construction of classical broadening theory of Rydberg line shapes.

ACKNOWLEDGMENTS

The work is supported by project INTAS 03-546348 and RFBR 06-02-16614.

REFERENCES

1. S.T. Belyaev, G.I. Budker, in *Plasma Physics and Nuclear Fusion Research vol.3,* edited by M.A. Leontovich, Moscow: Academy of Sciences of the USSR publisher, 1958, p.46.
2. A.B. Kukushkin, V.S Lisitsa, ***Sov. Phys. JETP*** **61**, 937 (1985).
3. L.A.Bureyeva, V.S. Lisitsa, *A Perturbed Atom,* Gordon and Breach, 2002.

NEUTRALS

Collision-Broadened Line Shapes: A Different Perspective via Moyal Quantum Mechanics

B. R. McQuarrie*, G. C. Tabisz[†¶] and T. A. Osborn[†]

* Division of Science and Mathematics, University of Minnesota at Morris, Morris, MN 56267, USA
[†] Department of Physics and Astronomy, University of Manitoba, Winnipeg, MB R3T 2N2, Canada
[¶]LERMA, l'Observatoire de Paris/Meudon, 5 Place Jules Janssen, Meudon, 92195, France

Abstract. Moyal quantum mechanics (MQM) is a statement of quantum mechanics completely set in classical phase space. A formalism that has been developed for describing the collisional broadening of spectral lines with the assistance of MQM is outlined. Progress in its application to the far infrared spectrum of HD-He at 77K is reported

INTRODUCTION

Who was Moyal and what is Moyal quantum mechanics? Jose Enrique (Jo) Moyal was born in Jerusalem in 1910 and educated in electrical engineering, mathematics, statistics and theoretical physics in Grenoble, Cambridge and Paris. After the Second World War, his took up his first academic appointment, in mathematical physics, at Queen's University, Belfast. Subsequently he served on the faculty in statistics at the University of Manchester, the Australian National University and Macquarrie University. He also spent 10 years at the Argonne national laboratory in the USA. He died tragically in a house fire in 1998.

During his career he made fundamental contributions to mathematics, statistics and physics. It is his work in quantum mechanics which concerns us here. In 1949, while at Belfast, he published a paper in the Proceedings of the Cambridge Philosophical Society [1], entitled "Quantum Mechanics as a Statistical Theory", in which he presented a complete statement of quantum theory set in classical phase space. His motivation stemmed from his perception that statistical concepts play an ambiguous role in quantum theory, being introduced only in a *post hoc* interpretation. His was an attempt to reformulate the principles of quantum mechanics in purely statistical terms in order to gain insight into the meaning of the theory.

His starting point was to consider two sets of observables, or operators, let's say the momenta $\mathbf{p}$ and coordinates $\mathbf{q}$, which commute amongst themselves but not with each other. Together they form a set of dynamical variables characterizing a given system. He obtains the phase space distribution F by looking for its Fourier transform, i.e. the mean of $\exp(i\tau\mathbf{p} + i\theta\mathbf{q})$, known in statistics as the characteristic function. The resulting phase space distribution is in fact the Wigner distribution.

CP874, *Spectral Line Shapes: 18th International Conference,*
edited by E. Oks and M. Pindzola

Moyal's effort to gain a deeper insight into the meaning of quantum mechanics has not had great influence. Nevertheless the formalism he constructed remains of interest because in it expectation values of operators can be found via asymptotic expansions in $\hbar$ about classical quantities.

Let us look at the basics of Moyal quantum mechanics (MQM) in a little more detail.

$$(\psi, \exp(i(\tau \boldsymbol{p} + \theta \boldsymbol{q})\psi)) = \int \exp(i(\tau p + \theta q)) F(p,q) dp dq \,. \tag{1}$$

The LHS of (1) is the expectation value of the exponential operator in state ψ. Another way of describing it is given on the RHS as the mean of the exponential function taken with the probability distribution $F(p,q)$. But this integral may also be regarded as the Fourier transform of the probability distribution function. Taking the inverse transform with the use of the Baker-Campbell-Hausdorf theorem to incorporate the non-commutivity aspects, one obtains

$$F(p,q) = (1/2\pi) \int \psi^* (q - (1/2)\hbar\tau) \exp(-i\tau p) \psi(q + (1/2)\hbar\tau) d\tau \tag{2}$$

which is the Wigner distribution [2].

The mathematical complexity of the formalism had given MQM a limited capability for solving realistic problems. In 1995 Osborn and Molzahn[3] reformulated MQM through a cluster graph expansion and thereby developed a practical method for solving the equations of motion of MQM. This formulation is computationally more accessible than the original Poisson Bracket form. The result is a description of time evolution, which in contrast to the WKB approximation to propagation, involves no singularity as $\hbar \to 0$, no multiple trajectories and no caustics. A remarkable and attractive feature of MQM is the simple structure assumed by the semiclassical expansion of the Heisenberg picture time evolution.

EQUATIONS OF MOYAL QUANTUM MECHANICS

The Moyal approach is based on the Wigner–Weyl transformation σ. Thus,

$$\sigma \boldsymbol{A} = \int dx \exp(-ip \cdot x/\hbar) \langle q + (x/2) | \boldsymbol{A} | q - (x/2) \rangle \equiv A_W(q,p) \,. \tag{3}$$

Here $\boldsymbol{A}$ is an operator in Hilbert space, A_W is a function in classical phase space where the subscript refers to *Weyl,* and the $\langle ..|..|..\rangle$ indicates a Dirac matrix element in coordinate space. A_W is said to be the symbol or image of $\boldsymbol{A}$.

It is found that the symbols of the momentum and coordinate operators are the classical momentum and position variables:

$$\begin{aligned}\sigma \boldsymbol{p}_j &= (p_j)_W = p_j \\ \sigma \boldsymbol{q}_j &= (q_j)_W = q_j\end{aligned} \quad . \tag{4}$$

Frequently, the symbol of the Heisenberg operator is its classical counterpart:

$$\sigma \boldsymbol{H} = H_W(\hbar, q, p) = H_{classical}(q, p) \ . \tag{5}$$

The Heisenberg evolution of the operator $\boldsymbol{A}$, namely $\boldsymbol{\Gamma}(t)\boldsymbol{A}=\boldsymbol{U}(t)^{\dagger}\boldsymbol{A}\boldsymbol{U}(t)$, where $\boldsymbol{U}(t)=exp(-i\boldsymbol{H}t/\hbar)$ is expressed by the Heisenberg-Weyl evolution operator Γ_t acting on A_W,

$$\Gamma_t A_W = \sigma \boldsymbol{\Gamma}(t)\sigma^{-1}\sigma \mathbf{A} = \sum_{n=0}^{\infty} \frac{\hbar^n}{n!} \gamma_t^{(n)}(A_W) \tag{6}$$

To calculate expectation values the symbol of the density operator is needed. For the state ψ, this is essentially the Wigner function $w_\psi(q,p) = w_\psi(z)$. Here z denotes the set of phase space variables (q,p).

$$w_\psi(z) \equiv \hbar^{-3}\sigma\boldsymbol{\rho} = \hbar^{-3}\int dx \exp(-ip\cdot x/\hbar) \times \langle q+(x/2)|\psi\rangle\langle\psi|q-(x/2)\rangle. \tag{7}$$

Then the expectation value of $\boldsymbol{A}$(t) is

$$\langle \mathbf{A}(t)\rangle_\psi = Tr\mathbf{A}(t) \quad = \int dz w_\psi(z)\Gamma_t A_W(z) = \sum_{n=0}^{\infty} \hbar^n \int dz w_\psi(z)\gamma_t^{(n)}(A_W(z)). \tag{8}$$

The Moyal Bracket $\{\ldots\}_M$ is the σ image of the quantum commutator.

$$\begin{aligned}\{X_W, Y_W\}_M &= \sigma\left(\frac{1}{i\hbar}[\mathbf{X},\mathbf{Y}]\right) = \frac{1}{i\hbar}(X_W * Y_W - Y_W * X_W) \\ &= \{X_W, Y_W\}(z) - \left(\frac{\hbar^2}{2}\right) B_{12}^3 \prec X_W, Y_W \succ (z) + \ldots\end{aligned} \tag{9}$$

In this equation it is seen that the image of operator multiplication is the star- or $*$-product, i.e. $\sigma\mathbf{XY} = X_W * Y_W$. The star-product may be expressed in several ways but perhaps its clearest statement is had by realizing that the product operation may be given by:

$$* \equiv \exp\left(\left(\frac{i\hbar}{2}\right)\left(\overleftarrow{\partial_q}\overrightarrow{\partial_p} - \overleftarrow{\partial_p}\overrightarrow{\partial_q}\right)\right) \quad . \tag{10}$$

The arrow pointing to the left indicates that the derivative is be taken on the function to the left side of the asterisk while an arrow pointing to the right indicates that the derivative is to be taken on the function to the right. This complex operation acts on functions over all of phase space introduces the aspect of non-commutivity into the formalism.

In the last line of (9), the leading term is the classical Poisson Bracket $\{\ldots\}$. The B_{12} is an extension of the Poisson Bracket operator [6] here applied three times to the first and second functions, denoted by the subscripts 1 and 2, within the tuple $\prec \ldots \succ$.

With these definitions, the quantum Heisenberg equation of motion

$$\frac{\partial}{\partial t}\boldsymbol{\Gamma}(t)\boldsymbol{A}(t) = \frac{1}{i\hbar}\left[\boldsymbol{\Gamma}(t)\boldsymbol{A}, \boldsymbol{H}\right]$$

has as its image simply:

$$\frac{\partial}{\partial t}\Gamma_t A_W = \{\Gamma_t A_W, H\}_M \,. \tag{11}$$

The leading or order $\hbar^0$ term in (11) has the following form:

$$\frac{\partial}{\partial t}\gamma_t^{(0)}(A_W) = \{\gamma_t^{(0)}(A_W, H_c\} \,. \tag{12}$$

H_c is the classical Hamiltonian. This is just the classical equation of motion in Poisson Bracket form. Thus $\gamma_t^{(0)}$ mirrors classical motion. If the coordinates and momenta for classical motion are designated by $g(t|z)$, then their evolution in time, the so-called classical flow, is given by Hamilton's equations,

$$\frac{\partial}{\partial t}g(t|z) = J\nabla H(g(t|z)) \,. \tag{13}$$

and

$$\gamma_t^{(0)}(A_W) = A_W(g(t|z)) \,. \tag{14}$$

Here J is the Poisson matrix [6].

While these equations of classical dynamics are familiar it must be remembered that what is being done here is quantum mechanics. This just the zeroth order approximation. Even at this level, when expectation values are taken, the effects of complex valued wave functions and interference phenomena are present. The non-commutative effects of the *-product are absent.

Higher order approximations such as those of order $\hbar^1$ and $\hbar^2$ [3,4,5] are more complicated. These require the calculation of the Jacobi field $\nabla g(t|z)$ along the trajectory. The Jacobi field describes the stability of the trajectory under small changes in initial conditions. The concept of a quantum trajectory arises from computing the Heisenberg-Weyl evolution of the classical phase space variables *(q,p)* or z [4].

SPECTRAL LINE SHAPE THEORY

Collision-broadened spectral line shapes contain information regarding the time-dependent dynamical and collisional processes occurring in the radiating medium. The intensity of the line may be expressed as [7]:

$$I(\omega) = \frac{1}{\pi} \mathrm{Re} \int_0^\infty dt \exp(i\omega t)(C(t))^N . \quad (15)$$

Here $C(t)$ is the one-perturber dipole autocorrelation function and N is the number of perturbers. The dipole autocorrelation function is generated by the dynamics of the Hamiltonian $\boldsymbol{H}$.

$$C(t) = \sum_{j=1}^{3} Tr[\boldsymbol{\mu}_j U(H;t)^\dagger \, \boldsymbol{\mu}_j U(H;t)\boldsymbol{\rho}] \quad (16)$$

In the strategy employed to apply the ideas of Moyal quantum mechanics to the line shape problem, two distinguished degrees of freedom are considered: the internal molecular coordinates and the relative radiator-perturber separation variables [5]. The Hamiltonian is split into three parts:

$$\boldsymbol{H} = \boldsymbol{H}_1 + \boldsymbol{H}_2 + \boldsymbol{H}_{12} .$$

$\boldsymbol{H}_1$ applies to the motion internal to the radiator. The relevant variables $\boldsymbol{p}$ and $\boldsymbol{q}$ are left in their Hilbert space form $\boldsymbol{z}$. $\boldsymbol{H}_2 + \boldsymbol{H}_{12}$ generate different parts of the radiator-perturber dynamics. We have taken $\boldsymbol{H}_2$ to contain the description of the isotropic part of the radiator-perturber interaction. The variables, designated by $(\boldsymbol{P},\boldsymbol{Q})$, are transformed to their semiclassical form (P,Q) or Z. $\boldsymbol{H}_{12}$ describes the anisotropic part of the radiator-perturber interaction and is a function of both z and Z. As a consequence of this approach a new entity, a mixed Weyl symbol had to be defined. It is a function of Hilbert space operators and functions and is given by

$$A_{MW}(z;Z) \equiv \left(\frac{1}{2\pi}\right)^{2d_1} \int_{R^{d_1}} du \int_{R^{d_2}} dz A_W(z;Z)\exp(iu\cdot(z-Z)). \tag{17}$$

The dimensions of the relevant subspaces are given by d_1 and d_2. Osborn [5] has extensively elaborated the properties and calculus of this symbol.

In the establishment of this theory the frequently used assumptions are applied. A dilute gas consisting of a single radiator embedded in N perturbers is considered. The binary collision approximation is made. Collisional broadening is assumed to dominate and Doppler and radiative broadening processes are ignored. The radiator and perturber subsystems are taken to statistically independent and as a consequence the density operator may be factored.

In the $H_1 + H_2$ interaction picture, the expression for $C_{mm}(t)$ then becomes

$$C_{mm}(t) = \sum_{j=1}^{3} Tr\left[(\widetilde{\boldsymbol{\Gamma}}(t)\boldsymbol{\mu}_j)U(H_1;-t)^{\dagger}\boldsymbol{\rho}_{mm}\boldsymbol{\mu}_j U(H_1;-t)\right] \tag{18}$$

where $\widetilde{\Gamma}$ (t) is the evolution operator in this interaction picture and the semiclassical expression for $\boldsymbol{\rho}_{mm}\boldsymbol{\mu}_j$ is

$$((\boldsymbol{\rho}_{mm}\boldsymbol{\mu}_j)(z) = \left|\Phi_m\right\rangle\left\langle\Phi_m\right|\boldsymbol{\mu}_j(\exp(-\beta\boldsymbol{H}_2))_W(z). \tag{19}$$

To perform the computation one must follow the evolving dipole moment over a large number of trajectories. Specifically one must calculate

$$X_{km}^{j}(t|z) = \left\langle\Phi_k\right|(\widetilde{\boldsymbol{\Gamma}}(t)\boldsymbol{\mu}_j)(z)\left|\Phi_m\right\rangle. \tag{20}$$

The autocorrelation function may then be written as

$$C_{mm}^{(qm)}(t) = \frac{1}{\hbar^3}\sum_{j=1}^{3}\sum_{k=1}^{\infty}\int_{R^6} dz \exp(-i(E_m - E_k)t/\hbar)X_{km}^{j}(t|z)X_{mk}^{j}(0|z)(\exp(-\beta H_2)_W(z). \tag{21}$$

APPROXIMATIONS FOR NUMERICAL EVALUATION

In order to evaluate numerically these expressions which are exact within the mixed Weyl symbol formalism, approximations are necessary. $\widetilde{\Gamma}$ (t) is approximated by its small $\hbar$ expansion. The matrix star product appearing in $C(t)$ is approximated by its $\hbar^0$ and $\hbar^1$ contributions. Further only a small number (N) of eigenvalues of $\boldsymbol{H}_1$, the radiator Hamiltonian, can be realistically considered.

Then (20) becomes

$$X_{sr}^{j}(t|z) \cong \mathrm{X}_{sr}^{0,j}(t|z) + \hbar \mathrm{X}_{sr}^{1,j}(t|z). \tag{22}$$

The autocorrelation function (21) in the semiclassical (sc) approximation is

$$C_{mm}^{(sc)}(t) = \frac{1}{\hbar^3}\sum_{j=1}^{3}\sum_{k=1}^{N}\int_{R^6} dz \exp(-i(E_m - E_k)t/\hbar)\left(\mathrm{X}_{km}^{0,j}(t|z) + \hbar \mathrm{X}_{km}^{1,j}(t|z)\right)\mathrm{X}_{mk}^{0,j}(0|z)\exp(-\beta H_2(z)). \tag{23}$$

The $\hbar^0$ contribution gives the usual classical path (cp) approximation used in line shape theory:

$$C_{mm}^{(cp)}(t) = \frac{1}{\hbar^3}\sum_{j=1}^{3}\sum_{k=1}^{N}\int_{R^6} dz \exp(-i(E_m - E_k)t/\hbar)\mathrm{X}_{km}^{0,j}(t|z)\mathrm{X}_{mk}^{0,j}(0|z)\exp(-\beta H_2(z)). \tag{24}$$

APPLICATION TO THE HD-He FAR-INFRARED SPECTRUM

The HD-He infrared spectrum shows interesting features due to interference between allowed and collision-induced features[8,9]. It is an ideal system for applying the ideas developed in this paper. First of all it is a light system and particularly at low temperature quantum mechanical effects on the collision dynamics should not be negligible. At 77K HD is essentially a two rotational energy level system. Therefore only a few energy levels are needed in the calculation of $C(t)$ (23). Much experimental data, as well as standard quantum mechanical calculations [10], on line broadening for this system are available [8,9].

The H_2-He intermolecular potential of Mulder et al.[11] was converted to use with the HD-He system by change of the origin coordinates from the center of symmetry of H_2 to the center of mass of HD. Results for the calculation for a single trajectory are given in the following figures. The trajectory is specified by $(g_1,g_2,g_3,g_4,g_5,g_6)$ where

g_1 to g_3 are the relative radiator-perturber position coordinates and g_4 to g_6 are the relative momenta. The trajectory starts at $t=0$ at (0.25,-16,0.05,0.01,0.035,0) where

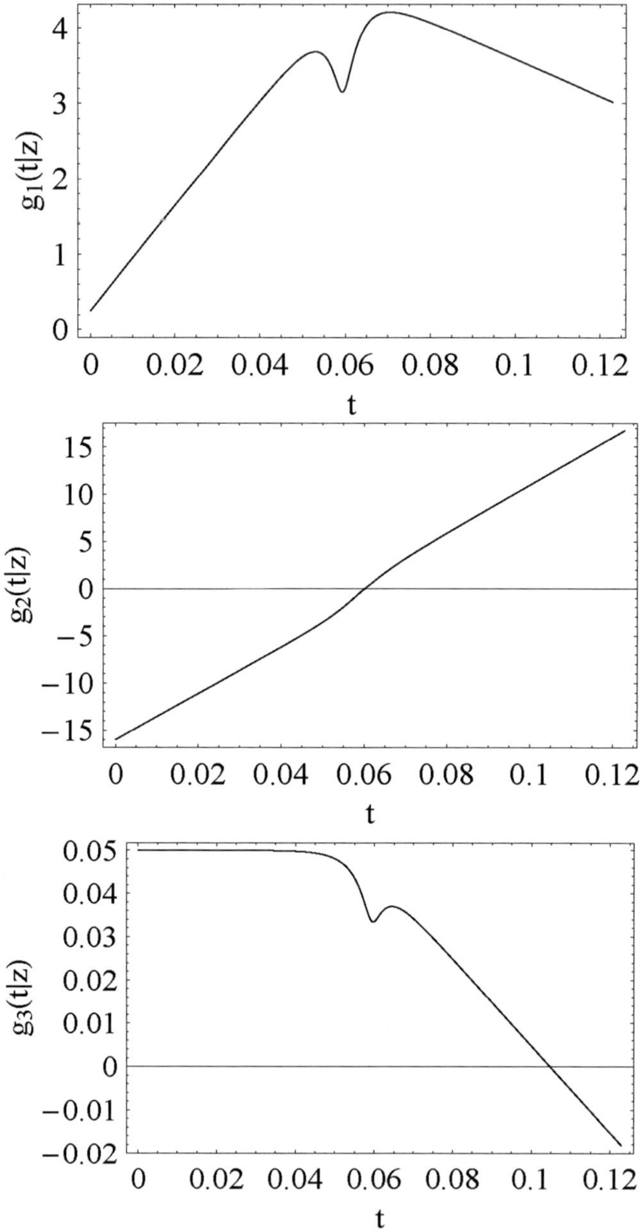

FIGURE 1. The coordinate components of position as a function of time for the sample trajectory.

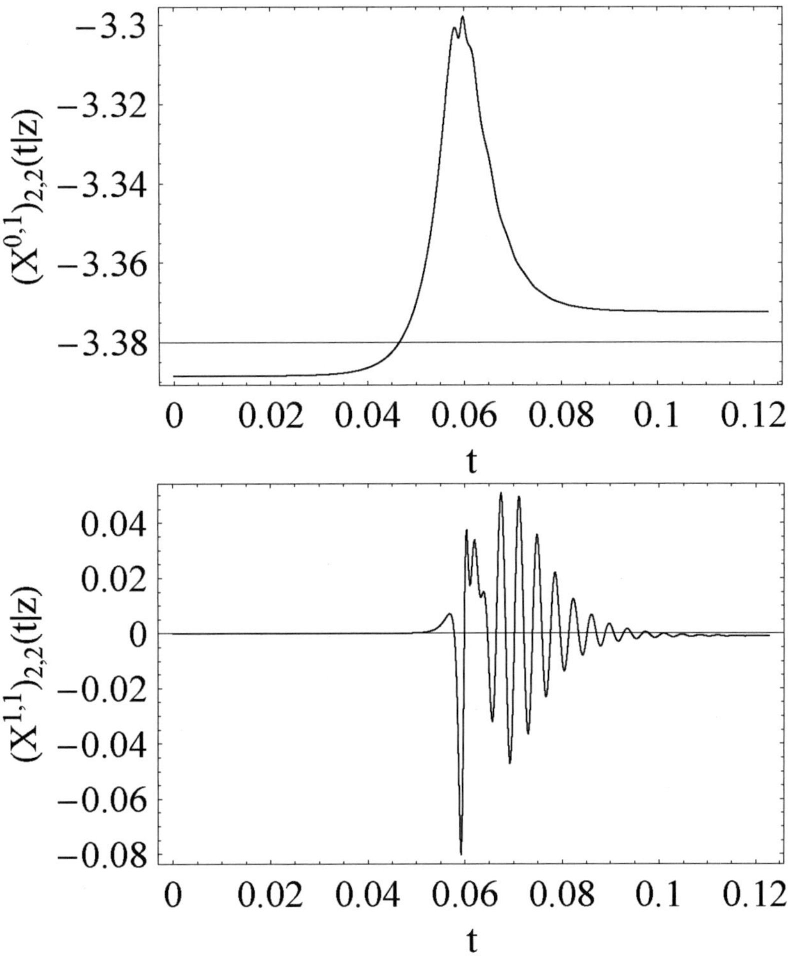

FIGURE 2. The time evolution over the sample trajectory of the real part of a diagonal matrix element of X. The top and bottom figures are respectively the order $\hbar^0$ and order $\hbar^1$ terms (Eq. 22).

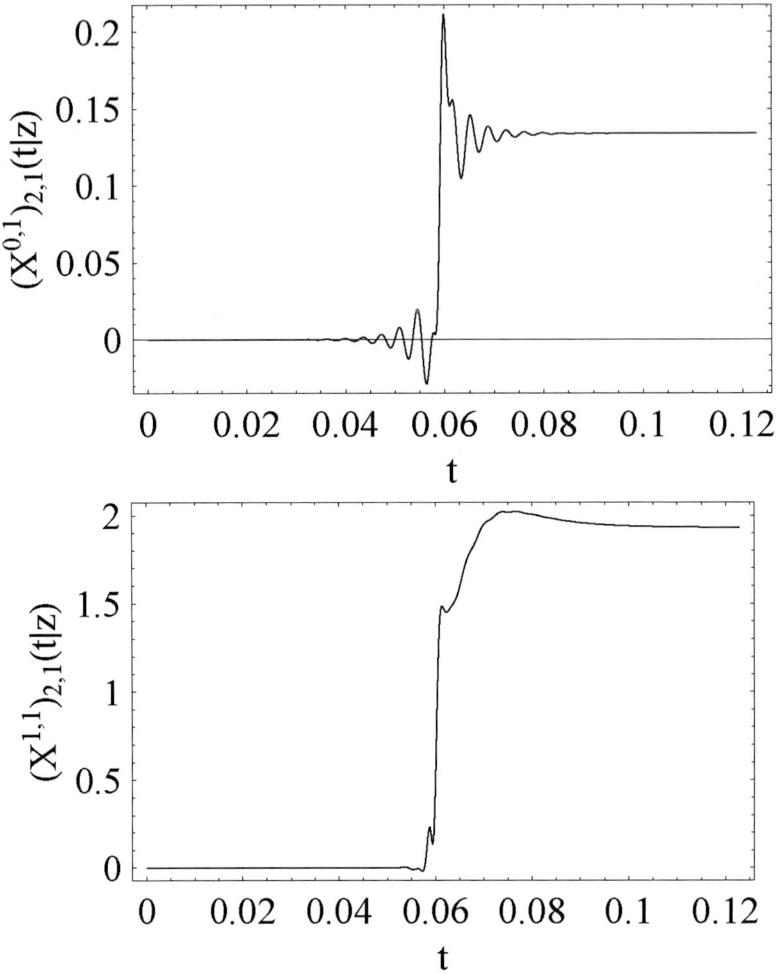

FIGURE 3. The time evolution over the sample trajectory of the real part of an off-diagonal matrix element of X. The top and bottom figures are respectively the order $\hbar^0$ and order $\hbar^1$ terms (Eq. 22).

time is expressed in units of 10^{-10} seconds, position in Ångstrom units and momentum in cm^{-1}-10^{-10}sec/Ångstrom. Figure 1 shows the trajectory. The point of closest approach occurs near t=0.06.

In Figs. 2 and 3 are shown examples of the evolution of the matrix elements of the X 's from which the autocorrelation function is constructed. The subscripts refer to rotational states *(J, M)* of the HD molecule in the ground vibrational-electronic state: *m=1 (J=0,M=0), m=2 (J=1,M=-1), m=3 (J=1,M=0), m=4 (J=1,M=+1).* The largest variation in both the leading (classical) term and the first quantum correction occurs in the region of closest approach. Far from this point these quantities are constant. For the classical path approximation, 54 dynamically evolving quantities must be followed along each trajectory. For the order $\hbar^1$ correction an additional 372 quantities must be followed. The X 's thus calculated are assembled to form the integrand in equation (23) for the autocorrelation function.

The present strategy is to attempt to calculate the 6-dimenional integral (23) by Monte Carlo techniques. At this writing it is not yet known whether this method is feasible or, if so, how many trajectories are necessary. Symmetry considerations can be introduced to simplify the calculation. The Fourier transform of the autocorrelation function will then be taken to obtain the spectral line profile.

SUMMARY

The possibility of employing the Moyal approach to quantum mechanics to problems in atomic scattering was demonstrated earlier[4]. A formalism for describing the collisional broadening of spectral lines with the assistance of this approach has been developed[5]. Calculations on the infrared spectrum of HD-He at low temperature are underway.

REFERENCES

1. J. E. Moyal Moyal, *Proc. Camb. Phil. Soc.* **45,** 99-124 (1949).
2. E. P. Wigner, *Phys. Rev.* **40,** 749-759 (1932).
3. T. A. Osborn and F. H. Molzahn, *Ann. Phys*(New York) **241**,79-127 (1995).
4. B. R. McQuarrie, T. A. Osborn and G. C. Tabisz, *Phys. Rev. A* **58**, 2944-2961 (1998).
5. T.A.Osborn, M.F.Kondrat'eva,G.C.Tabisz and B.R.McQuarrie, *J.Phys. A* **32**,4149-4169 (1999).
6. H. Goldstein, *Classical Mechanics 2nd edition*, Addison-Wesley, Reading,1980
7. H.Griem, *Plasma Spectroscopy*, McGraw-Hill, New York,1964.
8. P.G. Drakopoulos and G. C. Tabisz, *Phys. Rev. A* **36**,5566-5574(1987).
9. Z. Lu, G. C. Tabisz and L. Ulivi, *Phys. Rev. A* **47**,1159-1173 (1993).
10. M. Gustafsson and L. Frommhold, *Phys. Rev. A* **63**, 052514-1-052514-6 (2001).
11. F. Mulder, A. v.d.Avoird and P. E. S. Wormer, *Mol. Phys.* **37**, 159-180 (1979

Interaction–Induced Spectroscopy of H_2 in the Fullerenes

John Courtenay Lewis[1] and Roger M. Herman[2]

[1] *Department of Physics and Physical Oceanography, Memorial University of Newfoundland, St. John's NL, Canada A1B 3X7*

[2] *Department of Physics, Pennsylvania State University, University Park PA 16802*

Abstract. Carbon nanostructures of various sorts have been the subject of intensive research since their discoveries in the latter part of the 20th century [1]. Much of this research has been motivated by the intrinsic interest of these structures, though their potential as hydrogen storage media has also attracted attention [2]. It was realized [3,4] that the carbon–hydrogen interactions in these media would induce dipole moments which might lead to observable absorption of infrared spectra, and this work will be reviewed and extended in this paper. The fundamental vibration–rotation spectrum, of H_2 in a fcc C_{60} lattice (fullerite) at room temperature was first observed by S. A. FitzGerald and coworkers [3], who have subsequently extended their observations to near liquid nitrogen temperatures. Herman and Lewis have discussed the theoretical aspects of H_2 in carbon nanotube bundles and in fullerite [4,5,6]. We have developed a detailed theory for the spectrum of H_2 in fullerite. This theory assumes that the $H_2 - C$ potential can be accurately approximated by an exp–6 potential, the parameters of which are then obtained by fitting the line frequencies in FitzGerald's spectra. We have also obtained a model for the H_2 induced dipole moment based on the calculations of Frommhold and coworkers [7] on the induced dipole in $H_2 - He$. With one adjustable parameter this model gives a good account of the observed intensities.

In work to date the line width has been taken as an empirical parameter. However, the line width is in principal determinable from the $H_2 - C$ potential and induced dipole moment, together with the known properties of the phonon modes in fullerite. We conclude this paper with a discussion of the line width problem for H_2 in fullerite.

Keywords: interaction-induced spectroscopy; buckminsterfullerene;
PACS: 68.43.Pq, 78.30.Na, 33.20.Ea

INTRODUCTION

Since the discovery [1] of C_{60} by Kroto, Smalley, and coworkers in the mid-80s and the discovery of carbon tubules or nanotubes by S. Iijima in 1991 the chemical and physical properties of these all-carbon structures have been extensively studied by a wide range of methods.

Small molecules adsorbed in nanotubes and fullerenes are interesting for a variety of reasons, ranging from the advancement of our understanding of condensed phases

CP874, *Spectral Line Shapes: 18th International Conference*,
edited by E. Oks and M. Pindzola

(does H_2 adsorbed in nanotubes form a weakly bound anisotropic liquid at very low temperatures, for instance?) to achieving immediately practical applications. Pre-eminent among the latter is hydrogen storage. The search for an efficient low-pressure high-density storage system for molecular hydrogen has, in fact, motivated much of the work on hydrogen-adsorbing nanostructures.

The interaction-induced spectroscopy of the hydrogens in the gas, liquid, and solid phases has been discussed at previous ICSLSs. In this talk we discuss work from the last five years on the very interesting spectroscopy of H_2 in fullerene lattices. The IR spectra have been taken by S. A. FitzGerald of Oberlin College and his associates, and have been analyzed theoretically by T. Yildirim and associates at NIST, and by the present authors.

Buckminsterfullerenes

It is useful to review some of the properties of buckminsterfullerenes, and in particular properties of C_{60}. References [1] and [8] are particularly useful in this regard.

Buckminsterfullerene molecules are polyhedra of carbon atoms where each face is either a hexagon or a pentagon. They therefore have the same basic structure as "geodesic domes", albeit on a much smaller scale. Essential to their geometries is Euler's theorem for polyhedra, which asserts that

$$F+V=E+2,$$

where $F=$ the number of faces, $V=$ the number of vertices, and $E=$ the number of edges.

If we let $h=$ the number of hexagons and $p=$ the number of pentagons then

$$F=p+h, \quad 2E=5p+6h, \quad 3V=5p+6h$$

so that

$$\begin{aligned} 6(F+V-E) &= 12 \\ &= 6(p+h)+2(5p+6h)-3(5p+6h) \\ &= p. \end{aligned}$$

Hence there are exactly 12 pentagons in a fullerene molecule, regardless of its size. The smallest possible fullerene is C_{20}, known as the fullerane $C_{20}H_{20}$, which contains no hexagons. However, it is a rule that the presence of adjacent pentagons tends to destabilize the fullerene; the smallest fullerene with *no* adjacent pentagons is C_{60}, the celebrated "Buckyball", which is the most stable and easily produced of this class of molecule. The next largest fullerene satisfying the "no-adjacent-pentagons" rule is C_{70}, which is thermodynamically the most stable of the fullerenes. As of July 2006, C_{60} of 99.5% purity is available commercially at US$45 per gm and C_{70} of 99.0% purity at US$525 per gm.

C_{60} has also been detected occurring naturally in rock in Russia and in Colorado.

C_{60}

The basic properties of C_{60} are as follows:

- the exterior diameter of a C_{60} is about 10.34 Å;
- all 60 C are equivalent;
- there are two types of bond, a "single" bond of length 1.46 Å, which is significantly shorter than the C-C bond in alkanes or in diamond; and a "double" bond of length 1.39 Å;
- there is little resonance and an absence of aromatic characteristics;
- According to a standard reference on fullerene chemistry [8], "C_{60} is a fairly electronegative system. It behaves like an electron-poor conjugated polyolefin."
- Considerable charge separation is evident in C_{60}, with the "single" bonds positive and the "double" bonds negative [9]. The resultant electrostatic forces among C_{60} molecules result in rotational ordering at low temperatures.
- At room temperature its stable phase is fcc. This is known as "fullerite".

C_{70}

The second most common fullerene, C_{70}, includes among its properties that it has five distinct C sites and eight distinct C-C bonds; it looks like a rugby football; and its chemistry is similar to that of C_{60}. As for C_{60}, the stable room-temperature phase of C_{70} is fcc [9].

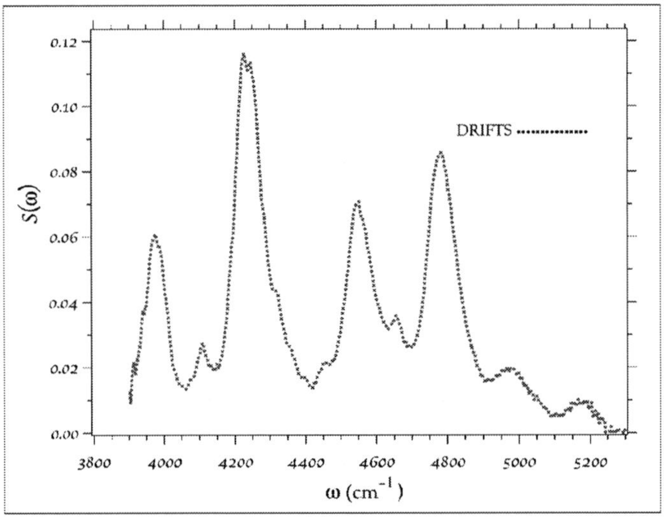

Figure 1. DRIFT spectrum from ref. [3] of the fundamental band of H_2 adsorbed to saturation in a fullerite lattice approximately at room temperature.

THE INDUCED FUNDAMENTAL BAND OF H_2 AT ROOM TEMPERATURE

When H_2 is adsorbed into a fullerite lattice, the interaction of the hydrogen molecules with the surrounding C_{60} lattice results in induced dipole moments and dipole transition moments. Observing the resultant absorption spectra is not straightforward, because macroscopic crystals of fullerite have not been grown, and samples are available only as powders. The fullerite lattice is, however, nearly transparent in the region of the hydrogen fundamental band, and the induced spectrum can be obtained [3] by the technique of Diffuse Reflectance Fourier Transform Spectroscopy (DRIFTS). A spectrum taken at room temperature is shown in Fig. 1.

The major features of the spectrum can be labeled in standard notation as follows:

$$\begin{array}{ll} Q(J)_R : J \to J, & n_z \to n_z + 1 \\ Q(J)_P : J \to J, & n_z \to n_z - 1 \\ S(J)_R : J \to J+2, & n_z \to n_z + 1 \\ S(J)_P : J \to J-2, & n_z \to n_z - 1 \\ O(J)_R : J \to J+2, & n_z \to n_z + 1 \\ O(J)_P : J \to J-2, & n_z \to n_z - 1 \end{array}$$

Here, "$n_z \to n_z \pm 1$" is shorthand for "$n_z \to n_z \pm 1$ and $\Delta n_x = 0$ and $\Delta n_y = 0$ or $n_x \to n_x + 1$ and $\Delta n_y = 0$ and $\Delta n_z = 0$ or $n_y \to n_y \pm 1$ and $\Delta n_z = 0$ and $\Delta n_x = 0$".

If a shift of approximately $-54\,\mathrm{cm}^{-1}$ from the corresponding free molecule vibration-rotation energies resulting from the differential perturbation of the $v = 0, 1$ levels is taken into account, the bands can readily be identified [C,F]. The larger $O(2)_R$ intensity will be obscured by the $Q(J)_P$ peaks, for all J, at $3950\,\mathrm{cm}^{-1}$. The tallest peak, at 4200 cm^{-1} arises from the $Q(J)_R$ transitions, for all J. This peak also contains, in its higher frequency wing, the $S(0)_P$ feature. The peak at $4550\,\mathrm{cm}^{-1}$ is mostly $S(1)_P$, also supplemented by the weaker $S(0)_R$. The tall peak at $4800\,\mathrm{cm}^{-1}$ is mostly $S(1)_R$, and obscures the $S(2)_P$ feature. The small peak at $5000\,\mathrm{cm}^{-1}$ is mostly $S(2)_R$, but has the $S(3)_P$ in its high frequency wing. The weak peak at $5200\,\mathrm{cm}^{-1}$ is almost entirely $S(3)_R$. The reason for the predominance of the $S(1)$ over the $S(0)$ features comes not only from the larger statistical weight of the $J = 1$ states over $J = 0$, but by virtue of the spin statistics, the $S = 1$ state that accompanies odd-J states being three times as frequently encountered, all other things equal, as the singlet spin state accompanying the even-J states.

In addition, there are three small relatively sharp features at approximately $4100\,\mathrm{cm}^{-1}$, $4450\,\mathrm{cm}^{-1}$, and $4650\,\mathrm{cm}^{-1}$. These "sharks-teeth" almost certainly reflect a breakdown of the octahedral symmetry of the fullerite, perhaps from defects or perhaps from partial rotational ordering of the C_{60} molecules. The various possibilities are discussed in ref. [6].

INTERMOLECULAR POTENTIALS AND ENERGY LEVELS

The C-H_2 Potential

Our work is based upon the C-H_2 potential, which we take to be of two forms:

$$U(r) = 4\varepsilon\left[\left(\frac{\sigma}{r}\right)^{12} - \left(\frac{\sigma}{r}\right)^{6}\right] \qquad \text{Lennard-Jones}$$

$$U(r) = 4\varepsilon\left[\exp\{-\beta(r-\sigma)\} - \left(\frac{\sigma}{r}\right)^{6}\right] \qquad \text{exp-6}$$

with

$$\beta = 11.2912 / \sigma.$$

The exp-6 potential with this parametrization for β has the same well depth ε, the same zero crossing radius σ and the same C_6 coefficient as the Lennard-Jones potential.

The C_{60}-H_2 Potential

If the 60 C atoms of a C_{60} molecule are smeared uniformly over a sphere of radius a then the above C-H_2 potentials yield the following potentials for the interaction of an H_2 molecule with a C_{60}:

$$U_{C_{60}\text{-}H_2}(R) = 4\varepsilon\frac{\sigma^2}{Ra}\left[3\left(\frac{\sigma}{R-a}\right)^{10} - \frac{15}{2}\left(\frac{\sigma}{R-a}\right)^{4} + \frac{15}{2}\left(\frac{\sigma}{R+a}\right)^{4}\right]$$

from the Lennard-Jones potential and

$$U_{C_{60}\text{-}H_2}(R) = 4\varepsilon\frac{\sigma^2}{Ra}\left[\frac{30(1+\beta(R-a))}{(\beta\sigma)^2}e^{-\beta(R-a-\sigma)} - \frac{15}{2}\left(\frac{\sigma}{R-a}\right)^{4} + \frac{15}{2}\left(\frac{\sigma}{R+a}\right)^{4}\right]$$

for the exp-6 potential. R is the displacement of the H_2 molecule from the center of a C_{60} molecule.

If the centers of the six C_{60} molecules bounding an octahedral cell enclosing an H_2 are located at $(\pm d, 0, 0)$, $(0, \pm d, 0)$, and $(0, 0, \pm d)$ then the total potential of the H_2, located at (x, y, z) relative to the equilibrium position of $(0, 0, 0)$ is

$$V(x,y,z) = U_{C_{60}\text{-}H_2}\left(\sqrt{(d+x)^2+y^2+z^2}\right) + U_{C_{60}\text{-}H_2}\left(\sqrt{(d-x)^2+y^2+z^2}\right)$$
$$+U_{C_{60}\text{-}H_2}\left(\sqrt{x^2+(d+y)^2+z^2}\right) + U_{C_{60}\text{-}H_2}\left(\sqrt{x^2+(d-y)^2+z^2}\right)$$
$$+U_{C_{60}\text{-}H_2}\left(\sqrt{x^2+y^2+(d+z)^2}\right) + U_{C_{60}\text{-}H_2}\left(\sqrt{x^2+y^2+(d-z)^2}\right).$$

Determination of the Energy Levels

We have found that the energy levels can be calculated approximately using a separable potential which is the sum of the three one-dimensional potentials across the faces of the fcc cell. If $V(x,y,z)$ is the exact nonseparable potential then

$$V(x,y,z) \simeq V_{1D}(x) + V_{1D}(y) + V_{1D}(z)$$

where

$$V_{1D}(x) = V(x,0,0) - \frac{1}{3}V(0,0,0).$$

The energy levels were determined in two ways. The first was a Discrete-Variable Representation (DVR) method in which a basis set of products of sines was used, the matrices of the position operators $\hat{x}, \hat{y}, \hat{z}$ were obtained and diagonalized, and their direct product was formed. The matrix of the potential $V(\hat{x}, \hat{y}, \hat{z})$ is diagonal in this representation and is easily calculated. Details are given in the Appendix. The result is then added to the matrix for the kinetic energy, and the whole is diagonalized to yield the energies. This procedure can be used equally for a separable and a non-separable potential.

The second method was to find the one-dimensional energies using the potential $V_{1D}(z)$ by a shoot-and-match procedure, and use these to obtain the energies for the separable three-dimensional potential. These energies served as a check on the DVR procedure, but also provided one-dimensional wave functions in useable form, from which the band intensities were calculated.

Table 1 compares energies for the separable potential with those for the non-separable potential to which it approximates. It is evident that the approximation of separability is satisfactory for the low-lying levels of greatest spectroscopic interest.

TABLE 1. exp-6 potential, σ = 3.22 Å, ε = 3.25meV

Energy level	Degen-eracy	Energy for $V(x,y,z)$ cm^{-1}	Energy for sum potential cm^{-1}	Principal states (n_x, n_y, n_z)
1	1	-926.7	-932.1	(0 0 0)
2	3	-830.9	-836.4	(0 0 1) and perms.
3	3	-739.4	-740.8	(0 1 1) and perms.
4	1	-714.7	-713.1	(0 0 2)* and perms.
5	2	-705.6	-713.1	same as above
6	1	-652.3	-645.1	(1 1 1)
7	3	-624.7	-617.5	(0 1 2) and perms.

8	3	-617.5	-617.5	same as above
9	3	-566.2	-569.9	(0 0 3) and perms.
10	3	-538.3	-521.8	(1 1 2) and perms.
11	1	-513.1	-494.2	(0 2 2)* and perms.
12	2	-501.8	-494.2	same as above
13	3	-489.7	-474.3	(0 1 3) and perms.
14	3	-474.2	-474.3	same as above
15	3	-427.2	-409.8	(0 0 4) and perms.
16	1	-419.7	-398.5	(1 2 2)* and perms.
17	2	-408.3	-398.5	same as above
18	1	-407.8	-378.6	(1 1 3)* and perms.
19	2	-394.1	-378.6	same as above
20	3	-374.5	-351.0	(0 2 3) and perms.
21	3	-368.6	-351.0	same as above
22	3	-325.0	-314.2	(0 1 4) and perms.
23	3	-324.6	-314.2	same as above
24	1	-318.8	-275.3	(2 2 2)
25	3	-308.2	-255.3	(1 2 3) and perms.
26	3	-284.5	-255.3	same as above

The separation states are linear combinations of the states in the fifth column. (*) indicates that specific linear combinations are required as first order approximations to the eigenstates for the nonseparable potential. Thus for the doubly degenerate level 5, the eigenfunctions for $V(x,y,z)$ can be represented to first order as

$$\frac{1}{\sqrt{6}}\left(2\left|002\right\rangle-\left|200\right\rangle-\left|020\right\rangle\right) \text{ and } \frac{1}{2}\left(\left|200\right\rangle-\left|020\right\rangle\right)$$

while level 4 to first order has the eigenfunction

$$\frac{1}{\sqrt{3}}\left(\left|002\right\rangle+\left|200\right\rangle+\left|020\right\rangle\right).$$

Levels 4 and 5 form one triply degenerate level in the separable-potential approximation.

The approximate separability of the potential is important, because the DVR method can certainly be used to calculate intensities, or indeed any other matrix element of functions of x, y, z, it is complicated to implement such calculations and we have not as yet been completely successful in doing so.

It should be emphasized that for realistic parameters σ, ε the potential $V(x,y,z)$ is neither very much like a box nor like an isotropic harmonic oscillator. In fact for typical values of these parameters it has a weak maximum at the center. Hence its approximate separability was not expected.

Our approach to calculating the energies was to vary σ, ε by hand, recalculating the spectrum for each new set of values. This procedure was speeded by the approximate separability property of the potential V, because it is of course much faster to obtain one-dimensional eigenvalues by a shoot-and-match procedure than to diagonalize the large matrix which is required by the DVR procedure. We started with values of σ, ε which were close to those for graphene-hydrogen [10]. The results were not satisfactory, as can be seen in Fig. 3. The graphene-hydrogen potential of Wang et al. [11] was similarly unsatisfactory [6]. Agreement was finally obtained, as shown in Fig. 2, by using a value of σ which is about 13% greater and 21% shallower than the graphene-hydrogen values of ref. [10].

We could also distinguish between the exp-6 potential and the Lennard-Jones 12-6 potential, as can be seen in Fig. 4. The Lennard-Jones potential has been found to be satisfactory in graphene-adsorbate studies, but for H_2 in fullerite it gives a spectrum which is too "boxy", i.e. the $V(x,y,z)$ which results is too "hard". The softer exponentially dependent repulsion of the exp-6 potential seems to be necessary for accurate energies.

Our final best C-H_2 potential was of exp-6 form with $\sigma = 3.22\,\text{Å}$, $\varepsilon = 3.25\,\text{meV}$.

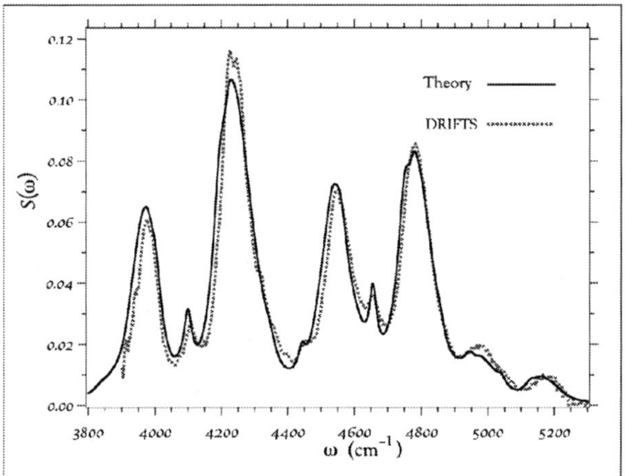

Figure 2. The DRIFTS from Fig. 1 together with our best theoretical spectrum, which was calculated for the exp-6 potential for which $\sigma = 3.22\,\text{Å}$ and $\varepsilon = 3.25\,\text{meV}$. The principal component Lorentzian line shapes have HWHH of $24\,\text{cm}^{-1}$. Other parameters are described in the text.

INDUCED DIPOLE MOMENTS

To calculate the intensities as opposed to the positions of the lines it is necessary to have some estimate for the dipole moment induced between the C_{60} molecules and the H_2 molecules. We define $\mathbf{s} = (s,\vartheta,\varphi)$ to be the orientation vector for the H_2, $\mathbf{r} = (r,\theta,\phi)$ to be the vector from a C to an H_2, and $\mathbf{R} = (R,\Theta,\Phi)$ to be the vector from the center of a C_{60} molecule to the center of an H_2 molecule. Then according to Poll and Hunt the vth spherical tensor component of the dipole moment induced between a C and an H_2 is given by

$$p_v(\mathbf{r},\mathbf{s}) = \frac{4\pi}{\sqrt{3}} \sum_{\Lambda LM} A_{\Lambda L}(r,s) C(\Lambda L1; M(v-M)) Y_{\Lambda M}(\vartheta,\varphi) Y_{L(v-M)}(\theta,\phi).$$

For vibrational $0 \rightarrow 1$ transitions this gives

$$p_0^{01}(\mathbf{r}) = \frac{4\pi}{\sqrt{3}} \sum_{\Lambda LM} B_{\Lambda L}(r) C(\Lambda L1; M, -M) Y_{\Lambda M}(\vartheta,\varphi) Y_{L,-M}(\theta,\phi)$$

where $p_0^{01}(\mathbf{r}) = \langle 0 | A_{\Lambda L}(r,s) | 1 \rangle$ is the matrix element of $p_0(\mathbf{r},\mathbf{s})$ between the ground and first vibrational levels.

From the symmetry of the H_2 molecule only even Λ and odd L contribute to the induced dipole moment. This implies a translational selection rule:

$$\Delta n_x = \Delta n_y = 0 \text{ and } \Delta n_z \pm 1$$

and permutations thereof.

It is expected that the most important contributions to p_0^{01} will be the B_{01} term, mainly from the overlap contribution, and the B_{23} term, which contains the dominant long range quadrupole induced contribution. However, we also include the induction contribution to B_{01}.

It is known that the overlap and induction contributions to the total dipole moment have an origin similar to the intermolecular force in distortion of the molecular charge clouds, and a similar functional dependence on the intermolecular separation, albeit with a somewhat greater range. It would be possible for us to calculate intensities with such a functional form and adjust parameters so as to obtain agreement with the intensities. However, we will endeavour to estimate those parameters *a priori*, or at any rate on the basis of detailed *ab initio* calculations of HD-H_2 by Borysow, Frommhold and Mayer [7], and on He-H_2 by Schaeffer and Kohler [12], rather than to interpret fitted parameters *a posteriori*.

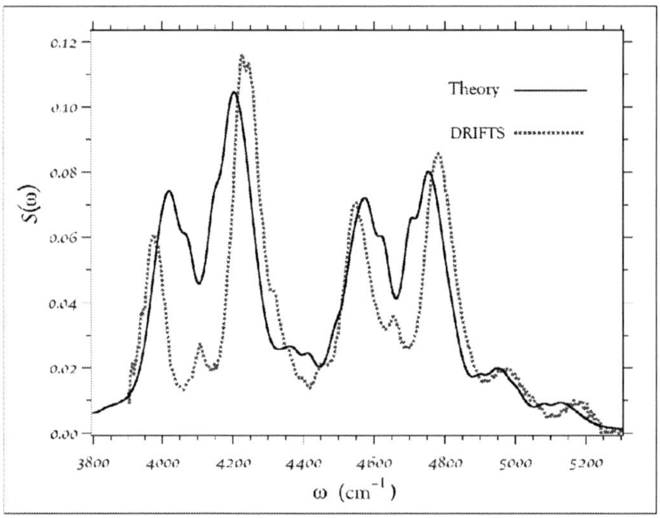

Figure 3. Theoretical spectrum using an exp-6 potential based on the Novaco [10] C-H_2 potential in graphene, for which $\sigma = 2.85\,\mathring{A}$ and $\varepsilon = 4.11\,\text{meV}$; with DRIFTS.

Long Range Behavior of the Induced Dipole Moment

At long range a bond polarizability model is used to relate $B_{23}^{01}(R)_{\mathrm{He-H_2}}$ to $B_{23}^{01}(R)_{\mathrm{C_{60}-H_2}}$. The result is

$$B_{23}^{01}(R)_{\mathrm{C_{60}-H_2}}\Big|_{\text{long range}} \simeq \frac{93.3}{R^4}$$

in atomic units. The long range (dispersion) part of B_{01} varies as R^{-7} and can be neglected in first approximation.

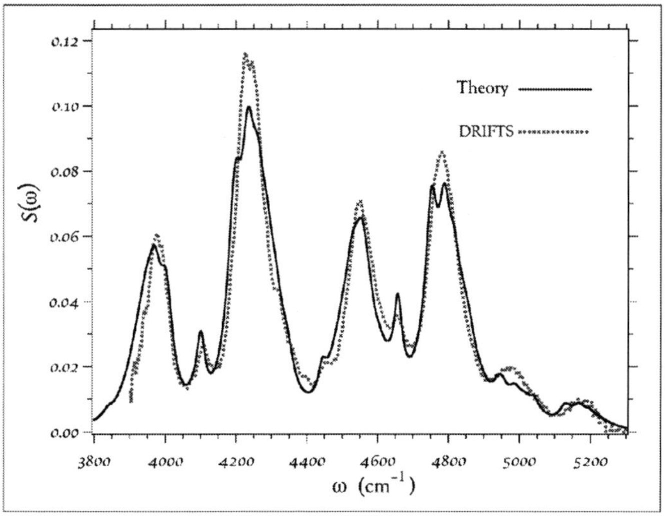

Figure 4. Theoretical spectrum using a Lennard-Jones interaction, with the same σ and ε as used for the exp-6 potential in Fig. 2.

Short Range Behavior of the Induced Dipole Moment

By graphical analysis of data from the afore-mentioned authors we find that

$$B_{01}^{01}(r)_{\mathrm{He-H_2}} \simeq 0.251\left[F_{\text{short range}}(r)_{\mathrm{He-H_2}}\right]^{0.75}$$

with numerical values in atomic units, for values of r around . We also find that

$$B_{23}^{01}(r) \simeq 0.24 B_{01}^{01}(r).$$

We conjecture that these relations hold for other atoms and small molecules as well as for He, provided that the bonding is largely physical rather than covalent, and provided that account is taken of differing charge deformability. The factor of 0.271 by itself cannot, be a general rule for relating the induced dipole moment and the 3/4th power of the intermolecular force. Both the intermolecular force and the induced dipole moment depend on the extent to which the atomic/molecular charge clouds are deformed by proximity of the colliding partner. However, this enters the intermolecular force as, in some sense, a *sum* of the deformabilities of the two collision partners, and the induced dipole moment as a *difference* of the deformabilities. Indeed, if the two collision partners are identical atoms, the induced dipole moment is zero. We conjecture that the reciprocal of the ionization potential I is an appropriate measure of the charge cloud deformability. Then, in general, instead of the factor 0.251 obtained for He-H_2, we expect that

$$0.251\left(\frac{1/I_{\mathrm{Y}}-1/I_{\mathrm{X}}}{1/I_{\mathrm{H}_2}-1/I_{\mathrm{He}}}\right)$$

will be the appropriate proportionality factor for colliding atoms and small molecules X and Y, in the absence of covalent bonding. Then for C-H_2

$$B_{01}^{01}(r)_{\mathrm{C-H}_2}\simeq 0.251\left(\frac{1/I_{\mathrm{H}_2}-1/I_{\mathrm{C}}}{1/I_{\mathrm{H}_2}-1/I_{\mathrm{He}}}\right)\left[F_{\text{short range}}(r)_{\mathrm{C-H}_2}\right]^{0.75}$$

should be a good representation of the short-range part of $B_{01}^{01}(r)_{\mathrm{C-H}_2}$. For the exp-6 potential the short-range part then yields

$$F_{\text{short range}}(R)_{\mathrm{C}_{60}\text{-}\mathrm{H}_2}=-\frac{120\varepsilon}{\beta^2 a}\frac{d}{dR}\left[\frac{1+\beta(R-a)}{R}e^{-\beta(R-a-\sigma)}\right]$$

which finally gives

$$B_{01}^{01}(R)_{\mathrm{C}_{60}\text{-}\mathrm{H}_2}=-3.55\gamma\left\{-\frac{\varepsilon}{\beta^2 a}e^{\beta\sigma}\frac{d}{dR}\left[\frac{1+\beta(R-a)}{R}e^{-\beta(R-a)}\right]\right\}^{0.75}.$$

Then the (23) – component is given by

$$B_{23}^{01}(R)_{\mathrm{C}_{60}\text{-}\mathrm{H}_2}=0.24B_{01}^{01}(R)_{\mathrm{C}_{60}\text{-}\mathrm{H}_2}+\frac{93.9}{R^4}.$$

The dimensionless parameter γ which enters both $B_{01}^{01}(R)_{\mathrm{C}_{60}\text{-}\mathrm{H}_2}$ and, implicitly, $B_{23}^{01}(R)_{\mathrm{C}_{60}\text{-}\mathrm{H}_2}$, is an adjustable parameter representing our imprecise knowledge of both the short range part and the long range part of the induced dipole moments. The ratio of the two tallest peaks in the spectrum is sensitive to it; we found that the value $\gamma = 3.1$ gave a good fit.

The calculation of intensities from these induced dipole moments is described in ref. [6].

Low-Temperature Spectra

S. A. FitzGerald and associates at Oberlin College have recently taken spectra at 80K and 10K [14,15]. We have applied our model to the 80K data. We obtain a good fit provided that:

- we assume that the lattice contracts by 1.4%;
- we take the temperature to be somewhat higher than 80K.

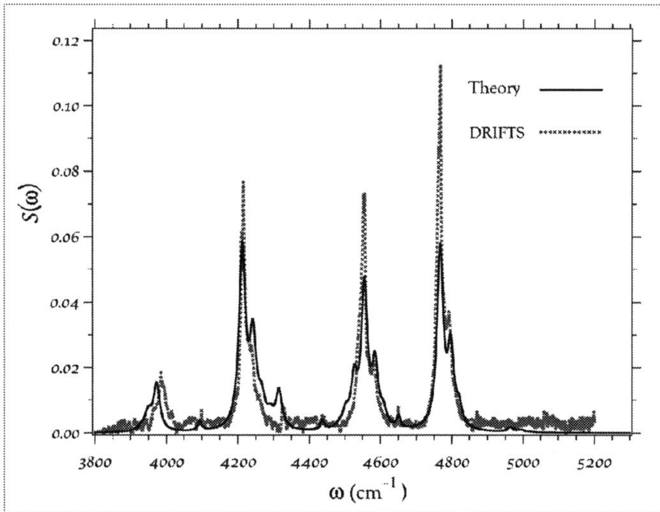

Figure 5. Observed (DRIFT) and theoretical spectra of H_2 in fullerite at approximately 80K. The DRIFTS was taken by S. A. FitzGerald and associates [14].

LINE WIDTHS AND LINE SHAPES

Whereas both energy levels and intensities are based on a priori estimates of relevant parameters, the line widths in our room temperature and 80K were chosen freely, to give the best fits with the experimental data. Our best spectrum, shown in Fig. 2, uses Lorentzian lines all of the same FWHH. We also used the same parameters with Gaussian line shapes, the results being shown in Fig. 6. Clearly the Gaussian line shapes give poorer agreement with the experimental data than do Lorentian lines, and we can therefore conclude that the broadening is largely homogeneous.

Given that the H_2-fullerite potential and induced dipole moment are now known, it should be possible to calculate the line broadening from first principles.

C_{70}

It should be relatively straightforward to obtain the induced spectrum of at least the fundamental band of H_2 in a C_{70} lattice. The stable room-temperature phase is fcc with complete orientational disorder [9]. Given the spheroidal shape of C_{70} and the presence of 5 inequivalent C atoms it is likely that the induced spectrum of H_2 in C_{70} would show considerably more homogeneous broadening than H_2 in C_{60}. At low temperatures, as the rotational motions freeze out, considerable inhomogeneous broadening would be expected.

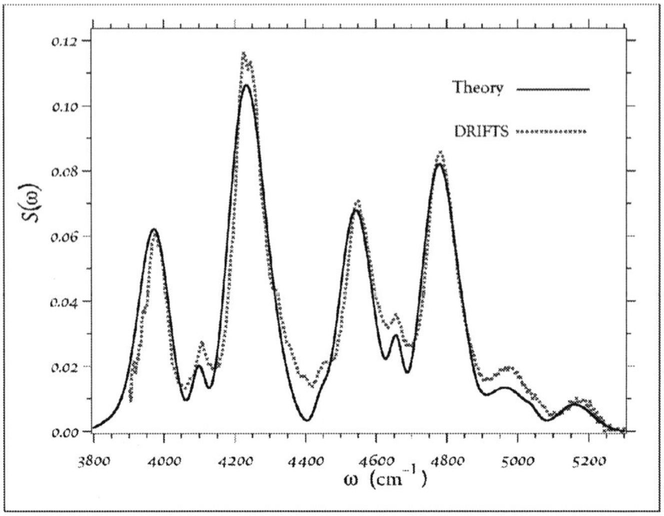

Figure 6. Theoretical room temperature spectrum using Gaussian line shapes. The width (and all other) parameters are the same as for the Lorentzian fit of Fig 2.

CONCLUSIONS

The first conclusion from the studies discussed herein is that the C-H_2 interaction in C_{60}, and probably in other small fullerenes, is substantially different from the C-H_2 interaction in graphene. This conclusions is in accord with present knowledge of the bonding in fullerenes [9] and of fullerene chemistry [8].

A more general conclusion is that DRIFT spectra, supplemented by neutron scattering spectra [13], contain much information on the C_{60}-H_2 interaction and likely on the dynamics of the fullerite lattice itself. This information can be obtained by careful analysis but without heroic extensions either to present experimental technique or to computational resources. The sensitivity which we have demonstrated of the theoretical spectra to small changes in parameters will lead to an excellent understanding of the C_{60}-H_2 interaction.

ACKNOWLEDGMENTS

We thank Stephen FitzGerald for permission to use his DRIFTS data in our figures, and for several extensive discussions about experiment and theory.

We acknowledge useful discussions on graphene and carbon nanostructures with M. W. Cole, P. C. Eklund, and V. Crespi.

RMH gratefully acknowledges receipt of a Fulbright Fellowship, with which part of this work was carried out, and thanks the Department of Physics and Physical Oceanography of the Memorial University of Newfoundland for hospitality during working visits.

JCL acknowledges support from the Natural Sciences and Engineering Research Council of Canada and thanks the Department of Physics of the Pennsylvania State University for its hospitality during several visits.

APPENDIX: A DISCRETE VARIABLE REPRESENTATION METHOD

We begin with a complete set of one-dimensional wave functions $\{\varphi_n(x)\}_n$. For the H_2-fullerite problem either sines (eigenfunctions for infinite rectangular wells), or simple harmonic oscillator wave functions, are feasible, though early on in our work we found that the sines seemed to give more rapid convergence than the simple harmonic wave functions. We denote the matrix element of an operator $\hat{O}$ by $\{\varphi_n|\hat{O}|\varphi_{n'}\}$. In particular the matrix element of the position operator $\hat{x}$ is $x_{nn'} = \{\varphi_n|\hat{x}|\varphi_{n'}\}$. This can be diagonalized by a unitary operator $\hat{u}$:

$$\psi_\lambda(x) = \sum_n u_{nl}\varphi_n(x)$$

$$\tilde{x}_\lambda \delta_{\lambda\lambda'} = \langle\psi_\lambda|\hat{x}|\psi_{\lambda'}\rangle = \sum_{nn'} u^*_{n\lambda}\{\varphi_n|\hat{x}|\varphi_{n'}\}u_{n'\lambda'} = \sum_{nn'} u^*_{n\lambda}x_{nn'}u_{n'\lambda'}.$$

The matrix element of the one-dimensional kinetic energy in the ψ basis is given by

$$\tilde{t}_{\lambda\lambda'} = \langle\psi_\lambda|\hat{t}|\psi_{\lambda'}\rangle = \sum_{nn'} u^*_{n\lambda}\{\varphi_n|\hat{t}|\varphi_{n'}\}u_{n'\lambda'} = \sum_{nn'} u^*_{n\lambda}t_{nn'}u_{n'\lambda'}.$$

The operator t is diagonal in the sines basis, though not in the harmonic oscillator basis, but we make no use of that fact.

If $\hat{I}$ is the one-dimensional identity operator then the operators for position and kinetic energy in three-dimensional space become

$$\hat{X} = \hat{x}\otimes\hat{I}\otimes\hat{I}, \quad \hat{Y} = \hat{I}\otimes\hat{y}\otimes\hat{I}, \quad \hat{Z} = \hat{I}\otimes\hat{I}\otimes\hat{z}$$

$$\hat{T} = \hat{t}_x\otimes\hat{I}\otimes\hat{I} + \hat{I}\otimes\hat{t}_y\otimes\hat{I} + \hat{I}\otimes\hat{I}\otimes\hat{t}_z$$

and the three-dimensional bases are, with some abuse of notation,

$$|n_x, n_y, n_z\} \equiv |N\} = \varphi_{n_x}(x)\varphi_{n_y}(y)\varphi_{n_z}(z)$$

$$|\lambda_x, \lambda_y, \lambda_z\rangle \equiv |\Lambda\rangle = \psi_{\lambda_x}(x)\psi_{\lambda_y}(y)\psi_{\lambda_z}(z)$$

with the unitary transformation $\hat{U}$ connecting the two bases:

$$|\Lambda\rangle = \sum_N U_{N\Lambda}|N\}$$

$$U_{N\Lambda} \equiv u_{n_x\lambda_x}u_{n_y\lambda_y}u_{n_z\lambda_z}$$

In the $|\Lambda\rangle$ basis the matrix elements of $\hat{X}$ are diagonal:

$$X_{\Lambda\Lambda'} = \tilde{x}_{\lambda_x}\delta_{\lambda_x\lambda'_x}\delta_{\lambda_y\lambda'_y}\delta_{\lambda_z\lambda'_z} \equiv \tilde{X}_\Lambda\delta_{\Lambda\Lambda'}$$

and similarly for $\hat{Y}$ and $\hat{Z}$; the matrix elements of the kinetic energy are given by

$$T_{\Lambda\Lambda'} = \tilde{t}_{\lambda_x\lambda_x'}\delta_{\lambda_y\lambda_y'}\delta_{\lambda_z\lambda_z'} + \delta_{\lambda_x\lambda_x'}\tilde{t}_{\lambda_y\lambda_y'}\delta_{\lambda_z\lambda_z'} + \delta_{\lambda_x\lambda_x'}\delta_{\lambda_y\lambda_y'}\tilde{t}_{\lambda_z\lambda_z'}.$$

It is straightforward and indeed almost trivial to calculate the matrix elements of the potential energy V in the $|\Lambda\rangle$ basis, because the position operators are diagonal; no integration is required, and little more than a function evaluation is necessary:

$$V_{\Lambda\Lambda'} = \delta_{\lambda_x\lambda_x'}\delta_{\lambda_y\lambda_y'}\delta_{\lambda_z\lambda_z'}V\left(\tilde{x}_{\lambda_x},\tilde{y}_{\lambda_y},\tilde{z}_{\lambda_z}\right) = \delta_{\Lambda\Lambda'}V\left(\tilde{x}_\Lambda,\tilde{y}_\Lambda,\tilde{z}_\Lambda\right).$$

The matrix elements of the three-dimensional translational Hamiltonian are given by $H_{\Lambda\Lambda'} = T_{\Lambda\Lambda'} + V_{\Lambda\Lambda'}$ and diagonalization of $H_{\Lambda\Lambda'}$ yields translational energy levels such as those shown in Table 1.

Evaluation of matrix elements of quantities dependent upon position requires back-transformation from the basis in which the Hamiltonian is diagonal, to the $|\Lambda\rangle$ basis.

REFERENCES

1. M. S. Dresselhaus, G. Dresselhaus, and P. C. Eklund, *Science of Fullerenes and Carbon Nanotubes*, Academic Press, 1996; and references therein.
2. B. K. Pradhan, A.R. Harutyunyan, D. Stojkovic, J. Fujiwara, and P. C. Eklund, *J. Mater. Res.* **17**, 2209 (2002).
3. S. A. FitzGerald, S. Forth, and M. Rinkoski, *Phys. Rev. B* **65**, 140302 (R) (2002).
4. R. M. Herman and J. C. Lewis, in *Spectral Line Shapes Vol. 12*, edited by C. Back, AIP Conference Proceedings no. 645, American Institute of Physics, Melville, NY, 2002, pp. 233–236.
5. R. M. Herman and J. C. Lewis, in *Proceedings of the 17th International Conference on Spectral Line Shapes*, edited by E. Dalimier, Frontier Group, 2004, pp. 369–371.
6. R. M. Herman and J. C. Lewis, *Phys. Rev. B* **73**, 155408 (2006).
7. A. Borysow, L. Frommhold, and W. Mayer, *J. Comp. Phys.* **88**, 4855 (1988).
8. A. Hirsch and M. Brettreich, *Fullerenes—Chemistry and Reactions*, WILEY-VCH, 2005.
9. J. Cioslowski, *Electronic Structure Calculations on Fullerenes and Their Derivatives*, Oxford University Press, 1995.
10. A. D. Novaco, Phys. Rev. Lett. **60**, 2058 (1988).
11. S. C. Wang, L. Sembetu, and C. W. Woo, J. Low Temp. Phys. **41**, 611 (1980).
12. J. Schaefer and W. E. Kohler, Physica A **129**, 469 (1985).
13. T. Yildirim and A. B. Harris, *Phys. Rev. B* **66**, 214301 (2002).
14. S. A. FitzGerald (private communication).
15. S. A. FitzGerald, H. O. H. Churchill, P. M. Korngut, C. B. Simmons, and Y. E. Strangas, *Phys. Rev. B* **73**, 155409 (2006).

Is There Anything Left to Do in the 100 Years Old Field of Pressure Broadening of Atomic Lines? (Abstract)

Kay Niemax

ISAS – Institute for Analytical Sciences, Bunsen-Kirchhoff-Strasse 1, 44139 Dortmund, Germany

The author presents his personal view on the status of pressure broadening of atomic lines, a research field founded by Lorentz exactly 100 years ago [1] and where significant progresses have been made in the thirties, forties and fifties of the last century. The fundamental work by Weisskopf, Holltsmark, Kuhn, Margenau, Jablonski, Baranger and Holstein was finally the basis of intensive research performed by many colleagues in the seventies when the understanding of line shapes became important in search of new laser media, e.g., excimer or exciplex systems. In the last years the interest in pressure broadening of atomic lines went down, mainly due to a lack of financial support by funding agencies providing money for fundamental research. An exception might be the interest in line shapes in ultra-cold gases.

However, there are several experimental observations in pressure broadening and shift of atomic lines still lacking understanding and proper theoretical description. Such observations are, for example, multi-perturber impact broadening and shift, the line shapes of bound- and continuum-state atomic pair absorption, the oscillations of the impact broadening and shift rates in alkali Rydberg series, and the peculiar multi-satellite series accompanying the self-broadened Cs principal series lines.

Very recently a special class of long range molecular Rydberg potentials of Rb*-Rb have been postulated [2,3] which show strong oscillatory behavior due to the scatter of the quasi-free optical electron on the Rb perturber. There was the assumption that the existence of these states can only be proved by laser excitation in an Rb Bose-Einstein condensate [2]. However, new line shape measurements in dense Rb vapor showed satellites in the far wings of the principle series lines whose positions are in agreement with the oscillatory potential curve extrema predicted [4]. On the other hand, the new calculations leading to the oscillatory behavior of the potentials seem to explain the peculiar multi-satellite series in the self-broadened Cs principal series lines found about thirty years ago.

REFERENCES

1. H.A. Lorentz, Proc. Amst. Acad. **18** (1906) 134.
2. C.H. Greene, A.S. Dickinson, H.S. Sadeghpour, Phys. Rev. Lett. **85** (2000) 2458.
3. A.A. Khuskivadze, M.I. Chibisov, I.I. Fabrikant, Phys. Rev A **66** (2002) 042508.
4. H. Crowell, E.L. Hamilton, C. Vadla, K. Niemax, C.H. Greene, DAMOP05 Meeting of the American Physical Society, Lincoln (Nebraska) and to be published.

CP874, *Spectral Line Shapes: 18th International Conference,*
edited by E. Oks and M. Pindzola

The Pressure Broadening and Shift of Spectral Lines Produced in an Intense Monochromatic Radiation Field (Abstract)

G. Peach[1] and F. Schuller[2]

[1]*Department of Physics and Astronomy, University College London, Gower Street, London WC1E 6BT, UK*

[2]*Laboratoire des Physique des Lasers, Institut Galile'e, UMR- C.N.R.S. 93430 Villetaneuse, France*

The study of the response of an atom to an intense narrow-band radiation field has become important ever since the advent of laser sources. When atoms that are surrounded by a buffer gas are absorbing or emitting radiation, the spectral line profile is both broadened and shifted, and if the radiation field is intense, additional effects are expected to occur. A general theory has been developed [1], to describe formally the combined effects of collisional and radiative perturbations on absorption and fluorescence profiles, but this theory has not been applied in practice. In [2], a unified theory of pressure broadening has been developed for atom-perturber interactions of the form $1/R^p$ where $p \geq 3$, which includes full averaging over a Maxwellian distribution of relative velocity. This theory reproduces the well-known results of the impact and static approximations in the low-density/high-temperature and high-density/low-temperature limits, and also predicts the shapes of profiles in the crucial intermediate region. The theories given in [1] and [2] have been used to describe coupled collisional and radiative processes, and preliminary results were reported at the 14th ICSLS in 1998 [3]. This combined theory has now been extended and has been applied to transitions in ultracold alkali-alkali and alkali-rare gas systems.

REFERENCES

1. F. Schuller and G. Nienhaus, *Physica* 121C, 281 (1983).
2. J. Peach, *J. Phys. B* **17**, 2599 (1984).
3. G. Peach and F. Schuller, Abstracts of 14th Int. Conf. Spectral Line Shapes, Penn State, University Park, 1998, p. 119.

CP874, *Spectral Line Shapes: 18th International Conference*, edited by E. Oks and M. Pindzola

Blue Satellite Bands and Photoassociation Spectra of Ultracold Cesium

Marin Pichler[1], Jainbing Qi[2], William C. Stwalley[3]
Robert Beuc[4], Goran Pichler[4]

[1] *Physics Department, Goucher College, Baltimore, Maryland 21204*
[2] *Department of Physics, Penn State University, Berks Reading, Pennsylvania 19610*
[3] *Department of Physics, University of Connecticut, Storrs Connecticut, 06269*
[4] *Institute of Physics, Zagreb, HR-10000, Croatia*

Abstract. Atomic line self-broadening was observed in ultracold cesium. Using resonant ionization detection we were able to photoassociate atoms into excited states of ultracold molecules below the *second* excited P_J atomic asymptotes $Cs(6S_{1/2}) + Cs(7P_{1/2})$ and $Cs(6S_{1/2}) + Cs(7P_{3/2})$. We were able to observe blue detuned satellite band features at 560 MHz and 800 MHz above the $7P_{1/2}$ and $7P_{3/2}$ atomic limits, respectively, as well as sharp hyperfine split features on the blue and red side of the atomic resonances. Such spectral features can be qualitatively supported by the local maxima in the long range molecular structure beyond 50 Å.

INTRODUCTION

The self-broadening of atomic lines at room and higher temperatures has been known to occur in many systems [1]. It arises due to interaction of like atoms. Until recently self-broadening has not been investigated in ultracold (T<1 mK) systems [2]. The blue and red satellite bands are associated with self-broadening. In ultracold systems though, the 0_g^- and 1_u pure long-range states [3, 4] may be regarded as "red" satellite bands. These states were observed in photoassociation (PA) spectra, corresponding to free-bound transitions red detuned from the atomic resonance [5]. Much research involving PA has been done recently in many different systems including production of ultracold molecules [6,7,8] and quantum degenerate gases [9]. Quantum degeneracy of molecules was not achieved through PA but by ramping a magnetic field near a Feshbach resonance (so-called magnetoassociation) in a degenerate Fermi system [10].

There have been prior attempts to observe blue detuned structure in PA experiments [11], but none were successful. Ultracold cesium was photoassociated experimentally below the $6P_J$ levels [12,13] and two-color PA to higher lying dissociation limits were performed in sodium [14], and potassium [15]. However, these two-color experiments involved two-step PA where the states dissociating to the first excited P asymptote served as the intermediate levels. In contrast our one-color experiments directly access the second excited P_J asymptotes. In this paper we present observation of blue satellite band features and sharp hyperfine-split photoassociative

CP874, *Spectral Line Shapes: 18th International Conference*,
edited by E. Oks and M. Pindzola

spectra on the red and blue wings of $7P_J$ lines. This work also presents the first experiment involving photoassociation near the second excited P level in any ultracold alkali-metal system.

The photoassociation spectroscopy below the highly excited asymptotes potentially offers the observation of many processes, such as autoionization, predissociation and the presence of ion-pair interaction. Weakly bound highly excited long-range molecules can be regarded as a negative energy analog to ground state-Rydberg state atomic collisions. The spectral structure of the highly excited states is also used as an intermediate step in two photon ionization detection of long range states dissociating to the lower $6P_J$ levels, some of which are not detected in trap-loss. Alternatively, the highly excited long range structure is necessary in performing multiple resonance experiments with three or more photons (e.g. all optical triple resonance-AOTR), where so-called dark states such as the 2_u ($6P_{3/2}$) state, which are forbidden in one step PA, can be investigated.

The experiments were focused on probing the spectral region near the $7P_J$ levels. Photoassociation here comes with a cost because the transition strengths are weak and transitions fall into the blue or ultraviolet spectral range. On the other hand, molecular structure is rich in long and pure long-range spectral features, which may offer a route to produce ultracold ground-state molecules in the lowest ro-vibrational levels.

Additional motivation comes from the connection to the proposed ultracold reaction $Cs(7P_J)$ + para-H_2 $\rightarrow$ $CsH(X\ ^1\Sigma_g^+$, v=0,J) + H [16], and ultracold reaction Cs_2^*+ para-H_2 $\rightarrow$ 2CsH, where Cs_2^* is formed below the $7P_J$ limits. In both cases the same spectral region is probed.

EXPERIMENT

The Cs magneto-optical trap (MOT), previously described in [13], is used for the PA near the $7P_J$ asymptotes in a similar fashion as it was used for the PA experiment below the $6P_J$ asymptotes. Trapped Cs atoms were illuminated with a PA-laser (probe) while monitoring the molecular ion signal. The λ=455.65 nm and λ=459.44 nm wavelengths are needed to excite the $7P_{3/2,1/2}$ transitions, respectively. A dye laser (Coherent 899 ring laser with Autoscan) running a Stilbene-3 laser dye was used as the PA-laser. To enhance the laser power, since these two wavelengths fall on the red edge of the dye gain profile, a Coumarin-102 output coupler was used. This configuration produced more output power (120-250 mW), and greatly improved scanning. The PA-laser beam was focused down to few hundred μm diameter in the cold cloud region. Further shrinking of the PA-laser had diminished the ion signal since a smaller fraction of the trapped atoms was addressed. To further maximize the PA-laser power, the laser beam was retroreflected and refocused back onto the cold cloud, almost doubling the available laser power, while preserving the maximum number of trapped atoms.

This became crucial for observing very weak spectral features. A separate laser resonant with the $6P_{3/2}$ level was used as a tracer beam for alignment over the MOT cloud. Laser scans were restricted to 0.3-1.0 cm^{-1} scan lengths, due to laser modehops and absolute wavelength calibration uncertainty, causing a scanning start frequency uncertainty of up to 0.22 cm^{-1} (reference cavity modehop) and 0.007 cm^{-1} (wavemeter uncertainty). To avoid these problems, the atomic $7P_J$ hyperfine levels were used for spectral calibration. We estimate the relative frequency readout was 5 MHz.

The 457 (or 459) nm PA laser is sufficient to ionize atoms in the cycling MOT transition, and causes a decay of the Cs trap. Therefore, it was necessary to alternately switch the PA-laser and the trapping laser beams. Switching of the MOT laser beams is accomplished with acousto-optic modulators (AOMs) at a 10 kHz rate and 60% on-40% off duty cycle. Fast switching of the trapping beams reduces the number and density of trapped atoms. To counter this effect, the MOT getter loading was increased.

Initially we have tried the trap-loss (fluorescence) detection. However, the predicted small PA-rates combined with the switching induced fluorescence instabilities, made trap-loss inadequate for the PA detection. Therefore the two-photon ion detection was implemented. The same light used for probing (photoassociation), is used for ionization of the long-range molecule (or atom in the case of atomic resonance). The first blue photon does the photoassociation, while the second (same color) photon ionizes the long-range molecule (or atom in the case of atomic resonance). The two-photon photoassociation/ionization was performed entirely during the time when the MOT lasers were off. Since the two blue photon ionization produces the ions very far into the continuum, the second blue photon was replaced with a infrared photon (~900 nm). This however did not improve the molecule detection, but produced an Autler-Townes splitting on atomic resonances. The atomic and molecular ions were accelerated and detected by a channeltron ion detector with -1750 to -2500 V operating in an analog mode. The detector gain of $\sim 10^7$-10^8 was further amplified by a current amplifier with a gain of 10^6-10^9 and sent to the data acquisition control unit, which recorded the laser scanning as well as detection signals. MOT fluorescence was monitored during PA scans for steady operation.

PHOTOASSOCIATION NEAR THE $7P_J$ LIMITS

The small formation rates result from the more than two orders of magnitude smaller atomic oscillator strengths for these second resonant transitions (6s →7p) compared to the principal resonant transitions (6s→6p). Nonetheless, the $7P_J$ photoassociation spectral region is very interesting because of the structure of the potentials, which can lead to observation of different spectral effects and potentially offers efficient formation of ultracold ground state molecules in the lowest ro-vibrational levels. One can improve the PA rates by increasing the PA laser intensity. This can be technically challenging especially in the case of generating spectrally narrow, continuous wave (cw) photons in the blue and ultraviolet (UV) spectral range. The detection presents another challenge as well, because due to the small formation

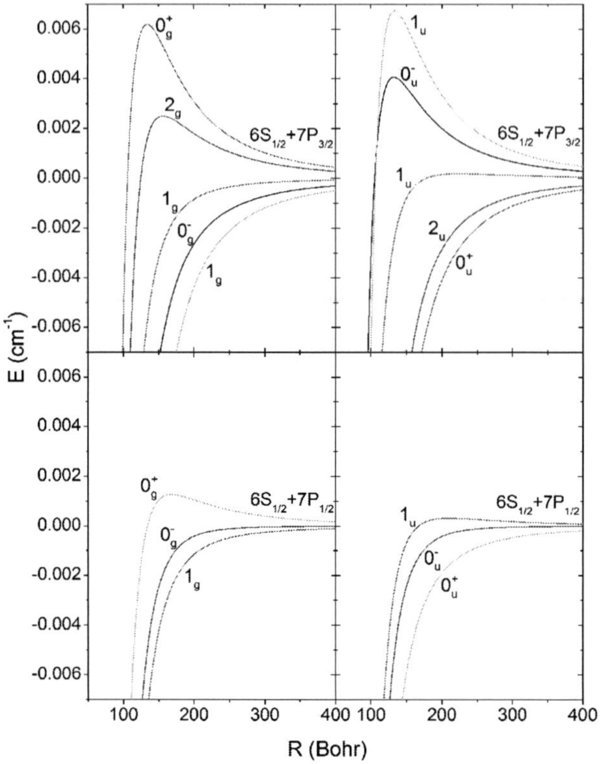

FIGURE 1. Calculated potential energy curves belonging to $6P_{1/2}$ + $7P_{1/2,3/2}$ asymptotes.

rates, one is forced to rely on ion detection, which, in the case of Cs with cw excitations, becomes difficult and insensitive at times.

The calculated highly excited Cs_2 potentials dissociating to $7P_2$ levels shown in Fig. 1 are used for quantitative analysis of the observed spectra. We have (initially) used C_3 and C_6 values of Marinecu and Dalgarno [17]. We found four states at the $7P$ limit (0_g^-, 1_g, 0_u^+ and 2_u) and two at the $7P_{1/2}$ limit 0_g^+ and 1_u) with local maxima (barriers) for large R >50 Å. However, all 16 states remain attractive inside 50 Å. By adding the ion-pair curve perturbations into the calculations (not shown here), the potential structure becomes even more dramatic for the 0_u^+ and 0_g^+ symmetries. This interaction results in a series of avoided crossings, inducing many long-range and pure long-range potential extremes. Spectroscopic features such as bound and quasibound states and interference effects are all associated with such complex long-range structure. Bound and quasibound states at large R provide large Franck-Condon overlap with the continua associated with two colliding ground state atoms.

The photoassociative reaction for cesium below the $7P_J$ levels is

$$Cs(6S_{1/2}, F=4)+Cs(6S_{1/2}, F=4)+h(\nu_0-\Delta_L) \rightarrow Cs_2(\Omega_{g,,u}, v, J, (7P_{1/2,3/2})) \quad (1)$$

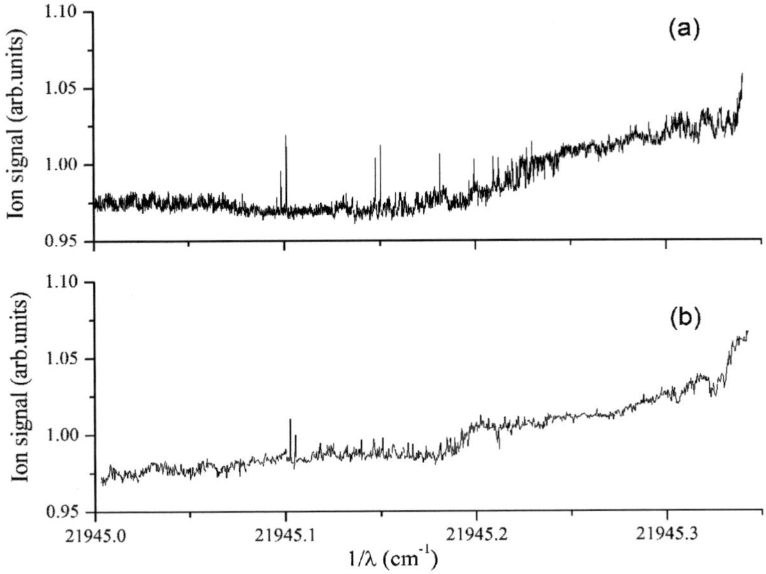

FIGURE 2. Photoassociation spectra below the $7P_{3/2}$ limit in the same region with higher scan acquisition rate (a) and lower acquisition rate (b).

where ν_0 is the frequency of the $6S_{1/2} \rightarrow 7P_{1/2,3/2}$ resonance line and Δ_L is the red detuning of the laser. The colliding ground state atom pair with 0_g^+, 0_u^-, or 1_u, symmetry, can be excited into any of the 0_g^+, 0_u^+, 0_g^-, 1_g, 1_u and 2_g ($7P_{1/2}$ or $7P_{3/2}$) attractive states, or, in the case of states with a pure long range barrier, into a quasibound level. Since all the states dissociating to the $7P_J$ asymptotes are attractive inside 50 Å, one can expect a higher overall density of states, compared to the $6P_J$ photoassociation, although the density of vibrational levels of an individual symmetry is probably still greater at the $6P_J$ asymptotes because of the larger C_3 coefficients.
The rate of photoassociation per atom can be written as

$$R_{PA}=A(i,f,\hat{e}_{PA})\,(3/2\pi)^{3/2}\,(h/2)\,n_{at}\,\lambda_{th}^{3}\,e^{-Er/kT}\,K^2\,S^2(\alpha,v) \tag{2}$$

where A is an angular factor of the dipole matrix element depending on initial and final state and the laser polarization, n_{at} is the atomic density, λ_{th} is thermal de Broglie wavelength, 2K is the atomic Rabi frequency ($K^2 = (\Gamma/2)(I/I_0)$), and $S= \langle\alpha,v\rangle$ is the overlap integral between the radial wavefunction of the initial continuum state α, and the bound vibrational state v. Due to the smaller oscillator strengths (entering in the factors A and K in Eq. 2) of 0.0113 and 0.0026 for the $7P_{1/2}$ and $7P_{1/2}$ transitions, respectively, the photoassociation rates are greatly reduced. Additionally, the ratio of $7P_{3/2,1/2}$ oscillator strengths is 4.3 which deviates from the predicted statistical value of ~2. This effect interpreted by Fermi in 1928 and recently discussed in [18] and references therein, will affect the PA spectrum. However, given the bound structure at large R ≥ 20 Å the overall PA rate may be slightly improved, especially for the case of the ion-pair curve perturbed potentials of 0_g^+ and 0_u^+symmetry.

The PA spectra below the $7P_{3/2}$ asymptote are shown in Fig. 2. The same region is scanned with different scanning parameters. The top scan is taken with a scan step of 2 MHz, while the bottom one with a 10 MHz step. Both scans are 10 GHz long. The ion signal is acquired by taking 500 readings and averaging in each scan step. The smallest scan step (the top scan) produces the best spectrum 20 GHz in the red wing of the $7P_{3/2}$ resonance line. A series of 7 doublet features is visible in the top spectrum, which represents a vibrational progression very close to the atomic asymptote. The blow-up of the strongest first feature is shown in Fig. 3, revealing a third line. The two additional lines belong to a different state. The spacing between each of the doublet (or triplet) features corresponds to the $7P_{3/2}$ hyperfine structure splitting of 66.5 MHz and 82.9 for the $F'=3\rightarrow4$ and $F'=4\rightarrow5$ intervals, respectively [19]. Therefore, the individual features of each doublet correspond to vibrational levels associated with the different hyperfine structure asymptotes, namely $6S_{1/2}$, $F=4$ + $7P_{3/2}$,$F=5$ and $6S_{1/2}$, $F=4$ + $7P_{3/2}$, $F=4$. Especially for the first triplet feature, the first faint line corresponds to the vibrational level associated with the $6S_{1/2}$, $F=4$ + $7P_{3/2}$,$F=3$ hyperfine structure asymptote. The match between the PA hyperfine splitting of the doublet (triplet) of lines and the exact atomic hyperfine structure splitting indicates that the PA transitions take place at very large internuclear distances, where the hyperfine molecular states do not mix significantly. This is in agreement with the fact that the spectrum is taken very close to the atomic resonance. The relative intensity of the individual lines in each doublet (or triplet) also corresponds to the relative hyperfine structure transition strengths. The higher frequency line in a doublet (triplet) represents the $F'=5$, and the

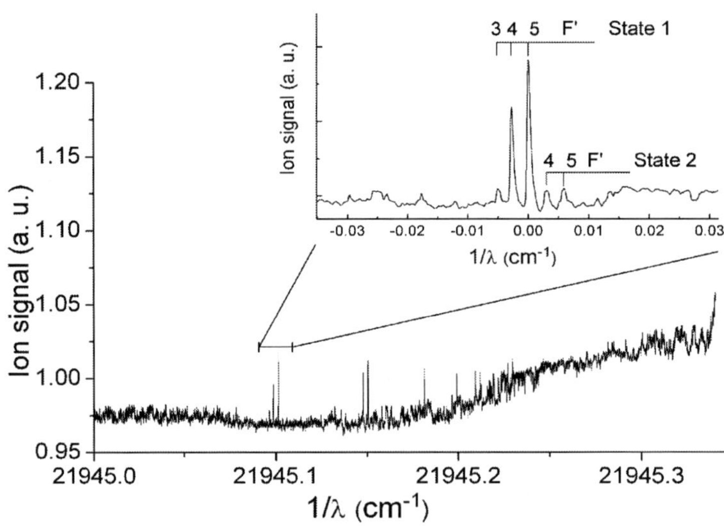

FIGURE 3. Photoassociation spectra below the $7P_{3/2}$ limit as in Fig. 1, and a blowout in the inset showing two states with resolvable hfs structure.

lower frequency one to $F'=4$ excited state hyperfine structure level. Since the normal Cs MOT contains mostly ground state atoms in $F=4$, the transition to $F'=5$ is stronger than the $F'=4$, which is still stronger than the $F'=3$ transition. This is supported by the relative line intensities shown in Fig. 3.

The loss of spectral features for detunings $\Delta > 40$ GHz suggests predissociation of levels, due to strong coupling with states dissociating to nearby asymptotes in the regions rich with avoided crossings, which arise because of the to ion-pair interaction. The predissociated features are presumably significantly broadened, and therefore too broad to be detected by ion detection. The rotational structure is very small ($B_v \sim 0.2$ MHz) for these detunings, compared even to the hyperfine structure, and therefore too small to be observed in the present spectra.

The PA spectra below the $7P_{1/2}$ asymptote yield fewer spectral features. This is not so surprising given the fact that the atomic oscillator strength is 4.3 times smaller. An example of PA spectra below the $7P_{1/2}$ asymptote is shown in Fig. 4(b). Only one doublet feature is visible in the red wing of the atomic resonance line. The spacing of the doublet lines again closely corresponds to the $7P_{1/2}$ atomic hyperfine splitting of 377 MHz for the $F'=3 \rightarrow 4$ interval [19]. The average $7P_{1/2}$ hfs splitting of the red detuned features from several scans of the same spectra is 385 MHz, which can be a result of hfs coupling.

THE BLUE DETUNED PHOTOASSOCIATION ABOVE THE $7P_{1/2}$ and $7P_{3/2}$ LIMITS

While scanning in the vicinity of the main atomic lines ($7P_{3/2}$ and $7P_{1/2}$) we have observed spectral lines, shown in Fig. 4, which are blue detuned or above the $6S_{1/2}$, $F=4$ + $7P_J$, F' asymptotes. We have observed one triplet and two doublet sets of sharp features just above the $7P_{3/2}$, F' limit. The line intervals again correspond to the $7P_{3/2}$ level hfs structure ($F'=3,4,5$ for the triplet and $F'=4,5$ for the two doublets). The hfs interval for the $F'=4,5$ has a slightly larger value of 100±5 MHz in the two doublet features, while the triplet hfs intervals of 70 MHz and 80 MHz for $F'=3,4,5$ are closer to the calculated values of 66.5 MHz and 82.9 MHz [19]. These small deviations can be attributed to coupling of the molecular hyperfine states. The $F'=5$ lines of the three multiplets are respectively 493, 1033 and 1596 MHz blue detuned from the ($6S_{1/2}$, $F=4$ + $7P_{3/2}$, $F'=5$) asymptote. The vibrational spacing seem to increase with increased blue detuning, which would be surprising if the levels belong to the same progression. Thus it is likely that the vibrational levels correspond to more than one electronic state. In addition, we observe that the first two sets of sharp hyperfine-split features sit on top of a less pronounced and much broader feature (hump), with an average detuning of $\Delta \sim +800$ MHz.

In contrast, the blue detuned spectrum above the ($6S_{1/2}$, $F=4$ + $7P_{1/2}$, F') asymptote shows only a single, but a much stronger and broader feature. This large blue detuned

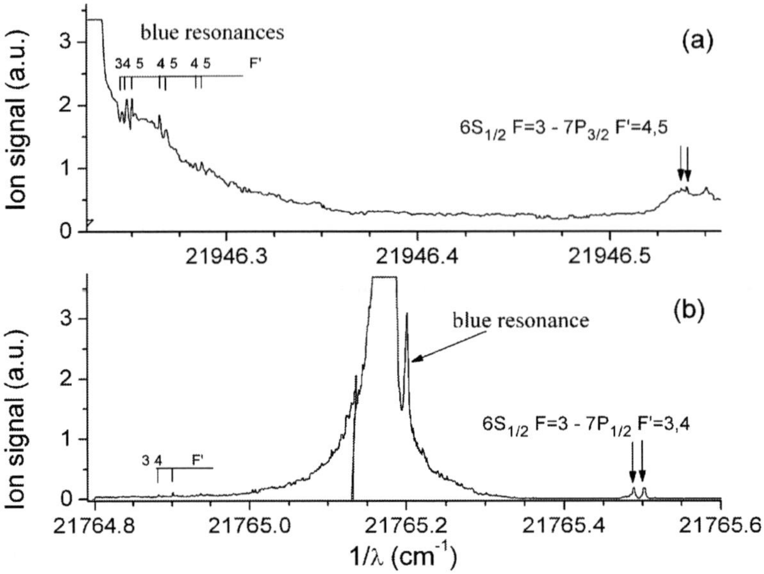

FIGURE 4. Blue satellite and photoassociation features near (a) the $7P_{3/2}$ level, and (b) the $7P_{1/2}$ level. The red wing also shows a bound state with hyperfine structure, which is also present in the sharp blue PA features of $7P_{3/2}$ level.

feature shown in Fig. 4(b) is $\Delta = +560$ MHz to the blue of the atomic line. Here we see no evidence of hyperfine structure.

The combination of competing attractive C_6 and repulsive C_3 long-range terms, especially for R≥50 Å [17] gives rise to pure long-range maxima, as visible in Fig. 1. Such structures can produce blue satellite bands and support quasibound long range states, as well as shape resonances. The blue satellite arises as a consequence of interference between two Condon points around the local maximum of the relevant difference potential curve. One would expect only one (strong) feature to occur at due to the averaging effect. We assign the large broad feature above the $7P_{1/2}$ limit in Fig. 3 (b) to be a blue satellite. The broad hump on the blue side of $7P_{3/2}$ line is also a blue satellite. It is likely that different long-range symmetries with slightly different barrier heights all blended together forming a single satellite band.

In Fig. 5 we show the results of calculations of wavefunctions around the $7P_{3/2}$ limit for 0_u^+ (Fig. 5 top) and 2_g (Fig. 5 bottom) states, which have the largest long-range barriers (Fig. 1). Close match of the calculated bound wavefunctions with the PA spectral lines is not evident. This is not surprising since the calculations are very sensitive to the details of the potential. Above the dissociation there are no quasibound levels, but there is phase effect on the free wavefunctions. This effect is seen even more dramatically the in Fig. 6 where transition strengths above the $7P_{3/2}$ limit are calculated (for 0_u^+ and 2_g symmetries). Blue satellites at somewhat smaller detuning

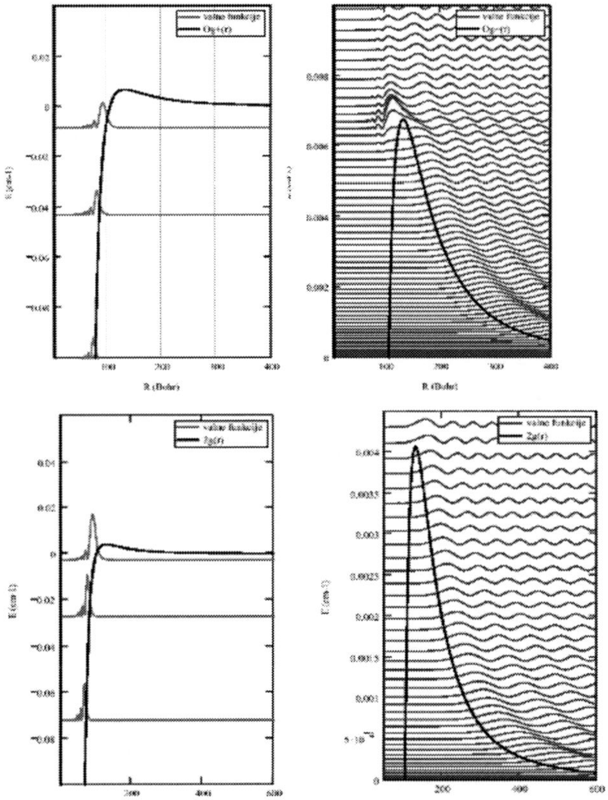

FIGURE 5. Calculated wavefunctions of bound and free states near the $7_{3/2}$ limit for 0_g^+ (top) and 2_g (bottom) states.

of Δ~ +250 MHz is reviled. This suggests slightly lower potential barriers used in calculations. For better simulation of blue detuned spectral features more precise long-range coefficients are needed, which take the fine- and hyperfine structure, the anomalous oscillator strength ratio and retardation effects into account.

There are several explanations for the sharp blue features (exhibiting hfs) above the $7P_{3/2}$ limit. The features are possibly bound states of 0_g^+ and 0_u^+ local minima that are mixing through an avoided crossing with states correlated with the $7P_{3/2}$ limit [20]. A rather unlikely explanation is that the features are quasibound states or shape resonances inside and below much higher barriers. Our calculations do not show transition lines with blue detunings comparable to the ones observed in the spectra, indicating that the barriers are too low to support quasibound states.

Finally, we argue that it is unlikely for the features to be bound states red detuned from $6S_{1/2}$, $F=3$ + $7P_{3/2}$, F' asymptote, corresponding to $f+f=3+4$ ground state collisions. Since, as mentioned earlier, our MOT contains very few atoms in the ground $F=3$ state, the initial collision most likely involves ground $f+f=4+4$ states.

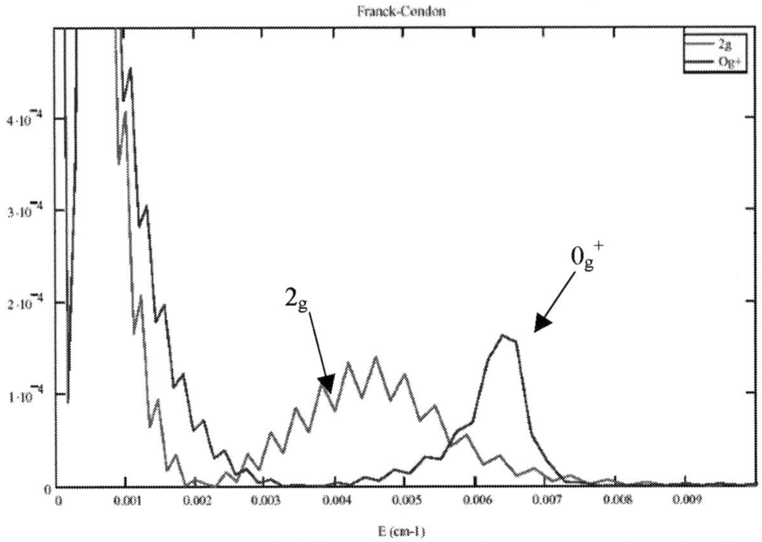

FIGURE 6. Transition probabilities for the blue-detuned region above the $7P_{3/2}$ limit.

Although both atomic resonances are visible for both the $7P_{1/2}$ and the $7P_{3/2}$ limits, the relative ion signal is more than 200 times larger for the $F=4$ resonances. Combined with already weak PA rates, we rule out the possibility that these resonances are bound states of the energetically higher asymptote associated with the $F=3$ ground state atom.

The long-range local maxima are produced with the smallest attractive C_6 and the largest repulsive C_3 on the blue side of the atomic resonance. To obtain C_n values the semi-classical LeRoy-Bernstein approach cannot be applied since the fine- and hyperfine energies are stronger than the predicted C_3 and C_6 terms. However, with the determined C_3 = 0.1501 a.u. from the oscillator strengths and averaged over $7P_{3/2}$ Hund's case *c* barrier states, we have derived the C_6 value from the position of the hump (blue satellite) that is 800 MHz blue detuned. Since the hump is associated with the local maximum of the potential, and assuming a local potential of the form $V(R) =(C_3/R^3)- (C_3/R^3)$, we find the value of $C_6 =4.626 \cdot 10^4$ a.u. Both values are effective long-range dispersion coefficient values.

CONCLUSION

A series of one step photoassociation experiments was performed in the ultracold cesium system above and below the $7P_{1/2}$ and $7P_{3/2}$ asymptotes. Photoassociation spectra were obtained in the region very close to the dissociation limit in the $7P_{3/2}$ case, whereas a few spectral features were observed below the $7P_{1/2}$ limit. This is consistent with the anomalous oscillator strength ratio for these levels, which is 4.3

and thus different from the statistical prediction value of 2. Resolved upper atomic state hyperfine structure is present in all observed features. Furthermore, comparable magnitude of the competing C_3 and C_6 long-range interactions in the probed spectral region, beyond 50 Å produce local potential maxima that support observed blue-detuned satellites. Sharp features that also show hyperfine structure, above the $7P_{3/2}$ limit were also observed. Such features are most likely result of mixing of 0_g^+ and 0_u^+ local minima (above the $7P_{3/2}$ limit) with states correlated with the $7P_{3/2}$ limit. Smaller photoassociation rates than for the $6P_J$ case prevent brother observation of bound and quasibound levels in the probed spectral region.

ACKNOWLEDGEMENTS

This work was partially supported by the National Science Foundation (U.S.A.), the Ministry of Science and Technology of the Republic of Croatia, the European Community Research Training Network Project on Ultracold Molecules and the Alexander von Humboldt Stiftung.

REFERENCES

1. K. Niemax and G. Pichler, J. Phys. B **8**, 179 (1975).
2. Marin Pichler et al. Phys. Rev A **73**, 021403(R) (2006).
3. M. Movre and G. Pichler, J. Phys. B **10**, 2631 (1977).
4. W. C. Stwalley, Y. H. Uang, and G. Pichler, Phys. Rev. Lett. **41**, 1164 (1978).
5. W. C. Stwalley and H. Wang, J. Mol. Spectrosc. **195**, 194 (1999).
6. J. T. Bahns, P. L. Gould, and W. C. Stwalley, Adv. At. Mol. Opt. Phys. **42**, 171 (2000).
7. F. Masnou-Seeuws and P. Pillet, Adv. At. Mol. Opt. Phys. **47**, 53 (2001).
8. K. M. Jones, E. Tiesinga, P. D. Lett, and P. S. Julienne, Rev. Mod. Phys. **78**, 483 (2006).
9. R. Wynar, R. S. Freeland, D. J. Han, C. Ryu, and D. J. Heinzen, Science **287**, 1016 (2000); J. M. Gerton, D. Strekalov, I. Prodan, and R. G. Hulet, Nature **408**, 692 (2000); C. McKenzie, et al. Phys. Rev. Lett. **88**, 120403 (2002); I. Prodan, M. Pichler, M. Junker, R. G. Hulet, and J. L. Bohn, Phys. Rev. Lett. **91**, 080402 (2003).
10. M. Greiner, C. Regal, and D. Jin, Nature (London) **424**, 47 (2003); S. Jochim et al. Science **302**, 2101 (2003); M. W. Zweirlein et al. Phys. Rev. Lett. **91**, 250401 (2003).
11. P.Pillet and H. Wang, private communication.
12. A. Fioretti, D. Comparat, A. Crubellier, O. Dulieu, F. Masnou-Seeuws, and P. Pillet, Phys. Rev. Lett. **80** 4402 (1998); D. Comparat et al. Eur. Phys. J. D **11**, 57 (2000).
13. M. Pichler, H. Chen, and W. C. Stwalley, J. Chem. Phys. **121**, 1796 (2004); M. Pichler, H. Chen, and W. C. Stwalley, J. Chem. Phys. **121**, 6779 (2004).
14. J. P. Shaffer, W. Chalupezak, and N. P. Bigelow, Phys. Rev. Lett. **83**, 3621 (1999); A. Amelink, K. M. Jones, P. D. Lett, P. van der Straten, and H. G. M. Heideman, Phys. Rev. A **61**, 042707 (2000).
15. H. Wang, P. L. Gould, and W. C. Stwalley, Phys. Rev. Lett. **78**, 4173 (1996).
16. M. Pichler, J. Qi, and W. C. Stwalley, Bull. Am. Phys. Soc. 46, H4 (2001).
17. M. Marinescu and A. Dalgarno, Phys. Rev. A **52**, 311 (1995).
18. J. Migdalek and Y.-K. Kim, J. Phys. B **31**, 1947 (1998).
19. E. Arimondo, M. Inguscio, and P. Violino, Rev. Mod. Phys. **49**, 31 (1977).
20. T. Ban et al., Chem. Phys. Lett. **345**, 423 (2001).

Dicke-Narrowed Line Shapes in CO-Ar: Measurements, Calculations, and a Revised Interpretation

R. Wehr[1], R. Ciuryło[2], A. Vitcu[1], F. Thibault[3], D. A. Shapiro[4], W.-K. Liu[5], F. R. W. McCourt[6], J. R. Drummond[1], and A. D. May[1]

[1]*Dept. of Physics, University of Toronto, Toronto, Canada M5S1A7*
[2]*Institut Fizyki ,Universytet Mikolaja Kopernika, Grudziadzka 5/7 87-100, Torun, Poland*
[3]*UMR 6627 du CNRS, Univ. de Rennes I, Campus de Beaulieu, 35042 Rennes Cedex, France*
[4]*Institute of Automation and Electrometry, Novosibirsk 630090, Russia*
[5]*Dept. of Physics and Astronomy, University of Waterloo, Waterloo, Canada N2L 3G1*
[6]*Dept. of Chemistry, University of Waterloo, Waterloo, Canada N2L 3G1*

Abstract. New line shape calculations for CO buffered by Ar are compared to high-resolution measurements from a difference-frequency laser spectrometer, over a range of thermodynamic conditions relevant to the atmosphere. The calculations are based on solving the quantum kinetic (i.e. transport/relaxation) equation for the molecules within the impact approximation, and rely on the commonly used MOLSCAT [1] and MOLCOL [2] codes to determine the speed-dependent collisional relaxation rate. Velocity-changing effects are treated classically using a rigid sphere potential. The comparison initially reveals that the experimental profiles exhibit only 10% to 30% of the expected Dicke narrowing, which leads us to reevaluate our understanding of the narrowing process. A more subtle aspect of the disagreement between theory and experiment draws our attention to an assumption implicit in the calculation of the collisional relaxation rate: the assumption of a Maxwellian form for the velocity dependence of the off-diagonal elements of the density matrix (i.e. the optical coherences). This assumption allows for an analytical simplification of the problem, but eliminates velocity-changing effects (so that they must be added back in using a supplementary classical calculation, which is based here on a rigid sphere interaction). We find that the removal of the above-mentioned assumption should allow for accurate and fully quantum mechanical (but numerical) line shape calculations for systems like CO-Ar on existing computers.

1. J. M. Hutson and S. Green, *Collaborative Computational Project no. 6 (CCP6)* (UK Science and Engineering Research Council) 1994.
2. D. R. Flower, G. Bourhis, and J.-M. Launay, *Comput. Phys. Commun.*, 131 (2000) 187.

Keywords: Spectral line shape; Quantum calculations; Infrared laser spectroscopy; Dicke narrowing, Collisional narrowing; Collisional broadening; Pressure broadening; Temperature dependence; Carbon monoxide; Inelastic collisions; Statistical correlation.
PACS: 33.70.Jg; 33.20.Ea; 33.20.Vq; 39.30.+w; 42.68.Ca

INTRODUCTION

Infrared absorption line shapes in the atmosphere are typically analyzed using semi-empirical models, the most common example being the Voigt profile. While there are many more complicated examples, accounting for collisional narrowing, speed-

CP874, *Spectral Line Shapes: 18th International Conference,*
edited by E. Oks and M. Pindzola

dependent broadening, and shifting, they are all limited ultimately by their reliance on empirical parameters. Using such models, it is possible to relate the line shape to some previously measured line shape, but it is not possible to predict the line shape from non-spectroscopic information, i.e. directly from the interaction potential, as is the objective of the line shape calculations described here. Owing to the complexity of the problem, there have only ever been a few attempts at a full line shape calculation [1-3], and none has been able to reproduce modern high-resolution spectra within their experimental uncertainties. Note that while there have been many purported calculations of line widths and shifts, these parameters depend on the line shape, and so it is impossible to calculate a width or shift directly in a rigorous way. Rather, the line profile must be calculated first and the width or shift measured from it.

The present study is restricted to isolated rovibrational absorption lines in the fundamental band of carbon monoxide buffered by argon (CO-Ar). We ignore the far wings of the lines and work within the binary impact approximation. CO-Ar was chosen for the present study for several reasons. First, dilutions of CO in Ar simulate the general atmospheric situation, in which infrared-active trace gases are diluted in nitrogen. Second, CO is a diatomic molecule, and thus its rovibrational absorption spectrum is simple, containing well isolated lines that can be measured and analyzed (nearly) independently of one another. Finally, the use of Ar (a structureless atom under atmospheric conditions) as the collision partner simplifies the calculations.

For atmospheric-type systems like CO-Ar, the principal processes contributing to the line shape are collisional broadening and Doppler broadening. However, at the level of accuracy provided by the present spectrometer, it is necessary to consider that the collisional broadening depends on the speed of the active molecule, and that the Doppler broadening is reduced by velocity-changing collisions. The latter effect is known as collisional, or Dicke, narrowing. Collisional line shifting is an order of magnitude weaker than the broadening in CO-Ar [4] and is neglected here (see, however, Ref. [5]). Line mixing is also weak below 1 atm, and the slight asymmetry it introduces into the experimental profiles can be accurately removed using a fitted parameter. Line mixing is thereby omitted from the calculations, in keeping with our focus on isolated lines.

There are a number of other, relatively minor, considerations that are nonetheless significant to the formation of the line shape when it is measured at high resolution. These include, for example: the homogenization of the speed-dependent collisional broadening by speed-changing collisions, the coupling of the collisional broadening and translational motion because they are both speed dependent, the influence of the CO:Ar mass ratio on the velocity memory, and a possible correlation between the velocity-changing and dephasing effects of collisions.[1] In principle, all of these considerations should be included in the present calculations by virtue of our approach, which is first to solve the quantum kinetic equation that describes the interaction of the molecules with each other and with the radiation field, and then to derive the line shape from the solution. However, we shall deliberately remove the correlation between velocity-changing and dephasing collisions from consideration by an approximation that will be discussed below.

[1] The word "dephasing" is used here in a generic sense and includes dephasing, reorienting, and state-changing processes.

In order to test our calculations, we measured the shapes of seven CO-Ar lines in the P branch of the fundamental band using a difference-frequency laser spectrometer. These line spectra were recorded at pressures of 0.025, 0.05, 0.1, 0.2, 0.5, and 1 atm, and at temperatures of 214, 236, 259, 296, and 324 K, thereby covering the thermodynamic conditions found in the Earth's troposphere and lower stratosphere. The particular lines studied were: P(1), P(2), P(5), P(7), P(10), P(13), and P(16). In this paper, we describe first the calculation of the shapes of these lines, and then the experimental apparatus used to measure them. Following that, we present our findings and a discussion of their implications.

CALCULATIONS

To obtain the transport/relaxation equation, we follow Ref. [6] and consider a transition between an initial state b and a final state a (in the case of photon emission) in the presence of a field represented by $E\exp[-\mathrm{i}(\omega t - \mathbf{k}\cdot\mathbf{r})]$, where ω is the frequency and $\mathbf{k}$ is the wavevector. We then write the resonant part ρ_r of the off-diagonal element of the density matrix for the active molecule as $\rho_r = -\mathrm{i}\varsigma(\omega,\mathbf{v})n_b\mu_{ba}E\exp[-\mathrm{i}(\omega t - \mathbf{k}\cdot\mathbf{r})]$, where n_b is the thermal equilibrium population of state b, and μ_{ba} is the matrix element of the electric dipole moment corresponding to a transition between states b and a. The population of the final state a is taken to be zero. Upon making the rotating wave approximation, we arrive at the following transport/relaxation equation [7]:

$$\varsigma_0(\mathbf{v}) = -\mathrm{i}(\omega - \omega_0 - \mathbf{k}\cdot\mathbf{v})\varsigma(\omega,\mathbf{v}) - \hat{S}\varsigma(\omega,\mathbf{v}). \tag{1}$$

Here $\varsigma_0(\mathbf{v})$ is the (typically Maxwellian) velocity distribution of the molecules in state b in the absence of the field, ω_0 is the resonant frequency, and $\hat{S}$ is the collision operator. The $\hat{S}$ term contains all the effects of intermolecular collisions, leading to Dicke narrowing and collisional broadening, while the other two terms are "free-streaming" terms. The $\mathbf{k}\cdot\mathbf{v}$ term, in particular, represents the Doppler shift that gives rise to Doppler broadening. The line shape $I(\omega)$ is obtained from $\varsigma(\omega,\mathbf{v})$ as [6]:

$$I(\omega) = \frac{1}{\pi}\mathrm{Re}\int \varsigma(\omega,\mathbf{v})\mathrm{d}^3\mathbf{v}. \tag{2}$$

In general, the effects of $\hat{S}$ may be difficult to calculate, or even to express in a useful form. One approach (and our approach) is to assume that $\hat{S}$ can be written as $\hat{S} = \hat{S}_{\mathrm{VC}} + \hat{S}_{\mathrm{D}}$, where $\hat{S}_{\mathrm{VC}}$ is the velocity-changing collision operator, which represents collisions that affect the translational motion, and $\hat{S}_{\mathrm{D}}$ is the dephasing collision operator, which represents collisions that affect the internal relaxation. By writing $\hat{S}$ in this form, one assumes that the effects of collisions on the translational motion and internal relaxation are statistically independent. Eq. (1) then becomes:

$$\varsigma_0(\mathbf{v}) = -\mathrm{i}(\omega - \omega_0 - \mathbf{k}\cdot\mathbf{v})\varsigma(\omega,\mathbf{v}) - \hat{S}_{\mathrm{VC}}\varsigma(\omega,\mathbf{v}) - \hat{S}_{\mathrm{D}}\varsigma(\omega,\mathbf{v}). \tag{3}$$

Nearly all line shape models effectively assume that the collision operator can be separated in this way, as they can be derived from Eq. (3) under various simplifying assumptions (see Ref. [8]).

In order to solve Eq. (3), it is converted into a matrix equation using an infinite and complete set of orthonormal basis functions (which is later truncated for numerical calculation). The details of this conversion are provided in Ref. [6]. The result is:

$$\mathbf{L}(\omega) = -\mathrm{i}(\omega - \omega_0)\mathbf{1} + \mathrm{i}\mathbf{K} - \mathbf{S}^f_{\mathrm{VC}} - \mathbf{S}^f_{\mathrm{D}}, \tag{4}$$

where $\mathbf{1}$ is the unit matrix, $\mathbf{K}$ is the Doppler shift matrix, and the $\mathbf{S}^f$ matrices arise from the velocity-changing and dephasing parts of the collision operator. Recipes for $\mathbf{K}$, $\mathbf{S}^f_{\mathrm{VC}}$, and $\mathbf{S}^f_{\mathrm{D}}$ are given in Ref. [7]. For the current study, $\mathbf{S}^f_{\mathrm{VC}}$ was calculated classically using a rigid sphere potential for the CO-Ar interaction, while $\mathbf{S}^f_{\mathrm{D}}$ was calculated from energy-dependent pressure-broadening cross-sections obtained using the MOLSCAT [9] and MOLCOL [10] quantum mechanical scattering codes, employing the *ab initio* potential energy surface of Toczyłowski and Cybulski [11]. The matrix $\mathbf{L}(\omega)$ was then used to solve the equation $\mathbf{b} = \mathbf{L}(\omega)\mathbf{c}(\omega)$ for the column vector $\mathbf{c}(\omega)$, given that $\mathbf{b}$ is a column vector whose first element is 1 and whose other elements are 0. Once $\mathbf{c}(\omega)$ is known, the line shape is given by:

$$I(\omega) = \mathrm{Re}\{c_0(\omega)/\pi\}, \tag{5}$$

where $c_0(\omega)$ is the first element of $\mathbf{c}(\omega)$.

More details on the calculations and the computational method can be found in Refs. [1,12] and especially Ref. [13]. Excepting the MOLSCAT and MOLCOL calculations, the calculations described here were performed on a 700 MHz desktop computer, and the calculation of a single line shape took about 10 seconds.

We conclude this section by noting that $\mathbf{S}^f_{\mathrm{VC}}$ and $\mathbf{S}^f_{\mathrm{D}}$ each contain a physically meaningful parameter that can be factored out of the matrix as a scalar. For $\mathbf{S}^f_{\mathrm{VC}}$, this parameter is the effective frequency of velocity-changing collisions, ν, while for $\mathbf{S}^f_{\mathrm{D}}$, it is the thermally-averaged collisional relaxation rate, Γ_0. There is no consensus in the literature (see Ref. [12]) as to whether ν should be equal to the dynamic friction coefficient, ν_{diff}, which can be calculated from the mass diffusion constant D via:

$$\nu_{\mathrm{diff}} = \frac{k_{\mathrm{B}}T}{mD}, \tag{6}$$

where k_{B} is the Boltzmann constant, T is the temperature, and m is the mass of the active molecule.[2] For the present calculations, we assumed initially that $\nu = \nu_{\mathrm{diff}}$ and later relaxed that assumption. As for the thermal average, Γ_0, it can be calculated as:

[2] The mass diffusion constant, D, can be related to the diameter, d, of the rigid sphere interaction [7]. However, to maintain contact with the literature, we use D rather than d to characterize the interaction leading to Dicke narrowing.

$$\Gamma_0 = \int_0^{\infty} f_M(\upsilon)\Gamma(\upsilon)\mathrm{d}\upsilon\,, \tag{7}$$

where υ is the active molecule speed, $f_M(\upsilon)$ is the Maxwellian speed distribution, and $\Gamma(\upsilon)$ is the speed-dependent collisional relaxation rate (obtained from the energy-dependent pressure-broadening cross-section). However, the accuracy with which Γ_0 can be calculated is only ±2% at present, which is the principal limitation on the accuracy of the overall calculation, and is too poor for the analysis of subtle line shape effects such as Dicke narrowing. Therefore, we elected to treat Γ_0 as a fitted parameter. We verify below that our fitted values of Γ_0 agree with our calculated values in their magnitude and their density dependence in all cases – with one important exception that we will discuss in some detail.

MEASUREMENTS

The spectrometer has been described in Refs. [13,14,15], but we repeat some pertinent details here. Two visible laser beams, one of which is tunable, are overlapped in an $LiIO_3$ crystal to generate an infrared (IR) beam at the difference frequency. This beam is split so that one branch passes through a temperature-controlled absorption cell, a second branch passes through a 15 cm long reference cell filled with low pressure CO at room temperature, and a third branch is used to monitor the incident intensity. The reference cell branch is only used to measure line shifts, which are not discussed here. Each branch of the beam is detected by a separate LN_2-cooled InSb detector, and the incident intensity signal is used to normalize the absorption cell signals and thereby remove absorption by laboratory air from the spectra. To detect the relatively weak laser beam against the laboratory's background IR signal, one of the visible laser beams is chopped by an acousto-optic modulator and the detected signals are processed by lock-in amplifiers. The spectrometer is depicted in Fig. 1. It has a frequency uncertainty of less than 1 MHz and a signal-to-noise ratio between 3000:1 and 10,000:1, depending mostly on the effective integration time employed (between 0.5 and 2 seconds).

The temperature-controlled, vacuum-jacketed absorption cell used for the current study was 50 cm long, and used a circulating heat transfer fluid and chiller to regulate the temperature. We were able to set the gas temperature to any value between roughly 210 K and 330 K, with a stability of ±0.1 K. The cell was fitted with CaF_2 windows mounted at Brewster's angle, and with five T-type thermocouples distributed along the length of the cell. Using a combination of the thermocouple measurements and spectroscopic temperature retrievals (following the method of Ref. [16]), we were able to determine the CO-Ar gas temperatures to within ±0.3 K. The gas pressures were recorded with an MKS Baratron capacitance manometer, which was calibrated by the manufacturer immediately before the study and found to be accurate to within 0.05% of the reading. The mixing ratio of CO to Ar was about 0.1%, and thus we can neglect the influence of self-broadening on the speed-dependence of the total broadening.

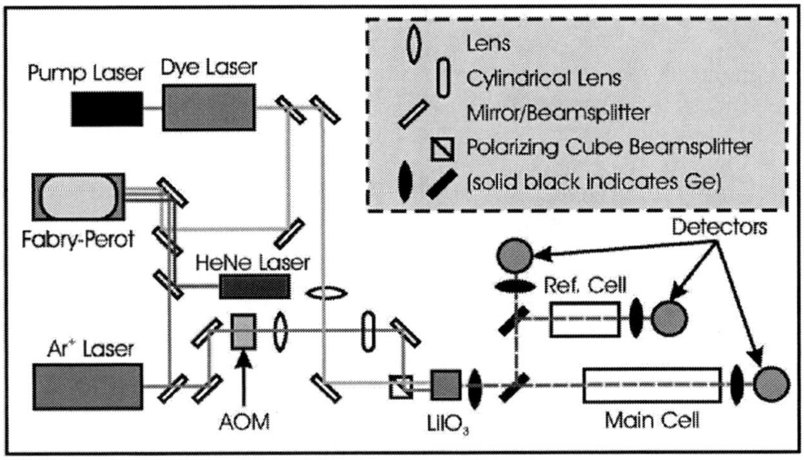

FIGURE 1. Schematic diagram of the difference-frequency laser spectrometer as viewed from above.

FINDINGS

At pressures above 0.2 atm, the line shape is insensitive to Dicke narrowing and is dominated by collisional broadening. For all temperatures and spectral lines in this pressure regime, our calculations (but with Γ_0 fitted) reproduce the measured line shapes within their experimental noise (i.e. within ±0.03% or better). As an example, the residual for the P(13) transition at 1.0 atm and 259 K is shown at the bottom of Fig. 2, underneath the corresponding absorption spectrum. The residual is defined as the difference between the calculated spectrum and the measured spectrum, and agreement between the two is confirmed by the fact that the residual is flat and equal to zero within the experimental noise. (While the spectra are recorded point by point, we have joined the points by straight line segments as a guide to the eye.) Note that there are a number of adjustable parameters used to fit our calculated line shapes to our experimental profiles and produce residuals such as the one shown in Fig. 2. Since our line shape calculation does not include the line strength (i.e. peak height) or the relative frequency of the line centre in a particular experimental scan, it is necessary to fit for those parameters. It is also necessary to fit for a linear baseline slope and intercept. None of these parameters influences the line shape. As mentioned earlier, weak line mixing does introduce a small asymmetry into CO-Ar lines under the conditions studied [1,17], but as we are interested in modelling an isolated line, we exclude line mixing from our calculation and account for the asymmetry with a final fitted parameter. All these fitted parameters are discussed further in Ref. [1]. The broadening and shifting coefficients obtained from our spectra in the > 0.2 atm regime are reported in Ref. [5].

The situation below 0.2 atm, where Dicke narrowing, Doppler broadening, and collisional broadening are all significant, is more problematic. In this pressure regime, when ν is fixed equal to ν_{diff} (as calculated using Eq. (6) and the empirical model of Bzowski et al. [18] for the mass diffusion constant D), we find for all lines and all temperatures that our line shape calculations fail to reproduce the measured line shapes within their experimental noise. That is true even though Γ_0 is fitted in a least-squares minimization of the residual. As an example, the residual for the P(7) line at 0.05 atm and 296 K is shown in Fig. 3(a), where the 'w' structure near line centre indicates a difference between the measured and calculated shapes. Note that while the maximum value of the residual is only about 1% of the peak absorption, it is well above the noise level as characterized by the fluctuations visible in the line wings.

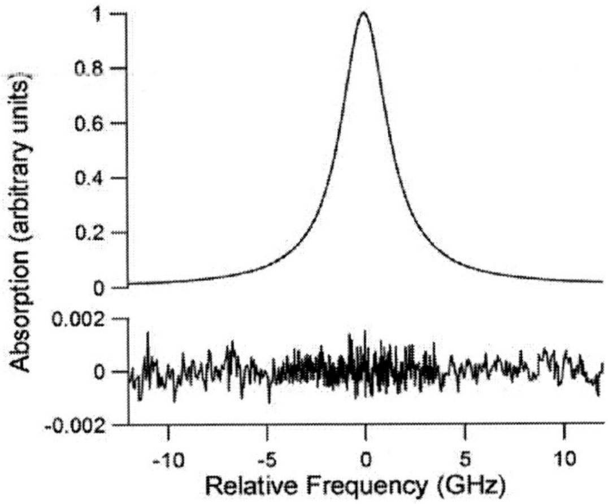

FIGURE 2. Measured spectrum of the P(13) line at 1.0 atm and 259 K (above), and the difference between the calculated spectrum and the measured spectrum (below). Note the expanded vertical scale in the lower plot. The arbitrary absorption units are the same for both plots, and are chosen such that the peak absorption is normalized to 1.

On the other hand, when ν is fitted, the calculations do reproduce the measurements within the experimental noise, as shown in Fig. 3(b). This same agreement is found for all spectral lines at all temperatures and pressures. Note that the increased noise near line centre in Fig. 3(b) should not be confused with structure in the residual. The noise increases because, for this scan, the CO:Ar mixing ratio was slightly higher than desirable, leading to a weak transmission signal at line centre.

The question therefore becomes: what are our fitted values of ν? Before presenting those values, we briefly check whether our fitted values of Γ_0 agree with our calculated values and whether they exhibit the correct physical variation with pressure. Fig. 4 shows a plot of the collisional broadening coefficient, $\gamma_0 = \Gamma_0/P$, versus pressure P for the P(2) line at 296 K. The results when ν is fixed equal to ν_{diff} and when ν is fitted are shown as hollow and solid circles, respectively. On physical grounds, γ_0 should be constant with pressure, as is the case in the $P > 0.2$ atm regime for both the solid and hollow circles. But a non-physical deviation (from a constant high-pressure value) is seen at low pressures in both cases. Fitting ν substantially reduces the deviation in γ_0, but does not eliminate it. The deviation in γ_0 when $\nu = \nu_{diff}$ was discussed in Ref. [1], where it was attributed to an overestimation of the narrowing (the least-squares fitting procedure compensates for an overestimation of the narrowing by increasing the broadening). That interpretation is supported here by the substantial reduction in the deviation when ν is fitted. However, we also see here that the interpretation of Ref. [1] is only partly correct, as γ_0 is not constant at low pressures even when ν is fitted.

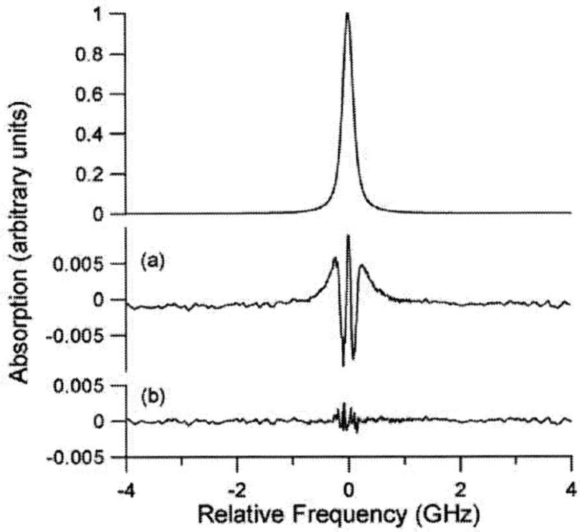

FIGURE 3. Measured spectrum of the P(7) line at 0.05 atm and 296 K, along with residuals (a) when ν is fixed equal to ν_{diff} and (b) when ν is fitted. Note the expanded vertical scale in the residual plots. The arbitrary absorption units are the same for all plots, and are chosen such that the peak absorption is normalized to 1.

The remaining deviation in γ_0 when ν is fitted cannot be attributed to any systematic error in the experiment, because its magnitude varies with temperature such that it is: (1) largest at 259 K, in the middle of the temperature range studied, and (2) negligible at 214 K. The first point makes a systematic error in the temperature or pressure measurement extremely unlikely, while the second point rules out any effect from the instrumental line shape. Thus we conclude that the low-pressure deviation in γ_0 results from some error in our calculation. We return to this point below. In any case, the low-pressure deviation in γ_0 does not affect our fitted values of ν, as will be shown shortly.

We now consider the fitted values of ν. Although there is disagreement in the literature as to the magnitude of ν, there is general agreement that it is proportional to the pressure (e.g. Refs. [19,20]), as expected of a collision rate. Within our experimental uncertainty, we also find ν to be proportional to the pressure, as shown in Fig. 5 for $P \leq 0.1$ atm. At $P \geq 0.2$ atm, the line shape becomes insensitive to ν, and the uncertainty in our retrieved values of ν is too large for those values to be useful. The solid line in Fig. 5 depicts ν_{diff}, which is three to ten times larger than ν, depending on the spectral line. Thus our measured spectra contain less Dicke narrowing than we would predict under the assumption that $\nu = \nu_{diff}$.

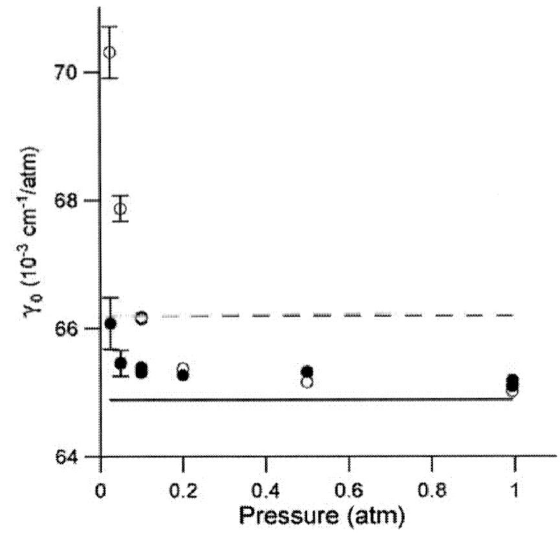

FIGURE 4. Broadening coefficient $\gamma_0 = \Gamma_0/P$ versus pressure P for the P(2) line at 296 K when ν is fixed equal to ν_{diff} (hollow circles) and when ν is fitted (solid circles). The solid horizontal line is the calculated value. The ±2% error in the calculated value is indicated in the positive direction by the dashed horizontal line. The 2σ error bars on the data points are too small to show for pressures above 0.1 atm.

To highlight the reduction of Dicke narrowing in our spectra, we plot in Fig. 6 (lower panel) the ratio ν/ν_{diff} versus rotational quantum number J. The solid triangles in Fig. 6 are our values obtained at 296 K and 0.05 atm. Note that normalizing by ν_{diff} removes the pressure dependence. In units of 10^{-3} cm^{-1}/atm, our calculated values of ν_{diff} are: 31.1 at 214 K, 28.6 at 236 K, 26.5 at 259 K, 23.8 at 296 K, and 22.2 at 324 K. These values should be accurate to within 2.5% [18]. We see from the top panel of Fig. 6 that the variation of Γ_0 with J mirrors that of ν/ν_{diff}, a fact which will be discussed below. As an aside, the hollow circles in Fig. 6 are values obtained when our calculated speed dependence for the collisional broadening is replaced by a fitted, quadratic speed dependence similar to that used in Ref. [14] and other papers. These erroneous values are qualitatively similar to those shown in Fig. 4(a) of Ref. [14], demonstrating the advantage of using a rigorous calculation for the form of the speed-dependent broadening.

In summary, for all spectral lines under all conditions, we find that the Dicke narrowing is reduced relative to what we would predict from the mass diffusion constant. This reduction is not substantially affected by using a soft or hard collision model in place of our rigid sphere calculation (results not shown). Nor is it affected by the low pressure deviation in γ_0 (or equivalently, in Γ_0), as evidenced by the following two facts: (1) unlike Γ_0, ν is proportional to the pressure; and (2) within the precision of the retrieval, the low-pressure values of ν/ν_{diff} at 214 K (where there is no significant deviation in γ_0) agree with the values at 259 K (where the deviation in γ_0 is most pronounced).

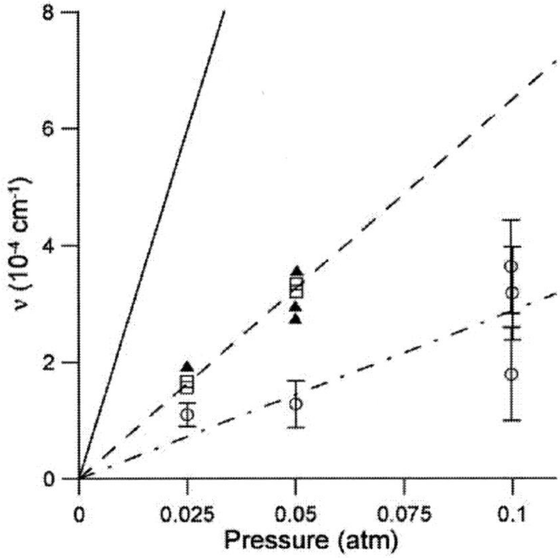

FIGURE 5. Pressure dependence of ν for the P(2) line (circles), the P(7) line (squares), and the P(13) line (triangles), at 296 K. The solid line indicates ν_{diff}, while the dashed line is a linear fit through the origin for P(7) and the dash-dotted line is a linear fit through the origin for P(2). For clarity, 1σ error bars are shown only on the P(2) data points.

DISCUSSION

We are left with two outstanding questions: (1) why do we find less Dicke narrowing in our measured spectra than we calculate from the mass diffusion constant? and (2) why is γ_0 not constant at low pressures even when ν is fitted? We begin with the first question. The model of Rautian and Sobel'man [21] (further developed in Refs. [22,23]), suggests that we overestimate the amount of Dicke narrowing because we separate the collision operator into independent velocity-changing and dephasing parts. By doing so, we neglect the fact that some velocity-changing collisions are really "VCD", or "velocity-changing and dephasing", collisions. Such collisions were also excluded from Dicke's original theory [24], which assumed that "the gas collisions should not influence the internal state of the radiator". It has not been established whether, or to what degree, VCD collisions should contribute to Dicke narrowing in CO-Ar.

To examine this issue, we discard the VC, D, and VCD nomenclature, which is based on intuitive notions of the effects of collisions rather than on a rigorous description of the collision process. Instead, we consider a rigorous decomposition of the collision operator $\hat{S}$, which is provided by Eqs. (2.142-2.148) of Ref. [25]. These equations relate $\hat{S}\rho$ (written there simply as **S**) to quantum mechanical scattering amplitudes by dividing it into a leaving term, $-\mathbf{S}^{(1)}$, and a returning term, $\mathbf{S}^{(2)}$. Adapting the notation of Ref. [25] slightly, $-\mathbf{S}^{(1)}$ is given by:

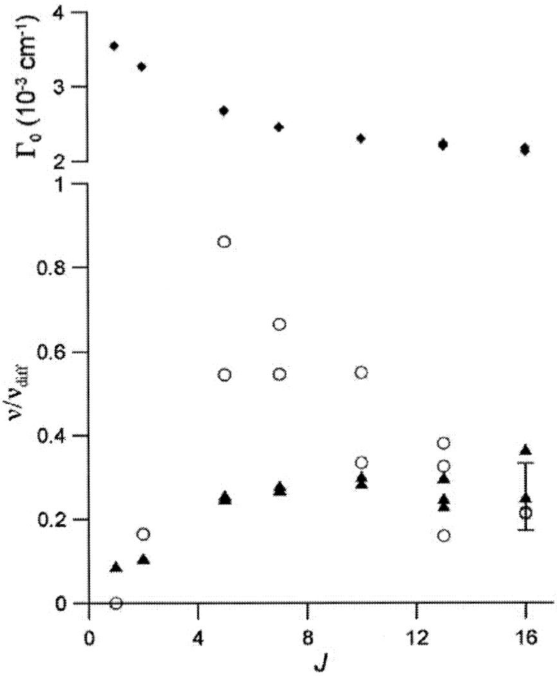

FIGURE 6. ν/ν_{diff} versus J at 0.05 atm and 296 K, as retrieved using our calculated speed dependence for $\Gamma(\upsilon)$ (solid triangles) and using a quadratic, fitted speed dependence instead (hollow circles). 1σ error bars are shown on one of the J = 16 data points. The relaxation rate Γ_0 for the same spectra is shown on the scale above, for comparison. The error bars on Γ_0 are too small to show.

$$-\mathbf{S}^{(1)}(ba,\mathbf{v}) = -\sum_{dc} \nu(ba|dc,\mathbf{v})\rho_{dc}(\mathbf{v}), \tag{8}$$

where a and b are the lower and upper states, respectively, of the rovibrational transition corresponding to the spectral line of interest; c and d are those states for some other transition; and $\nu(ba|dc,\mathbf{v})$ represents the collisional transfer of coherence from the transition ba and the velocity $\mathbf{v}$ to the transition dc and all other velocities $\mathbf{v}'$. Note that $\nu(ba|dc,\mathbf{v})$ is more general than the effective frequency of velocity-changing collisions ν in that it includes transfer between components of the optical coherence as well as between velocities. The leaving term $-\mathbf{S}^{(1)}$ therefore involves leaving both to other velocities and to other components of the optical coherence (i.e. to other spectral lines), although only the latter aspect is clear from Eq. (8). The leaving to other velocities can be seen by applying the optical theorem [25, pp. 61 *ff.*] to the forward-scattering amplitudes. The returning term $\mathbf{S}^{(2)}$ is analogous to $-\mathbf{S}^{(1)}$ in that it involves returning both from other velocities and from other components of the optical coherence, but both of these aspects are explicit in the returning term:

$$\mathbf{S}^{(2)}(ba,\mathbf{v}) = \sum_{dc}\int_{\mathbf{v}'} A(ba\mathbf{v}|dc\mathbf{v}')\rho_{dc}(\mathbf{v}')d\mathbf{v}', \tag{9}$$

where A(ba**v**|dc**v**') represents the collisional transfer of coherence from the transition *dc* and the velocity **v**' to the transition *ba* and the velocity **v**.

Inspection of Eq. (9) reveals that, in the case of rotationally inelastic collisions (i.e. $dc \neq ba$)[3], the returning from other velocities occurs in the same terms as the returning from other components of the optical coherence. For an isolated line, those terms can be neglected (because, by the definition of an isolated line, ρ_{dc} is negligible at frequencies near the transition *ba* for all $dc \neq ba$), and so the velocity exchange associated with rotationally inelastic collisions cannot homogenize the Doppler broadening. Thus the structure of the collision operator determines that *only rotationally elastic collisions contribute to the Dicke narrowing of an isolated line.* The velocity changes associated with rotationally inelastic collisions still occur, but they do not influence the spectrum. Accordingly, a velocity-changing collision rate based on mass diffusion (i.e. ν_{diff}) will overestimate the effective (elastic) rate ν. Furthermore, since rotationally inelastic collisions are the dominant mechanism for collisional broadening in CO-Ar at atmospheric temperatures [26], we expect large values of Γ_0 to be associated with small values of ν – a "correlation", evident in Fig. 6. The abstention of rotationally inelastic collisions from Dicke narrowing is implied in Ref. [27], in which the returning in velocity is absent in the case of an isolated line, and more recently in Ref. [28], where Dicke narrowing due to rotationally inelastic collisions reappears in the presence of line mixing.

We now turn to our second outstanding question: why is γ_0 not constant at low pressures even when ν is fitted? Since the deviation occurs at low pressure, and increases with decreasing pressure, it is unlikely to result either from an error in the speed-dependent collisional broadening or from an overestimation of speed-class exchange. Errors in either of these two aspects of the calculation ought to become more significant with increasing pressure. The deviation is even less likely to result from an error in the Doppler broadening, which is well understood and trivial to calculate. Our use of a rigid sphere potential to calculate the Dicke narrowing is absolved of responsibility by noting that the deviation is not appreciably affected by replacing our rigid sphere model with a hard or soft collision model (results not shown). So we hypothesize that the deviation occurs because we have oversimplified the collision operator by separating it as in Eq. (3), and because fitting ν is not sufficient to restore the lost information.

In considering this point, it is worth examining the approach by which the dephasing part of the collision operator, $\hat{S}_D$, is calculated. The scattering amplitudes in a binary collision can be calculated using either the MOLSCAT or the MOLCOL software package. In principle, these amplitudes are sufficient to determine all the effects of collisions on the line shape; however, progressing from scattering amplitudes to the line shape is difficult because of the integral nature of the $\mathbf{S}^{(2)}$ term in Eq. (9), with its $\rho(\mathbf{v}')$ dependence (note the prime). That dependence is a feature

[3] By assuming that *a,b,c* and *d* represent different levels, we are ignoring spatial degeneracy (*m* sublevels). We thus avoid the added complexity of *m*-changing collisions or pure reorientation effects and the problem of complete mixing of lines that are degenerate in the absence of a magnetic field.

common to all Boltzmann-like equations and is the principal reason why analytical solutions have not been found (with the sole exception of Maxwell molecules [29, Ch. 5]). To avoid this problem, the available post-processor codes calculate pressure-broadening cross-sections from the scattering amplitudes based on the Ben-Reuven [30] Shafer-Gordon [31] theory, which gives the line shape only in the high pressure limit (i.e. neglecting Doppler broadening and the effects of velocity-changing collisions on the spectrum). The result of this theory can be obtained from our kinetic theory by employing the ansatz:

$$\rho_{dc}(\omega,\mathbf{v}') = \overline{\rho}_{dc}(\omega) n_{\text{act}} f_{\text{M}}(\mathbf{v}') \tag{10}$$

for some velocity-independent function $\overline{\rho}_{dc}(\omega)$, where n_{act} is the number density of the active molecules and $f_{\text{M}}(\mathbf{v}')$ is the Maxwellian velocity distribution. Substituting Eq. (10) into Eqs. (8) and (9) produces the collisional change

$$\begin{aligned}\int_{\mathbf{v}} \hat{S}\rho_{ba}(\omega,\mathbf{v})\mathrm{d}\mathbf{v} &= \int_{\mathbf{v}} [-\mathbf{S}^{(1)}(ba,\mathbf{v}) + \mathbf{S}^{(2)}(ba,\mathbf{v})]\mathrm{d}\mathbf{v} \\ &= -n_{\text{act}} \sum_{dc} \int_{\mathbf{v}} \left[\nu(ba|dc,\mathbf{v}) - \int_{\mathbf{v}'} A(ba\mathbf{v}'|dc\mathbf{v})\mathrm{d}\mathbf{v}' \right] \overline{\rho}_{dc}(\omega) f_{\text{M}}(\mathbf{v})\mathrm{d}\mathbf{v},\end{aligned} \tag{11}$$

where in the last term, we have interchanged $\mathbf{v}$ with $\mathbf{v}'$ in order to bring out the common factor $f_{\text{M}}(\mathbf{v})$. For an isolated line, we only have $dc = ba$; in that case, by identifying $\rho_{ba}(\omega,\mathbf{v})$ with $\varsigma(\omega,\mathbf{v})$ in Eq. (1), using the ansatz of Eq. (10), and ignoring the $\mathbf{k}\cdot\mathbf{v}$ term (which is negligible compared to the collision term in the high-pressure limit), we obtain a Lorentzian line with a width given by Eq. (7), in which

$$\Gamma(\upsilon) = \nu(ba|ba,\mathbf{v}) - \int_{\mathbf{v}'} A(ba\mathbf{v}'|ba\mathbf{v})\mathrm{d}\mathbf{v}'. \tag{12}$$

More generally, we can define the dephasing contribution as:

$$\mathbf{S}_{\text{D}}(ba,\mathbf{v}) = -\sum_{dc} \left[\nu(ba|dc,\mathbf{v}) - \int_{\mathbf{v}'} A(ba\mathbf{v}'|dc\mathbf{v})\mathrm{d}\mathbf{v}' \right] \rho_{dc}(\mathbf{v}), \tag{13}$$

which can be computed from an integral over the relative velocity of a binary collision (see Eq. (24) of Ref. [20]) after the appropriate scattering amplitudes have been obtained from MOLSCAT or MOLCOL.[4] To see how this contribution relates to the total collision operator, $\int_{\mathbf{v}'} A(ba\mathbf{v}'|dc\mathbf{v})\rho_{dc}(\mathbf{v})\mathrm{d}\mathbf{v}'$ can now be added to and subtracted from the full colllisional contributions of Eqs. (8) and (9), resulting in:

[4] Here $\mathbf{S}_{\text{D}}$ should not be confused with the corresponding matrix in the section "Calculations".

$$\mathbf{S}(ba,\mathbf{v}) = \mathbf{S}_{\mathrm{D}}(ba,\mathbf{v}) + \sum_{dc}\int_{\mathbf{v}'}\left[A(ba\mathbf{v}|dc\mathbf{v}')\rho_{dc}(\mathbf{v}') - A(ba\mathbf{v}'|dc\mathbf{v})\rho_{dc}(\mathbf{v})\right]d\mathbf{v}'. \quad (14)$$

The term involving the sum and the integral in Eq. (14) is approximated by our $\hat{S}_{\mathrm{VC}}$, which is responsible for Dicke narrowing and whose contribution is calculated here using the classical billiard-ball model. A similar division of $\hat{S}$ has also been carried out in the classical description of Rautian and Sobel'man [21]. Such a separation is not unique, and is useful only if one has reliable models for $\hat{S}_{\mathrm{D}}$ and $\hat{S}_{\mathrm{VC}}$.

The preceding paragraphs are meant not only to explain the low-pressure deviation in our fitted values of γ_0, but also to suggest an approach to performing fully quantum mechanical line shape calculations in the near future. The MOLSCAT and MOLCOL codes already calculate all the quantum mechanical scattering amplitudes required to determine the full collision operator; the limitation has been in obtaining a solution for the off-diagonal density matrix elements responsible for the spectral line shape, i.e. $\rho_{ba}(\omega,\mathbf{v})$. Using today's computers, it should be possible to do so numerically, abandoning Eq. (10). In that case, it would not be necessary to patch Dicke narrowing back into the collision operator with a classical term $\hat{S}_{\mathrm{VC}}$, nor to fit the velocity-changing collision rate ν. The scattering calculations would yield the complete collision operator, including all effects implied by the labels VC, D, and VCD, and the resulting quantum kinetic equation would yield the correct shape for an isolated spectral line. There appears to have been only one attempt along this line [2].

CONCLUSIONS

We have demonstrated that theoretical line shapes obtained by solving the quantum kinetic equation for CO-Ar can reproduce the most accurately measured line shapes available to within their experimental noise, provided that the line shapes are insensitive to the amount of Dicke narrowing present (i.e. for pressures above 0.2 atm). In that case, the line shapes can be calculated *ab initio* except for the thermally averaged relaxation rate Γ_0, which must be fitted to overcome the present limit on the accuracy with which it can be calculated. For pressures at which the line shape is sensitive to the amount of Dicke narrowing (i.e. below 0.2 atm), a comparison between our measured and quantum kinetic line shapes has shown that the magnitude of the Dicke narrowing in CO-Ar is 10% to 30% of that predicted from mass diffusion. Furthermore, an examination of the structure of the collision operator has revealed that the reduced narrowing is probably due to inelastic collisions, which do not cause Dicke narrowing for an isolated line. Finally, we have noted that current quantum mechanical calculations of pressure-broadening cross-sections assume that the velocity dependence of the off-diagonal elements of the density matrix is Maxwellian – an assumption that removes velocity-changing effects from the line shape problem. By abandoning that assumption and solving the problem numerically, it should be possible to calculate the complete line shape quantum mechanically, starting only from the potential energy surface for the CO-Ar interaction.

ACKNOWLEDGMENTS

This work was supported by the Natural Sciences and Engineering Research Council of Canada. The figures and portions of the text herein were originally published by the authors in the Journal of Molecular Spectroscopy [5,12].

REFERENCES

1. R. Wehr, A. Vitcu, R. Ciuryło, F. Thibault, J. R. Drummond, and A. D. May, *Phys. Rev. A* **66**, 062502 (2002).
2. R. Blackmore, S. Green, and L. Monchick, *J. Chem. Phys.* **91** (7), 3846-3853 (1989).
3. L. Demeio, S. Green, and L. Monchick, *J. Chem. Phys.* **102** (23), 9160-9166 (1995).
4. C. Luo, R. Wehr, J. R. Drummond, A. D. May, F. Thibault, J. Boissoles, J.-M. Launay, C. Boulet, J.-P. Bouanich, and J.-M. Hartmann, *J. Chem. Phys.* **115** (5), 2198-2206 (2001).
5. R. Wehr, A. Vitcu, J. R. Drummond, A. D. May, and F. Thibault, *J. Mol. Spectrosc.* **235**, 69-76 (2006).
6. D. A. Shapiro, R. Ciuryło, J. R. Drummond, and A. D. May, *Phys. Rev. A* **65**, 012501 (2001).
7. R. Ciuryło, D. A. Shapiro, J. R. Drummond, and A. D. May, *Phys. Rev. A* **65**, 012502 (2001).
8. A. D. May, *Phys. Rev. A* **59** (5), 3495-3505 (1999).
9. J. M. Hutson and S. Green, *Collaborative Computational Project no. 6 (CCP6)*, UK Science and Engineering Research Council (1994).
10. D. R. Flower, G. Bourhis, and J.-M. Launay, *Comput. Phys. Commun.* **131**, 187-201 (2000).
11. R. R. Toczyłowski and S. M. Cybulski, *J. Chem. Phys.* **112** (10), 4604-4612 (2000).
12. R. Wehr, R. Ciuryło, A. Vitcu, J. R. Drummond, A. D. May, and F. Thibault, *J. Mol. Spectrosc.* **235**, 54-68 (2006).
13. R. Wehr, Dicke-narrowed spectral lines in carbon monoxide buffered by argon, Ph.D. thesis, University of Toronto (2005).
14. P. Duggan, P. M. Sinclair, R. Berman, A. D. May, and J. R. Drummond, *J. Mol. Spectrosc.* **186**, 90-98 (1997).
15. P. Duggan, P. M. Sinclair, M. P. L. Flohic, J. W. Forsman, R. Berman, A. D. May, and J. R. Drummond, *Phys. Rev. A* **48** (3), 2077-2083 (1993).
16. R. Wehr, E. McKernan, A. Vitcu, R. Ciuryło, and J. R. Drummond, *Appl. Opt.* **42** (33), 6595-6604 (2003).
17. R. Berman, P. M. Sinclair, A. D. May, and J. R. Drummond, *J. Mol. Spectrosc.* **198**, 283-290 (1999).
18. J. Bzowski, J. Kestin, E. A. Mason, and F. J. Uribe, *J. Phys. Chem. Ref. Data* **19** (5), 1179-1231 (1990).
19. P. Duggan, P. M. Sinclair, A. D. May, and J. R. Drummond, *Phys. Rev. A* **51** (1), 218-224 (1995).
20. A. S. Pine, *J. Quant. Spectrosc. Radiat. Transfer* **62**, 397-423 (1999).
21. S. G. Rautian and I. I. Sobel'man, *Sov. Phys. Uspekhi* **9** (5), 701-716 (1967).
22. A. S. Pine and R. Ciuryło, *J. Mol. Spectrosc.* **208**, 180-187 (2001).
23. F. Chaussard, X. Michaut, R. Saint-Loup, H. Berger, P. Joubert, B. Lance, J. Bonamy, and D. Robert, *J. Chem. Phys.* **112** (1), 158-166 (2000).
24. R. H. Dicke, *Phys. Rev.* **89** (2), 472-473 (1953).
25. S. G. Rautian and A. M. Shalagin, *Kinetic Problems of Non-Linear Spectroscopy*, New York: Elsevier Science Pub. Co., 1991, p. 43.
26. F. Thibault, R. Z. Martinez, J. L. Domenech, D. Bermejo, and J.-P. Bouanich, *J. Chem. Phys.* **117** (6), 2523-2531 (2002).
27. R. Blackmore, *J. Chem. Phys.* **86** (7), 4188-4197 (1987).
28. V. P. Kochanov, *J. Quant. Spectrosc. Radiat. Transfer* **66**, 313-325 (2000).
29. C. Truesdell, *Indiana Univ. Math. J.* **5** (1), 55-128 (1956).
30. A. Ben-Reuven, *Phys. Rev.* **145**, 7 (1966).
31. R. Shafer and R. G. Gordon, *J. Chem. Phys.* **58**, 5422 (1973).

Surface Interactions in Extremely-Thin Vapour Cell Spectroscopy: Coherent Dicke Narrowing of Atomic Transitions (Abstract)

Martial Ducloy

Laboratoire de Physique des Lasers, UMR 7538 du CNRS, Institut Galilée
Université Paris 13, 99 Av. JB Clément, 93430 Villetaneuse, France

Atom-wall interactions [1] are a key feature of atom dynamics in thin gas cells. For instance, for room-temperature alkali atoms with a residence time on the order of a few μs and a μm-thick cell, atoms are spending a thousand more time adsorbed on the wall than in free flight between the walls. In optical transmission spectroscopy, for vapour cells short enough to make the atomic free path anisotropic, the transient build-up of the resonant interaction with light is responsible for a specific enhancement of the response of the slowest atoms (*i.e.*, atoms moving quasi-parallel to the walls): this provides the principle of a novel method for Doppler-free spectroscopy, applicable to a variety of situations (velocity-dependent optical pumping, linear absorption, two-photon transition, etc.). In the linear response regime, the free-flight coherent dynamics of the atom-light resonant interaction naturally leads to Dicke narrowing of the lineshape [2], which has been observed in the optical domain with its periodical decay and revival of the sub-Doppler optical resonance [3]. Extensions of Dicke coherent narrowing to multiphoton (two-photon or Raman-type) resonances have been recently analysed [4]. For smaller cell thickness, van der Waals (vW) atom-surface interactions deeply alter the resonance lineshape. Nanometre-thin vapour cells allow one to monitor vW interactions in the yet unexplored 50-150 nm range [5]. On the other hand, Raman transitions between ground sublevels, or dark resonances [6], should in the principle allow one to explore long range Casimir interactions.

REFERENCES

1. D. Bloch and M. Ducloy, "*Atom-wall interaction*", in "*Advances in Atomic, Molecular and Optical Physics*" (B. Bederson and H. Walther eds, *Academic Press, San Diego*, 2005) **50**, 91.
2. R. H. Romer and R. H. Dicke, Phys. Rev. **99**, 532 (1955).
3. G. Dutier *et al*, Europhys. Lett., **63**, 35 (2003) ; D. Sarkisyan *et al*, Phys. Rev. A **69**, 065802 (2004).
4. G. Dutier *et al*, Phys. Rev. A **72**, 040501(R) (2005).
5. I. Hamdi *et al*, Laser Physics **15**, 987 (2005), and to be published.
6. H. Failache, A. Lezama, D. Bloch and M. Ducloy, "Coherent spectroscopy in sub-millimetre cells", communication presented to ECAMP VIII (Rennes, France, 2004).

CP874, *Spectral Line Shapes: 18th International Conference*,
edited by E. Oks and M. Pindzola

On the Calculation of Vector Collision—Sequence Interference in Collision-Induced Absorption

John Courtenay Lewis

Department of Physics and Physical Oceanography, Memorial University of Newfoundland St. John's, NL Canada A1B 3X7

Abstract. A theory of the intercollisional interference effect [1] in collision-induced absorption was published over thirty years ago [2]. However, owing to computational difficulties it has only been applied to simple models of the intermolecular potential and induced dipole moment. In this paper we reformulate the expressions for the interference dip, and detail procedures for the numerical evaluation of these kinetic--theoretical expressions for realistic intermolecular interactions in the classical limit.

Keywords: collision-induced absorption; vector collision-sequence interference
PACS: 33.80.-b, 33.20.Ea, 33.70.-w, 33.90.+h

ISOLATED ELASTIC BINARY COLLISIONS

We will follow the notation of Chapman and Cowling [3] closely though not exactly, and otherwise the notation of ref. [2].

With a spherically symmetric interaction potential, eight variables are needed to specify a collision. These may be

$$Q = \{b, \varepsilon, \mathbf{c}_1, \mathbf{c}_2\}$$

where subscripts 1 and 2 refer to atoms of two species so labelled; but more useful are

$$Q = \{\chi, \varepsilon, \mathbf{c}_1, \mathbf{c}_2\}, \quad Q = \{\mathbf{k}, \mathbf{c}_1, \mathbf{c}_2\} \text{ and } Q = \left\{\mathbf{c}_1, c_1', \xi, \varphi, g, \varphi_g\right\}.$$

Derived Quantities

Several derived variables are important in the treatment of intercollisional interference: the change in velocity of molecule 1, $\mathbf{C} = \mathbf{c}_1' - \mathbf{c}_1$; the velocities $\mathbf{g}_{21} = \mathbf{c}_2 - \mathbf{c}_1$; $\mathbf{g}_{21}' = \mathbf{c}_2' - \mathbf{c}_1'$; $g = |\mathbf{g}_{21}| = |\mathbf{g}_{21}'|$; and the angles $\xi = \angle \mathbf{c}_1', \mathbf{c}_1$; $\eta = \angle \mathbf{C}, \mathbf{c}_1$; $\eta' = \angle \mathbf{C}, \mathbf{c}_1'$; and $\chi = \angle \mathbf{g}_{21}', \mathbf{g}_{21}$.

The interference dip needs only the zero-frequency limit of the Fourier transform of the induced dipole moment. We define

$$\bar{\mu}(Q) = \int_{-\infty}^{\infty} \mu\left(\mathbf{x}_1(t), \mathbf{x}_2(t)\right)_Q \, dt$$

CP874, *Spectral Line Shapes: 18th International Conference*, edited by E. Oks and M. Pindzola

In fact because $\mu(t)$ is symmetric about the apse of the binary collision we obtain

$$\bar{\mu}(Q) = \bar{\mu}(g,C)\,\hat{\mathbf{C}}\,.$$

CALCULATING THE DIP SPECTRUM

The expression for the intercollisional interference dip can be written in terms of four collisional aaverages. First is $K = \langle \bar{\mu}^2 \rangle_Q$, which gives the intensity at zero frequency in the absence of interference. Second is the mean projection of the integrated induced dipole moment on the direction of the velocity before collision,

$$A(c_1) = \langle \bar{\mu}(g,C)\cdot\hat{\mathbf{c}}_1 \rangle_{\xi,\varphi,g,\varphi_g} = \int_0^\infty dc_1' \int_0^\pi d\xi \sin\xi \int_0^\infty \bar{\mu}(g,C)\cos\eta \, P\left(g,\xi,c_1' \middle| c_1\right)$$

The third, $B\left(c_1'\right)$, is the mean projection of the integrated induced dipole moment on the direction of the velocity after collision, wih expression similar to that of A. The final function does not depend on the integrated induced dipole moment, but is a measure of the persistence of velocity during a binary collision:

$$\Delta\left(c_1' \middle| c_1\right) = \left\langle \hat{\mathbf{c}}_1' \cdot \hat{\mathbf{c}}_1 \right\rangle_{\xi,\varphi,g,\varphi_g} = \int_0^\pi d\xi \sin\xi \cos\xi \, P\left(\xi, c_1' \middle| c_1\right)$$

The interference dip spectrum can then be written as

$$w(\omega) = K + 2\,\mathrm{Re}\int_0^\infty dc_1' \, \nu_c\left(c_1'\right) \Big/ \left(\nu_c\left(c_1'\right) + \iota\omega\right) B\left(c_1'\right) G_c\left(\omega, c_1'\right)$$

where $G_c(\omega, c_1)$ is the solution of the Fredholm integral equation

$$G_c(\omega, c_1) = A(c_1) + \int_0^\infty dc_1' \, \nu_c\left(c_1'\right) \Big/ \left(\nu_c\left(c_1'\right) + \iota\omega\right) G_c\left(\omega, c_1'\right)$$

Until recently this had been applied in detail only to hard spheres, in most detail in a Lorentz gas.

Developments

Recent analyses indicates that the most important of the collisional probabilities is

$$P(C, c_1) = \frac{8\pi^3 \gamma_1 \gamma_2}{N_{12} a_2} c_1 e^{-ac_1^2} \int_0^\infty dg \int_0^{b_{\max}} \mathrm{db}\,\mathrm{b}\, g^2 e^{-a_2 g^2} \left(1 - e^{-2a_2 g c_1}\right) \delta\left(V - 2M_2 g \left|\sin\tfrac{\chi(g,b)}{2}\right|\right)$$

where $a_i = \dfrac{m_i}{2k_B T}$, $a = \dfrac{m_1 + m_2}{2k_B T}$, $M_2 = \dfrac{m_2}{m_1 + m_2}$, $\gamma_i = n_i \left(\dfrac{a_i}{\pi}\right)^{3/2}$, and the total collision frequency is N_{12}. Then the persistence of velocity kernel can be written as

$$\Delta\left(c_1' \middle| c_1\right) = \int_{\left|c_1 - c_1'\right|}^{c_1 + c_1'} dC \frac{c_1^2 + c_1'^2 - C^2}{2c_1 c_1'} P(C, c_1) = \frac{8\pi^3 \gamma_1 \gamma_2}{N_{12} a_2} \frac{e^{-ac_1^2}}{c_1'} \left\{\Im_1\left(c_1', c_1\right) - \Im_2\left(c_1', c_1\right)\right\}$$

$$\Im_1\left(c_1', c_1\right) = \left(c_1'^2 + c_1^2\right) \int_0^\infty dg\, g^2 e^{-a_2 g^2} \sinh(a_2 g c_1) \int_0^{b_{\max}} db\, b\, \Theta(b, g)$$

$$\Im_2\left(c_1',c_1\right)=4M_2^2\int_0^\infty dg\,g^4e^{-a_2g^2}\sinh\left(a_2gc_1\right)\int_0^{b_{\max}}db\,b\sin^2\frac{\chi(g,b)}{2}\Theta(b,g).$$

$$\Theta(b,g)=1 \text{ if } \left|c_1-c_1'\right|\le 2M_2g\left|\sin\frac{\chi(g,b)}{2}\right|\le c_1+c_1' \text{ and 0 otherwise.}$$

We have recently found that by symmetrising the kernel D and modifying the A and B functions a more economical and computationally feasible expression for the line shape function can be found. We set

$$\tilde{\Delta}\left(c_1'\middle|c_1\right)=P\left(c_1'\right)^{-1/2}P\left(c_1\right)^{-1/2}\int_0^\pi d\xi\,\sin\xi\cos\xi\,P\left(\xi,c_1',c_1\right)$$

and

$$\tilde{G}(\omega,c_1)=-P(c_1)^{1/2}G_c(\omega,c_1) \text{ and } \tilde{B}(c_1)=B(c_1)P(c_1)^{-1/2}=-A(c_1)P(c_1)^{1/2}.$$

Then

$$\tilde{G}(\omega,c_1)=\tilde{B}(c_1)+\int_0^\infty dc_1'\,v_c\left(c_1'\right)\Big/\left(v_c\left(c_1'\right)+i\omega\right)\tilde{\Delta}\left(c_1'\middle|c_1\right)\,\tilde{G}(\omega,c_1)$$

and the equation for the interference line shape then becomes

$$w(\omega)=K-2\,\mathrm{Re}\int_0^\infty dc_1\,v_c(c_1)\big/\left(v_c(c_1)+i\omega\right)\tilde{B}(c_1)\,\tilde{G}(\omega,c_1).$$

To obtain an expression for the function $\tilde{B}(c_1)$, define

$$\tilde{S}_1(a)\equiv\int_0^\infty d\vartheta\sin^2\vartheta\,e^{-a\cos\vartheta} \text{ and } \tilde{S}_2(a)\equiv\int_0^\infty d\vartheta\sin\vartheta\cos\vartheta\,e^{-a\cos\vartheta}$$

$$\tilde{N}(c_1)=\int_0^\infty dg\,g^3e^{-m_2g^2/k_BT}\left\{\tilde{S}_1(m_2gc_1/k_BT)\tilde{M}_1(g)+\tilde{S}_2(m_2gc_1/k_BT)\tilde{M}_2(g)\right\}$$

$$\tilde{M}_1(g)=\int_0^{b_{\max}}db\,b\,\overline{\mu}(g,b)\cos\frac{\chi(g,b)}{2} \text{ and } \tilde{M}_2(g)=\int_0^{b_{\max}}db\,b\,\overline{\mu}(g,b)\sin\frac{\chi(g,b)}{2}.$$

Then

$$\tilde{B}(c_1)=\exp\left[-(m_1/2+m_2)c_1^2/2k_BT\right]\tilde{N}(c_1).$$

Computational Schema

Calculate χ and $\overline{\mu}$ from g and b; then calculate the functions $\tilde{M}_1$ and $\tilde{M}_2$ and the function $\tilde{N}$. Then calculate $\tilde{B}(c_1)$. Then solve the discretised integral equation for $\tilde{G}$ for each desired value of ω (this is a relatively easy matrix inversion problem) and calculate the interference spectrum $w(\omega)$. Each step is straightforward and no step requires more than one-dimensional integration.

REFERENCES

1. Lewis, J. C., and Herman, R. M., *Phys. Rev. A* **68**, 032703 (2003); and references therein.
2. Lewis, J. C., *Can. J. Phys* **50**, 2881 (1972).
3. Chapman, S., and and Cowling, T. G., *The Mathematical Theory of Non-uniform Gases*, Cambridge University Press 1964.

Collisional Line Broadening, Line Mixing, and Rotational Relaxation Rates at Low Temperatures: ab initio Studies and Comparision with Experiments (Abstract)

Franck Thibault

Laboratoire de Physique des Atomes, Laser, Molecules et Surfaces, Universite de Rennes I, Campus de Beaulieu, 35042 Rennes Cedex, France

To model complex band spectra of an optically active molecule in a perturber bath one may have to resort to the Energy Corrected Sudden (ECS) Approximation. For instance, the so-called Infinite Order Sudden Approximation has been used to compute cross sections for collisions between ammonia in the ground rotational level and Helium, Hydrogen or Argon. The entire relaxation matrix was then generated within the ECS framework to simulate absorption in the ν_1 and ν_2 bands. Examples for NH_3-HE mixtures will show that this method opens promising perspectives for the remote sensing study of planetary atmospheres.

The availability of increasingly powerful computing resources now makes possible quantitative predictions of the band-shape evolution with the perturber density for simple molecules using fully ab initio methods. Close coupling (CC) calculations were carried out for the first time for the Q branch Raman spectra of the fundamental band of CO in low-temperature helium baths. Both diagonal (line width and shift) and off-diagonal (line couplings) generalized cross sections were computed starting with accurate ab initio potential energy surfaces. As shown, results are in good agreement with Q-branch Raman data.

For CO-Ar mixtures, pressure broadening cross sections are also calculated very accurately, allowing the extraction of collisional line widths of isolated lines for both IR and Raman spectra over a wide range of temperature. These CC values provide benchmarks for approximate quantum or (semi-) classical methods. The relative failure of the coupled state approximation will be discussed for this system.

In addition to the interest of ab initio calculations for several fields such as planetology, lineshape parameters may also help to determine semi-empirical PESs. The Rg-C_2H_2 PESs are currently under construction. The example of the Ar-braodening lines of acetylene is presented.

Most recent PESs are constructed to agree with experimental spectra of complexes. These PESs are thus expected to be very accurate in the region of their potential well however, comparisons of low temperature (say, below 10 K) theoretical pressure broadening and shifting parameters with measurements obtained using the collisional cooling technique are till puzzling. In fact large discrepancies are found for several systems, such a CO in HE which will be presented in this talk. Such substantial

CP874, *Spectral Line Shapes: 18th International Conference,*
edited by E. Oks and M. Pindzola

disagreement is not observed in state-to-state rotational relaxation rates at low temperatures, despite the fact that the rates are sensitive to the same part of the PES as the pressure broadening coefficients and that the latter are related to the former. Indeed, quite satisfactory results down to a few Kelvin have been obtained for rotational relaxation of N_2 in He, para-H_2 colliding pairs. Preliminary results for H2 interacting with He are also encouraging. Illustrative results are presented.

Density Matrix Treatment for the Solid HD Spectrum in the Vicinity of the $R_1(0)$ Feature

R. M. Herman

Department of Physics, The Pennsylvania State University, 104 Davey Laboratory, University Park, PA 16802, USA

Abstract. The spectroscopic structure of the somewhat complicated $R_1(1)$ feature and its surroundings in solid HD [1] represents a long-standing mystery. The complexity arises from the interference between the direct transition and the surrounding Q_1-R phonon band. In this approach, we first establish a density matrix formalism to describe overlapping single transitions of different intrinsic line widths and intensities, coupled through a static interaction. This approach is then generalized to the interaction of a single narrow transition embedded in a band of broad transitions. Results are compared with the spectrum of ref. [1].

Keywords: solid hydrogen deuteride, rotation-translation spectra.
PACS: 78.90+t, 33.70Jg.

INTRODUCTION

While there have been several studies of the HD $R_1(0)$ feature (v,J = 0,0 → 1,1) and the overlapping $Q_1(0)_R$ phonon band (v,J = 0,0 → 1,0, together with a single phonon excitation) [1-3] none has successfully treated the pronounced interference features including a dramatic dip around 5 cm^{-1} below and an unexpected peak around 14 cm^{-1} above the sharp $R_1(0)$ peak. See Fig. 1 which shows the spectrum of ref. [1] and is quite close to that of ref. [2]. Both the sharp $Q_1(1)$ and $R_1(0)$ features in the solid are vibrationally downshifted by 7.3 cm^{-1} from their free molecule counterparts, while being spaced by 85 cm^{-1} as they are in the free molecule. In the present problem a static interaction couples states which are differently stochastically perturbed. In this regard the usual situation of collisionally-coupled line mixing is not in force, and the author is not aware of any previous similar treatment for this type of system.

THEORETICAL CONSIDERATIONS

To attack the problem we begin with a simple phase shift analysis of the random perturbations, then generalize the result to include other stochastic effects. Consider a single lower state 0, and a pair of upper states 1, 2 coupled by a static perturbation V. In the impact limit, each of the density matrix elements ρ_{10} and ρ_{20} vary with time dependence $\exp(-i\omega t)$, so that state 1 cannot easily distinguish whether it was populated by direct excitation from state 0 or, by V, from state 2. Thus relevant relaxation properties must reflect the excitation pathway, and not simply the two states

CP874, *Spectral Line Shapes: 18th International Conference*, edited by E. Oks and M. Pindzola

indicated in the specification of the density matrix element. One therefore must separate ρ_{10} into a sum of a directly excited part $\rho_{10}{}^{d}$ coming from the almost 100% populated state 0, and an indirectly excited part $\rho_{10}{}^{i}$ coming from state 2. We write the corresponding Liouville equations (with all energies as circular frequencies)

$$i\dot{\rho}_{10}{}^{d} = \left(\omega_{10} + v_{10}(t)\right)\rho_{10}{}^{d} - C_{10}\exp(-i\omega t) \tag{1}$$

$$i\dot{\rho}_{10}{}^{i} = \left(\omega_{10} + v_{12}(t)\right)\rho_{10}{}^{i} + V_{12}\left(\rho_{20}{}^{d} + \rho_{20}{}^{i}\right) \tag{2}$$

with $v_{10}(t) = \left(V_{11}(t) - V_{00}(t)\right)_{\text{coll}}$, $v_{12}(t) = \left(V_{11}(t) - V_{22}(t)\right)_{\text{coll}}$, and $C_{10} = \frac{1}{2}Ep_{10}$.

Corresponding equations exist for $\rho_{20}{}^{d}$ and $\rho_{20}{}^{i}$. To solve Eq. (1) in the impact regime, write

$$\rho_{10}{}^{d} = \sigma_{10}{}^{d}\exp\left[-i\left(\omega_{10}t + \int_{-\infty}^{t} v_{10}(t')dt'\right)\right] \tag{3}$$

with the result

$$\dot{\sigma}_{10}{}^{d} = iC_{10}\exp\left[i(\omega_{10} - \omega)t + i\int_{-\infty}^{t} v_{10}(t')dt'\right] \tag{4}$$

This equation can be converted into a difference equation connecting time intervals large compared with a collision duration, and then be directly integrated as in conventional impact theory with the result, (when one gets back to $\rho_{10}{}^{d}$),

$$\rho_{10}{}^{d} = iC_{10}\exp(-i\omega t)/\left(\gamma_{10}/2 + i(\omega_{10} + \delta_{10} - \omega)\right) \tag{5}$$

with γ_{10} and δ_{10} being the standard linewidth and shift for the 10 transition by itself, leading to the expected Lorentz profile in the absence of V_{12}. This result and similar ones for other components, could have been obtained with the equivalent equations

$$\left(\gamma_{10}/2 + i(\omega_{10} - \omega)\right)\rho_{10}{}^{d} = iC_{10}\exp(-i\omega t) \tag{6}$$

$$\left(\gamma_{12}/2 + i(\omega_{10} - \omega)\right)\rho_{10}{}^{i} = -iV_{12}\left(\rho_{20}{}^{d} + \rho_{20}{}^{i}\right) \tag{7}$$

$$\left(\gamma_{20}/2 + i(\omega_{20} - \omega)\right)\rho_{20}{}^{d} = iC_{20}\exp(-i\omega t) \tag{8}$$

$$\left(\gamma_{21}/2 + i(\omega_{20} - \omega)\right)\rho_{20}{}^{i} = -iV_{21}\left(\rho_{10}{}^{d} + \rho_{10}{}^{i}\right) \tag{9}$$

with line shift parameters neglected and the stochastic effects simply represented by effective γ-parameters throughout. This is often done in other relaxation analyses, where these parameters now reflect all sources of broadening. Here $\gamma_{12} = \gamma_{21}$, while possibly differing from γ_{10} and γ_{20}.

An interesting (and necessary) feature of Eqs. (6-9) is that in the event that the γ's are equal and small, the usual results for intensities and line positions of the coupled states 1,2 as being directly excited from the 0 state are obtainable for all values of V_{12} and $(\omega_{20} - \omega_{10})$. One now proceeds by allowing state 2 to be replaced by a continuum of states k, and on the r.h.s. of Eq. (7), one sums over all states k. Substitution of the solution of Eq. (8) and the form for Eq. (9) into Eq. (7) and combining with Eq. (6), one can get a closed form solution for $\left(\rho_{10}{}^{d} + \rho_{10}{}^{i}\right)$. Then substituting that result into Eq. (9), $\rho_{20}{}^{i}$ can be established. The results can be written as a (rather complicated) function of the noninterfered phonon spectrum, the unperturbed $R_1(0)$ Lorentzian and an interaction-perturbed $R_1(0)$ Lorentzian, together with their (Kramers-Kronig) dispersion counterparts.

RESULTS

The results of the present treatment are shown in Fig. 1, together with the spectrum of ref. [1]. The best guess for a smoothed, noninterfering HD phonon spectrum, consistent with the known D_2 phonon spectrum [1] is also shown. This is then used in the calculation of the total response, as indicated above

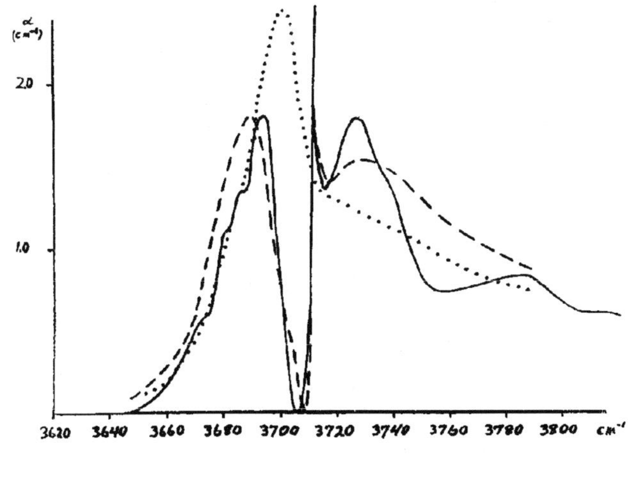

FIGURE 1. The observed [1] total spectrum ——, the proposed smoothed uninterfered phonon spectrum inferred from that of D_2 [1] • • • •, and resulting predicted total contour – – – for parameters C_{10}= 6.0, $(V_{1k}/C_{10}C_{1k})$ = 0.060, γ_{10} = 1.5 and $\gamma_{1k} = \gamma_{k1}$ = 12.

These results are preliminary, and more study on both the methodology and correct noninterfered phonon spectrum are needed.

ACKNOWLEDGEMENT

The author is pleased to acknowledge helpful discussions with R. H. Tipping and with J. C. Lewis.

REFERENCES

1. A. Crane, and H. P. Gush, *Can. J. Phys.* **44**, 373-398 (1976).
2. A. R. W. McKellar, and M. J. Clouter, *Can. J. Phys* **68**, 422-427 (1990).
3. J. D. Poll, R. H. Tipping, Sang Young Lee, Sung-ik Lee, Tae W. Noh and J. R. Gaines, *Phys. Rev. B* **39**, 11372-11377 (1989), and references therein.

Isotope Effects Associated With Optical Collisions In The Cd-Ar System

P. Masłowski, J. Domysławska, K. Bielska, R. Ciuryło, D. Lisak, R. S. Trawiński, J. Szudy

Instytut Fizyki, Uniwersytet Mikołaja Kopernika, ul. Grudziądzka 5/7, 87-100 Toruń, Poland

Abstract. Results of a systematic study of profiles of the hfs components of the intercombination 326.1 nm line of the even-odd ^{113}Cd isotope perturbed by argon performed by means of a laser-induced fluorescence method are reported. We have found that the values of total line asymmetries and Lorentzian widths of both hfs line components are different.

Keywords: Spectral line shapes, hyperfine structure
PACS: 32.70.-n, 31.50.+w, 34.20.

INTRODUCTION

In this paper we present results of experiments on the influence of argon on the shape of intercombination $5^1S_0 - 5\,^3P_1$ line (326.1 nm) of the even-odd ^{113}Cd isotope. Preliminary results of these experiments [1, 2] done at one value (81 Torr) of Ar-pressure indicated that the observed asymmetry of the 326.1 nm ^{113}Cd line is caused both by the collision duration (CD) and the line-mixing (LM) effects. To our knowledge this was the first observation of the LM asymmetry for atomic lines perturbed by neutrals. Such an asymmetry which originates from the coupling of the adjacent lines is a typical feature of spectral profiles associated with atomic spectra perturbed by electrons and ions in plasmas as well as for infra-red lines of molecules. In the course of the present study we used the laser-induced fluorescence (LIF) method to perform systematic measurements of argon-broadened profiles of the 326.1 nm ^{113}Cd line in the pressure range up to 200 Torr.

Beyond the impact limit, i.e. in the case when the finite duration of collisions must be taken into account the collisional part of the resulting line shape $I_{CD}(\nu)$ is represented as the sum of ordinary Lorentzian shape and a dispersion term [3-6], whose magnitude is characterised by an asymmetry parameter χ related to the collision-duration time. Using the Baranger [7] theory of collisional effects on overlapping lines Rosenkranz [8] has shown that in the first-order approximation spectral line profiles that take into account mixing of adjacent lines can be expressed in terms of sums of $I_{LM}(\nu)$ profiles which are defined as sum of Lorentzian shape and a dispersion term in the way formally identical to $I_{CD}(\nu)$ profiles. However, in this case the magnitude of the dispersion term which accounts for the interference due to the

CP874, *Spectral Line Shapes: 18th International Conference,*
edited by E. Oks and M. Pindzola

line mixing is characterised not by χ but by another asymmetry parameter Y related to the relaxation matrix.

The intercombination $5^1S_0 - 5\,^3P_1$ (326.1 nm) line of the even-odd ^{113}Cd isotope consists of two hfs components: F′ = 1/2 - F′′ = 1/2 (weaker) and F′= 3/2 - F′′ = 1/2 (stronger). We can thus expect that for this line the observed asymmetry is due to simultaneous occurence of the mixing of the two hfs components and the collision duration effect. Following Refs. [9, 10] in the case of weak coupling of adjacent lines the collisional part of the resulting line shape $I_{LM\text{-}CD}(\nu)$, in which both the LM and CD effects are taken into account, can be written in a form analogous both to $I_{CD}(\nu)$ and to the Rosenkranz formula for $I_{LM}(\nu)$, i.e. as the sum of Lorentzian and dispersion profiles. The total asymmetry parameter M is then given by the following approximate relation: $M \approx \chi + Y$, i.e. the sum of parameters describing the asymmetry due to CD and LM effects.

RESULTS

Measurements of the shape of ^{113}Cd 326.1 nm line were performed by means of LIF method using a modified version of a digital laser spectrometer with actively stabilized Coherent 899-21 ring dye laser described elsewhere [1, 11]. Our data analysis was simillar to those described in [1, 2]. We have used a least-squares algorithm for nonlinear parameters due to Marquardt [12] to perform the best-fit procedure. Figure 1 shows the values of the best-fit total asymmetry parameter M determined for stronger (plot - a) and weaker (plot - b) hfs components of the 326.1 nm line.

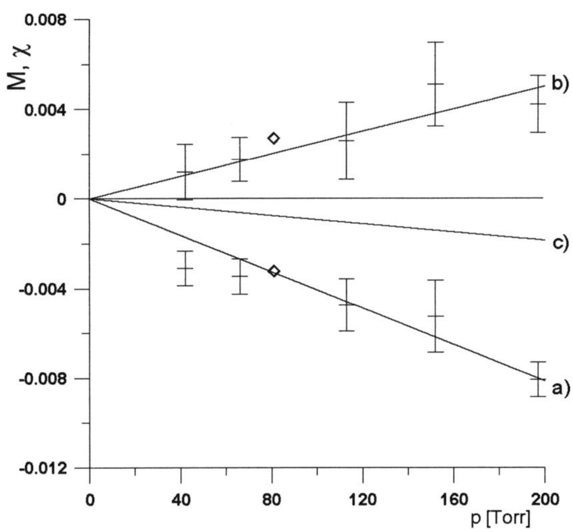

FIGURE 1. The values of total M ((a) and (b) plots) and collision-duration χ (plot - c) asymmetry parameters as the function of argon pressure - p. Rhombs - total asymmetry values obtained in earlier work [1, 2] for p = 81 Torr.

We can see that these asymmetries are of different signs and absolute values. From the slopes of fitted straight lines we determined for the total asymmetry pressure coefficients m = M / p, where p is the perturbing gas pressure, the following values: $m_{(strong)}$ = (- 4.06 ± 0.20) $*10^{-5}$ Torr $^{-1}$ for the stronger line component and $m_{(weak)}$ = (2.51± 0.23)$*10^{-5}$ Torr $^{-1}$ for the weaker component. For the comparison in Figure 1 we also plotted the values of collision-duration asymmetry parameters χ (plot c) for the even-even isotope ^{114}Cd perturbed by argon determined in our earlier work [6]. The slope of the line c gives rise to the CD-asymmetry pressure coefficient the value χ / p = (- 0.92 ± 0.15) $*10^{-5}$ Torr $^{-1}$.

From the line shape analysis we have also found that the values of collisional (Lorentzian) widths - γ of both hfs components are different: $\gamma_{(weak)}$ / $\gamma_{(strong)}$ = 1.029 ± 0.002.

ACKNOWLEDGMENTS

This work was supported from funds for science in years 2005-2008 as a research project.

REFERENCES

1. A . Bielski, R. Ciuryło, J. Domysławska, D. Lisak, P. Masłowski, J. Szudy and R. S. Trawiński, *Acta. Phys. Pol. A* **105**, 329 (2004).
2. A. Bielski, R. Ciuryło, J. Domysławska, D. Lisak, P. Masłowski, J. Szudy and R. S. Trawiński, *Spectral Line Shapes vol. 13* ed. by E. Dalimier, pp. 88, Frontier Group, Paris (2004).
3. M. Harris, E. L. Lewis, D. McHugh, I. Shannon, *J. Phys. B* **17**, L661 (1984); **19**, 3207 (1986).
4. J. Szudy, W. E. Baylis, *Phys. Rep.* **266**, 127 (1996).
5. A. Bielski, R. Ciuryło, J. Domysławska, D. Lisak, R. S. Trawiński, J. Szudy, *Phys. Rev. A* **62**, 032511 (2001).
6. A. Bielski, D. Lisak, R. S. Trawiński, *Eur. Phys. J. D* **14**, 27 (2001).
7. M. Baranger, *Phys. Rev.* **111**, 494 (1958).
8. P. W. Rosenkranz, *IEEE Trans. Anten. Prop.* **23**, 498 (1975).
9. R. Ciuryło, J. Szudy, *Phys. Rev. A* **63**, 042714 (2001).
10. W. F. Wang, J. M. Sirota, *J. Chem. Phys.* **116**, 532 (2002).
11. A. Bielski, R. Ciuryło, J. Domysławska, D. Lisak, R. S. Trawiński and J. Wolnikowski, *Acta Phys. Pol.* **97**, 1003 (2000).
12. D. W. Marquardt, *J. Soc. Industr. Appl. Math.* **11**, 431 (1963).

Spectroscopic Study of the Negative Corona Discharge in the Liquid ^{4}He

Z.L. Li[1], N. Bonifaci[1], A. Denat[1], V.M. Atrazhev[2] and V.A. Shakhatov[3]

[1]*Laboratoire d'Electrostatique et de Materiaux Diélectriques, CNRS, BP 166 38042 Grenoble Cedex 9, France*
[2]*Institute for High Energy Densities, RAS, Moscow, Russia*
[3]*A.V.Topchiev of Petrochemical Synthesis Institute, RAS, Moscow, Russia*

Abstract: The corona discharge plasma in liquid ^{4}He has been studied by spectroscopy as a function of the pressure between 0.1MPa and 10.4MPa. The rotational temperature was approximately equal to 350K at the pressure of 0.1MPa. With increasing pressure the blue shift and the broadening of the atomic helium line at 706.5nm ($2p^3P$-$3s^3S$) and 728.1nm ($2p^1P$-$3s^1S$) and the bands of the singlet $d^1\Sigma_u^+ \rightarrow b^1\Pi_g$ transition at 639nm and triplet $d^3\Sigma_u^+ \rightarrow b^3\Pi_g$ transition at 659nm have been observed.

Keywords: liquid helium, discharge, shifts.

INTRODUCTION

It is well-known if negative high voltage is applied to corona-forming electrode multiplication of avalanches is primarily mechanism of electron production in corona discharge in gaseous (GHe) and liquid (LHe) helium. Atoms in electronically-excited states are populated by step electron impact in the vicinity of electrode. Molecules of He_2 in electronically-excited states are mainly produced by three-particle recombination. The excited species can produce strong light emission in LHe. This is an essential prerequisite to investigate by spectroscopy a variety of species in the excited states, which are interaction with a dense surrounding medium. We present results of measurements of gas temperature, shift and broadening of the atomic lines and molecular bands of helium in the GHe and LHe located in vicinity of electrode.

EXPERIMENT AND RESULTS

The corona discharge initiated in the point-plane geometry under negative high voltage (up to 20kV) has been studied [1, 2]. The radii r_p of tungsten tips is of 0.45 and 2.5μm. The emission spectrum from the ionization region located in the vicinity of the corona-forming electrode is collected on the entrance slit of a spectrograph through a sapphire window in the cell. The experimental set-up has been described the more details in [2]. In the visible range from 500 to 1080nm, all but six spectral components of spectra can be identified as emission from the excited states of He_2. Of these, two spectral components are interpreted as emission from the excited states of He. Table1 gives the identified and resolvable atomic lines and molecular bands observed at T=4.2K and P=0.1MPa. The

CP874, *Spectral Line Shapes: 18th International Conference,* edited by E. Oks and M. Pindzola feature of the spectra taken from the LHe lies in the fact that atomic lines and molecular bands are substantially broader and shifted in ultraviolet spectral region by a few wavenumbers relative to these observed in the GHe. Actually, as showed in Fig.1 and 2, the resolvable atomic line at 706.4nm can just be observed up to 4MPa, while we monitor the molecular band at 639nm at P=1.68MPa. The blue shift for the line and band is evident. No lines and bands are observed in spectrum at P>4.0MPa.The pressure shift of the lines and bands given in Fig.1-2 is linear. Value of the pressure shift determined by slope is equal to 2.47nm/MPa. Compared with our results, the line shift at 706.4nm caused electron-heavy particle collisions [3] is also showed in Fig.2. Our result is near Soley's result 3.034nm/MPA at the pressure lower than 1MPa. And even it is a trifle over at P>1MPa.

TABLE 1. Molecular bands and atomic lines observed in LHe

Atomic line (nm)		Band head (nm)	
$3s^1S \rightarrow 2p^1P$	728.1	$f^3\Delta_u \rightarrow b^3\Pi_g$	577.4
$3s^3S \rightarrow 2p^3P$	706.5	$f^3\Pi_u \rightarrow b^3\Pi_g$	587.8
		$d^3\Sigma_u^+ \rightarrow b^3\Pi_g$	639.6
		$D^1\Sigma_u^+ \rightarrow B^1\Pi_g$	659.5
		$C^1\Sigma_g^+ \rightarrow A^1\Sigma_u^+$	912.8
		$c^3\Sigma_g^+ \rightarrow a^3\Sigma_u^+$	918.3

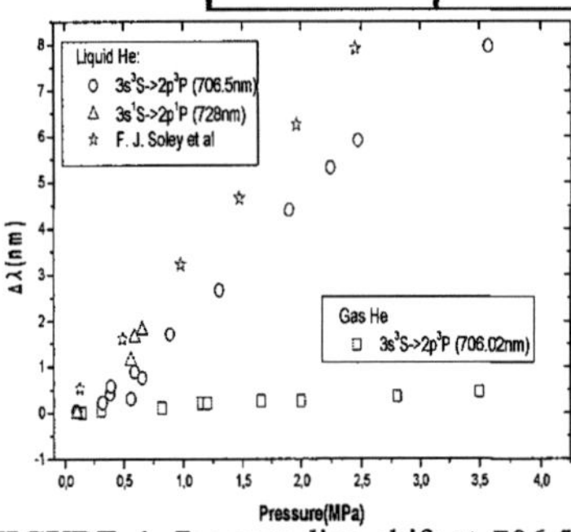

FIGURE 1. Pressure line shift at 706.5 nm (o) and 728 nm (Δ) in LHe and at 706.5nm (□) in GHe.

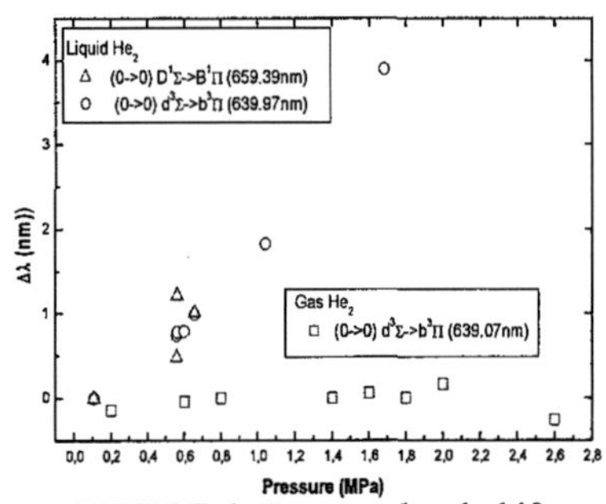

FIGURE 2. Pressure band shift at 659 nm (Δ) and 639 nm (o) in LHe and at 660 nm (□) in GHe

ANALYSIS

The atomic line pressure shifts (at 706.4nm) in LHe are larger than those in GHe (Fig.1 and 2). But no band pressure shift is observed in GHe. The difference follows that the density in LHe is very higher than that in GHe. The Stark effect can not give explanation observed the blue shift which takes place in LHe and also in GHe. For example, the atomic line shift (at 706.4 nm) is the red shift by the stark effect [4]. But the blue shift is not only observed by us, it is also obtained by the others [3] who explained theoretically by the bubble model [5, 6]. The theoretical value predicted by the bubble model is about 3.85nm/MPa [5]. It is close to our.

The rotational temperature T_r of He_2 can be determined from an analysis of the rotational structure of He_2 bands. We suppose that T_r corresponds to the gas temperature T_{th} of the medium in the corona zone [7]. It is due to the strong coupling between translational and rotational degree freedom of He_2. T_r is defined from the comparison of measured and synthetic spectra of the singlet $d^1\Sigma_u^+ \rightarrow b^1\Pi_g$ and the triplet $d^3\Sigma_u^+ \rightarrow b^3\Pi_g$ transitions. Distribution over rotational levels in excited $d^1\Sigma_u^+$ and $d^3\Sigma_u^+$ states were characterized by Boltzmann function. The line strength takes from [8] and the transition is considered between v'=0 and v''=0 for electronic states. Simulation taking into account of contribution of the band (v'=1, v"=1) is also being carried out. Until now, simulated spectra were not perturbed by this contribution if the vibrational temperature was taken less than 800 K. Fitted rotational temperature is in the range of 350±25K.

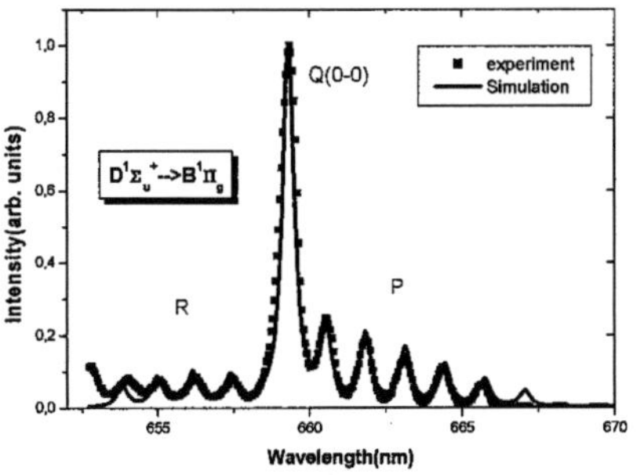

FIGURE 3. Comparison of experimental (points) and calculated (solid line) spectra for transitions $d^1\Sigma_u^+ \rightarrow b^1\Pi_g$ in LHe at P=0.1MPa

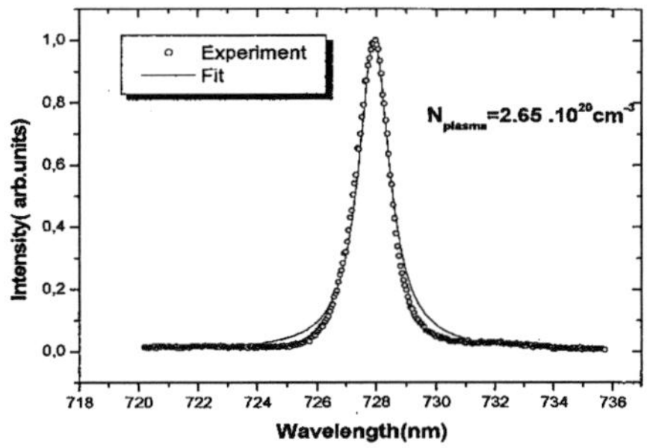

FIGURE 4. Simulation of $3s^1S \rightarrow 2p^1P$ in LHe at P=0.1MPa

It should be noted that the calculated rotational structure of the systems qualitatively agree with that, which observed in the emission spectra excited by electron beam [9] and synchrotron radiation [10] in LHe. T_{th} in the ionization zone ($T_{th} \cong T_r$) is much higher than T=4.2 K far from the corona zone in LHe. Plasma density number N_{plasma} can be evaluated from the resonant broadening of atomic lines of HeI [11]. An analysis of line at 728.53nm (Fig.4) gives: $N_{plasma} \sim 10^{22}$cm^{-3} at P=0.1MPa. Perhaps, the rotational temperature of He_2 molecule is not connected with gas temperature because the molecules are created not by direct electron impact excitation of the molecule in the ground state (there is not stable ground state of He_2 molecule) but as a result of triple-body reactions of excited and non-excited He atoms.

The work was supported by grant of Program of Basic Investigations of RAS № 20, and NATO grant PST.CL G-979649. V. Atrazhev and V. Shakhatov thank LEMD-CNRS for hospitality.

REFERENCE

1. A V. A. Goncharov and V. I. Levitov *Izvestiya Academii Nauk, Energetika i Transport (in Russian)* 12 134 1975;
2. Z. Li, N. Bonifaci, A. Denat, V.M. Atrazhev *IEEE 15th Intern. Conf. on Dielect. Liquids, ICDL-2005, Coimbra, Portugal,* 275 2005.
3. F. J. Soley and W. A. Fitzsimmons, *Phys. Rev. Lett.* 32 988 1974.
4. H.R.Griem, "Spectral line broadening by plasmas", *Academic Press, INC*, 1974.
5. A.P. Hickman, W.Street and N.F. Lane, *Phys. Rev. B* 14, 4205-8, 1975
6. W.Street, A.P. Hickman and N.F. Lane, *Chem. Phys. Lett.* 28,31,1974
7. M. Phillips, *J. Phys. D: Appl. Phys.,* 8, 507 1975.
8. I.Kovacs. "Rotational structure in the spectra of diatomic molecules", *Adam Hilfer LTD, London*
9. W. S. Dennis, E. Durbin, Jr., W. A. Fitzsimmons, O. Heybey, and G. K. Walters, *Phys. Rev. Lett.* 23 1083 1969.
10. K. von Haeften, T.Laarmann, H.Wabnitz, T.Moller, *Phys. Rev. Lett.* 88 233401 2002.
11. G. Peach, *Advance Phys.* 30 367 1981

SYMPOSIUM

SPECTROSCOPIC DIAGNOSTICS FOR MAGNETIC FUSION

Spectral Line Emission from Light Elements

M. S. Pindzola*, S. D. Loch*, F. Robicheaux*, C. P. Ballance†, D. C. Griffin†, N. R. Badnell‡, M. G. O'Mullane‡, H. P. Summers‡, M. Fogle◇, D. R. Schultz◇, J. P. Colgan• and C. J. Fontes•

* *Department of Physics, Auburn University, Auburn, AL*
† *Department of Physics, Rollins College, Winter Park, FL*
‡ *Department of Physics, University of Strathclyde, Glasgow, UK*
◇ *Physics Division, Oak Ridge National Laboratory, Oak Ridge, TN*
• *Theoretical Division, Los Alamos National Laboratory, Los Alamos, NM*

Abstract. We review a three step process in which fundamental atomic collision physics is linked to spectral line emission observations in controlled fusion experiments.

Keywords: atomic, molecular, collisions
PACS: 34

INTRODUCTION

A three step process may be used to accurately understand spectral line emission features and overall radiative power loss in plasmas containing light elements. A first stage fundamental collision database is assembled which contains electron-impact excitation, ionization, and recombination rate coefficients for all atomic levels of all ion stages of a particular element. Computationally intensive non-perturbative R-matrix with pseudo-states (RMPS) and time-dependent close-coupling (TDCC) methods are combined with perturbative distorted-wave (DW) methods to calculate electron-ion collision processes in the light element isonuclear sequences. A second stage effective collision database is then assembled which contains generalized collisional-radiative, photon emissivity, and radiative power loss coefficients using the ADAS collisional-radiative package[1]. The reduction, from the large all excited levels database of stage one to the relatively small ground and metastable levels database of stage two, greatly reduces the computational effort needed in the third stage plasma modelling of light element transport and spectral line emission, such as the current studies on Li and Be at DIII-D and JET. These tokamak studies[2] are testing critical design issues for ITER.

CP874, *Spectral Line Shapes: 18th International Conference,*
edited by E. Oks and M. Pindzola

FUNDAMENTAL ATOMIC COLLISIONS

Electron Excitation of He

The R-matrix method[3] is based on expanding a time-independent wavefunction in atomic eigenfunctions:

$$\Psi(\vec{x}_1, ...\vec{x}_{N+1}) = \mathcal{A} \sum_{i,j} a_{ijk} \phi(\vec{x}_1, ...\vec{x}_N, \hat{r}_{N+1}\sigma_{N+1}) \frac{P_j(r_{N+1})}{r_{N+1}} + \sum_i b_{ik} \Phi(\vec{x}_1, ...\vec{x}_{N+1}) , \quad (1)$$

where $\mathcal{A}$ is an antisymmetrization operator, ϕ are N-electron target states coupled to the angular and spin functions of the scattered electron, Φ are $(N + 1)$-electron bound eigenfunctions needed for completeness, and $P(r)$ are continuum basis orbitals. The coefficients a_{ijk} and b_{ik} are determined by diagonalization of the $(N + 1)$-electron Hamiltonian. Recently, standard R-matrix and R-matrix with pseudo-states calculations were carried out for the electron-impact excitation of the neutral He atom[4]. Excitation rate coefficients for the $1s^2\,{}^1S \rightarrow 1s4s\,{}^3S$ and the $1s^2\,{}^1S \rightarrow 1s4s\,{}^1S$ transitions are shown in Figure 1. A substantial reduction in the rate coefficient is found for both transitions in going from a 29 LS-term standard R-matrix calculation to a 101 LS-term RMPS calculation.

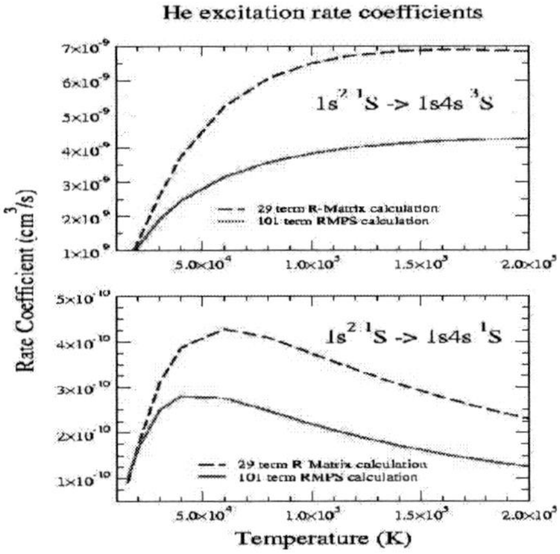

FIGURE 1. Electron-impact excitation rate coefficients for Helium.

Electron Ionization of H

For two-electron systems, the time-dependent close-coupling method[5, 6] is based on expanding a time-dependent wavefunction in coupled spherical harmonics:

$$\Psi(\vec{r}_1, \vec{r}_2, t) = \sum_{l_1, l_2} \frac{P^L_{l_1 l_2}(r_1, r_2, t)}{r_1 r_2} W[(l_1, l_2)L] \ , \tag{2}$$

where L is the total angular momentum. From projection onto the time-dependent Schrodinger equation, a set of time-dependent close-coupled partial differential equations are used to solve for the $P^L_{l_1 l_2}(r_1, r_2, t)$ radial functions.

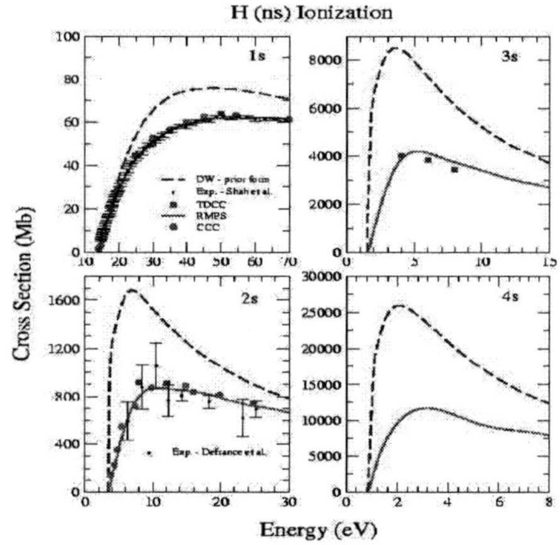

FIGURE 2. Electron-impact ionization cross sections for Hydrogen.

Recently, time-dependent close-coupling, R-matrix with pseudo-states, and distorted-wave calculations were carried out for the electron-impact ionization of the neutral H atom[7]. Ionization cross sections for the $1s$, $2s$, $3s$, and $4s$ states are shown in Figure 2. The RMPS calculations (solid curves) are found to be in excellent agreement with the TDCC calculations (solid squares). On the other hand, the DW calculations (dashed curves) are found to become increasingly inaccurate as one moves to the ionization of the more highly excited states.

Electron Double Ionization of H$^-$

For three-electron systems, the time-dependent close-coupling method[8] is based on expanding a time-dependent wavefunction in coupled spherical harmonics:

$$\Psi(\vec{r}_1, \vec{r}_2, \vec{r}_3, t) = \sum_{l_1, l_2, L, l_3} \frac{P^{\mathcal{L}}_{l_1 l_2 L l_3}(r_1, r_2, r_3, t)}{r_1 r_2 r_3} W[((l_1, l_2)L, l_3)\mathcal{L}] \ , \tag{3}$$

where $\mathcal{L}$ is the total angular momentum. From projection again onto the time-dependent Schrodinger equation, a set of time-dependent close-coupled partial differential equations are used to solve for the $P^{\mathcal{L}}_{l_1 l_2 L l_3}(r_1, r_2, r_3, t)$ radial functions.

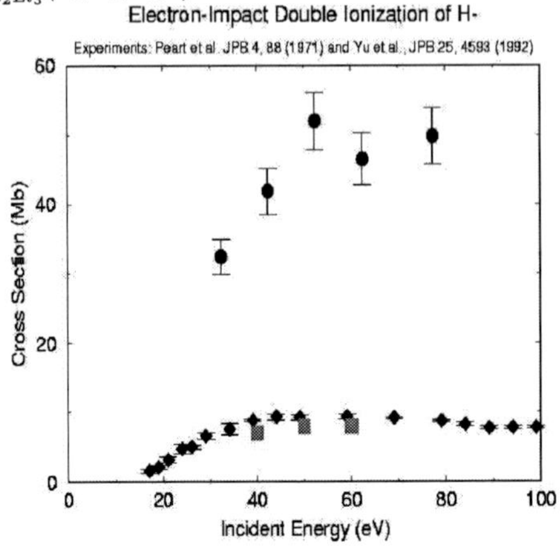

FIGURE 3. Electron-impact double ionization cross sections for H^-.

Recently, time-dependent close-coupling calculations were carried out for the electron-impact double ionization of the H^- atomic ion[9]. The TDCC calculations (solid squares) shown in Figure 3 are found to be more than a factor of five below the experimental measurements of Peart et al.[10] and in excellent agreement with the experimental measurements of Yu et al.[11].

Electron Recombination of O^{4+}

Recently, experimental measurements using the CRYRING storage ring and perturbative distorted-wave calculations were carried out for electron recombination of several Be-like atomic ions[12]. Cross sections for O^{4+} are shown in Figure 4. Theory and experiment are generally in good agreement for the many dielectronic recombination resonances leading up to the $2s2p\ ^1P$ excitation threshold at 19.7 eV. In addition, several trielectronic recombination resonances are seen in the spectrum, including an especially strong resonance at 60 meV.

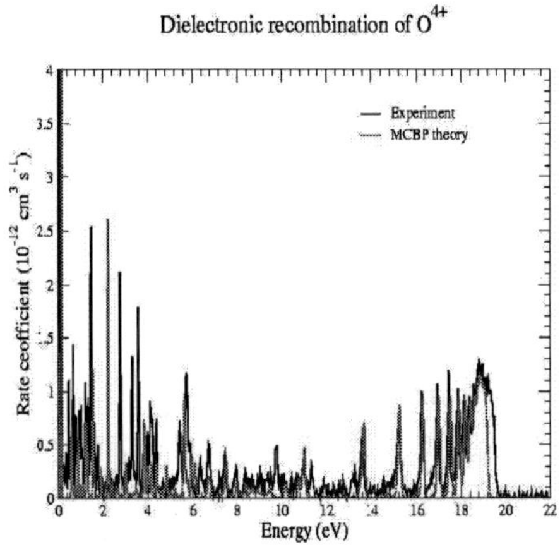

FIGURE 4. Electron-impact recombination cross sections for O^{4+}.

GENERALIZED COLLISIONAL-RADIATIVE COEFFICIENTS

The first stage fundamental atomic collision data presented in Section II can be used directly to calculate term populations via the collisional-radiative equations:

$$\begin{aligned}\frac{dN_j^z}{dt} &= -\sum_{j'<j} A_{j\to j'} N_j^z - N_e \sum_{j'\neq j} q_{j\to j'} N_j^z \\ &+\sum_{j'>j} A_{j'\to j} N_{j'}^z + N_e \sum_{j'\neq j} q_{j'\to j} N_{j'}^z \\ &-N_e \sum_k \mathcal{S}_{j\to k} N_j^z + N_e \sum_k \mathcal{R}_{k\to j} N_k^{z+1} \ , \qquad (4)\end{aligned}$$

where $A_{j\to j'}$ are radiative decay rates, $\mathcal{S}_{j\to k}$ are electron-impact ionization rates, $\mathcal{R}_{k\to j}$ are electron-impact recombination rates, and N_e is the electron density.

For many plasma conditions, the ground and metastable term populations evolve on a much slower timescale than the excited term populations, allowing the time derivative of the excited term populations to be set to zero. In the quasi-static equilibrium approximation[1], the ground and metastable term populations can be calculated via the collisional-radiative equations:

$$\frac{dN_\beta^z}{dt} = -N_e \sum_{\beta'\neq\beta} (X_{\beta\to\beta'} + Q_{\beta\to\beta'}) N_\beta^z$$

$$+N_e\sum_{\beta'\neq\beta}(X_{\beta'\to\beta}+Q_{\beta'\to\beta})N^z_{\beta'}$$
$$-N_e\sum_{\gamma}S_{\beta\to\gamma}N^z_{\beta}+N_e\sum_{\alpha}S_{\alpha\to\beta}N^{z-1}_{\alpha}$$
$$+N_e\sum_{\gamma}R_{\gamma\to\beta}N^{z+1}_{\gamma}-N_e\sum_{\alpha}R_{\beta\to\alpha}N^z_{\beta}\ . \tag{5}$$

The cross-coupling rate coefficients are given by:

$$X_{\beta\to\beta'}=(\mathcal{C}^z_{\beta'\beta}-\sum_j\mathcal{C}^z_{\beta'j}\sum_{j'}(\mathcal{C}^z_{jj'})^{-1}\mathcal{C}^z_{j'\beta})/N_e, \tag{6}$$

and the parent cross-coupling rate coefficients are given by:

$$Q_{\beta\to\beta'}=N_e\sum_i\mathcal{S}_{i\to\beta'}\sum_{i'}(\mathcal{C}^{z-1}_{ii'})^{-1}\mathcal{R}_{\beta\to i'}\ , \tag{7}$$

where the collisional-radiative matrix elements for ion stage z are given by:

$$\mathcal{C}^z_{jj}=\sum_{j'<j}A_{j\to j'}+N_e\sum_{j'\neq j}q_{j\to j'}, \tag{8}$$

$$\mathcal{C}^z_{jj'}=-A_{j'\to j}-N_e q_{j'\to j}\qquad for\quad j'>j, \tag{9}$$
$$\mathcal{C}^z_{jj'}=-N_e q_{j'\to j}\qquad for\quad j'<j. \tag{10}$$

The generalized ionization rate coefficients are given by:

$$S_{\beta\to\gamma}=\mathcal{S}_{\beta\to\gamma}-\sum_j\mathcal{S}_{j\to\gamma}\sum_{j'}(\mathcal{C}^z_{jj'})^{-1}\mathcal{C}^z_{j'\beta}, \tag{11}$$

$$S_{\alpha\to\beta}=\mathcal{S}_{\alpha\to\beta}-\sum_i\mathcal{S}_{i\to\beta}\sum_{i'}(\mathcal{C}^{z-1}_{ii'})^{-1}\mathcal{C}^{z-1}_{i'\alpha}. \tag{12}$$

The generalized recombination rate coefficients are given by:

$$R_{\gamma\to\beta}=\mathcal{R}_{\gamma\to\beta}-\sum_j\mathcal{C}^z_{\beta j}\sum_{j'}(\mathcal{C}^z_{jj'})^{-1}\mathcal{R}_{\gamma\to j'}, \tag{13}$$

$$R_{\beta\to\alpha}=\mathcal{R}_{\beta\to\alpha}-\sum_i\mathcal{C}^{z-1}_{\alpha i}\sum_{i'}(\mathcal{C}^{z-1}_{ii'})^{-1}\mathcal{R}_{\beta\to i'}. \tag{14}$$

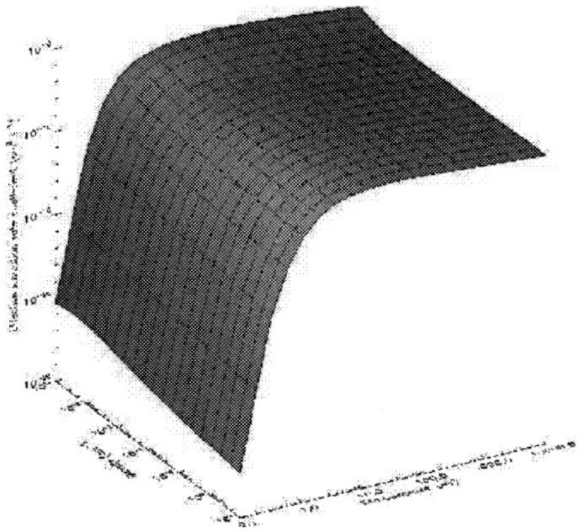

FIGURE 5. Generalized ionization rate coefficient for Li.

Our second stage effective collision database contains the generalized collisional-radiative (GCR) coefficients. These GCR coefficients are a function of both temperature and density. Recently, the GCR coefficients for all ion stages of Li were calculated[13] based on previous RMPS, TDCC, and DW calculations for the fundamental atomic collision database. The generalized ionization coefficient for Li is shown in Figure 5, while the generalized recombination coefficient for Li^+ is shown in Figure 6. We note that it is these GCR coefficients that are most useful in plasma transport codes, since they contain the effects of the excited states on the ground and metastables, without the need for a large number of coefficients to be archived.

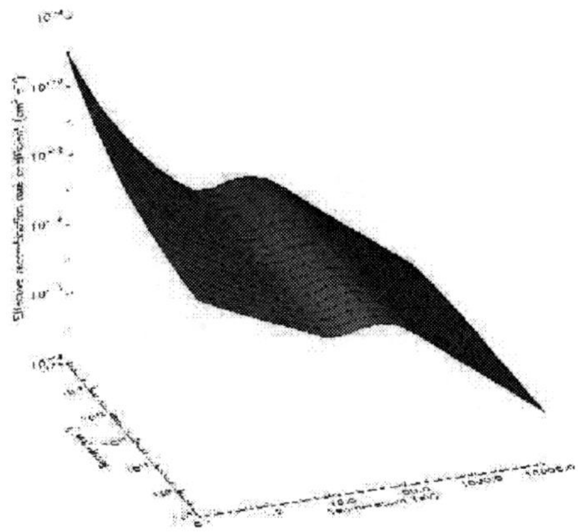

FIGURE 6. Generalized recombination rate coefficient for Li^+.

SPECTRAL LINE EMISSION

The radiated power loss function from a single excited term, due to all possible bound-bound spectral line emissions, is given by:

$$RPL_j^z = \sum_{j'<j} \frac{A^z_{j\to j'} \Delta E^z_{jj'} N^z_j}{N_e N_{tot}} , \tag{15}$$

where $\Delta E^z_{jj'}$ is the transition energy. On the other hand, the total radiated power loss function from an ionization stage, due to spectral line emission, is given by:

$$RPL^z = \sum_{\beta} \frac{P^z_{exc,\beta} N^z_\beta}{N_{tot}} + \sum_{\gamma} \frac{P^z_{rec,\gamma} N^{z+1}_\gamma}{N_{tot}} . \tag{16}$$

Note that to use this last equation, one needs to know the fractional abundances of each of the ion stages, either from an ionization balance calculation, or from an impurity transport calculation. The generalized excitation power loss coefficients are given by:

$$P^z_{exc,\beta} = \sum_j \sum_{j'<j} \frac{A^z_{j\to j'} \Delta E^z_{jj'}}{N_e} \left[- \sum_{j''} \mathcal{C}^{-1}_{jj''} \mathcal{C}_{j''\beta} \right] , \tag{17}$$

and the generalized recombination power loss coefficients are given by:

$$P^z_{rec,\gamma} = \sum_j \sum_{j'<j} \frac{A^z_{j\to j'} \Delta E^z_{jj'}}{N_e} \left[N_e \sum_{j''} \mathcal{C}^{-1}_{jj''} \mathcal{R}_{j''\gamma} \right] . \tag{18}$$

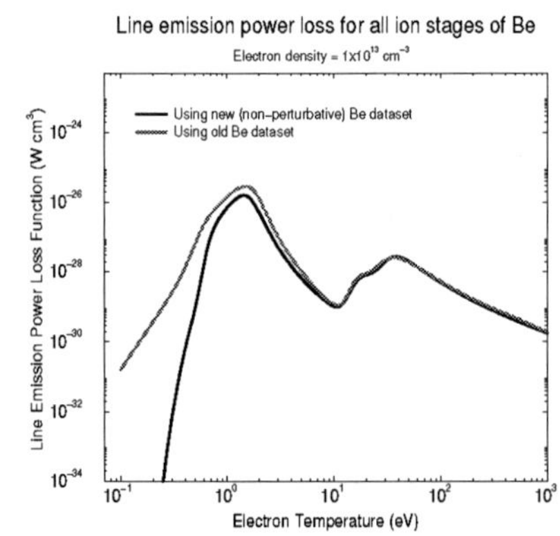

FIGURE 7. Generalized excitation power loss coefficient for Be.

Recently, the GCR ionization, recombination, and power loss coefficients for all ion stages of Be were calculated[14] based on RMPS, TDCC, and DW calculations for the fundamental atomic collision database. The total radiated power loss function for all ion stages of Be at $N_e = 1.0 \times 10^{13}$ cm^{-3} is shown in Figure 7. The ion stage abundances $\frac{N_\beta^z}{N_{tot}}$ needed in Eq.(16) are determined by solving Eq.(5), after setting $\frac{dN_\beta^z}{dt} = 0$. An older ADAS dataset yields a radiated power loss function that is always higher than the new ADAS dataset, especially at low temperatures. The peaks in the curves around 1-2 eV are due to Be and Be^+, while the peaks in the curves around 10-20 eV are due to Be^{2+} and Be^{3+}.

SUMMARY

In the future, we plan to complete RMPS calculations for the electron-impact excitation of C, C^+, and C^{4+}, which when combined with earlier work on C^{2+}[15], C^{3+}[16], and C^{5+}[17], will help us complete a second stage GCR database for all ion stages of C.

ACKNOWLEDGEMENTS

This work was supported in part by grants from the U.S. Department of Energy to Auburn University, Rollins College, Oak Ridge National Laboratory, and Los Alamos National Laboratory. Computational work was carried out at the National Energy Research Scientific Computing Center in Oakland, California, and at the National Center for Computational Sciences in Oak Ridge, Tennessee.

References

[1] H. P. Summers and M. G. O'Mullane, "The Atomic Data and Analysis Structure (ADAS)", in *Nuclear Fusion Research*, eds. R. E. H. Clark and D. H. Reiter, (Springer, 2005), p 399.

[2] D. G. Whyte, T. E. Evans, C. P. C. Wong, W. P. West, R. Bastasz, J. P. Allain, and J. N. Brooks, Fusion Engineering and Design **72**, 133 (2004).

[3] P. G. Burke and K. A. Berrington, *Atomic and Molecular Processes: an R-Matrix Approach*, (IOP Press, 1993).

[4] C. P. Ballance, D. C. Griffin, S. D. Loch, R. F. Boivin, and M. S. Pindzola, Phys. Rev. A **74**, 012719 (2006).

[5] M. S. Pindzola and D. R. Schultz, Phys. Rev. A **53**, 1525 (1996).

[6] M. S. Pindzola and F. Robicheaux, Phys. Rev. A **54**, 2142 (1996).

[7] D. C. Griffin, C. P. Ballance, M. S. Pindzola, F. Robicheaux, S. D. Loch, J. A. Ludlow, M. C. Witthoeft, J. Colgan, C. J. Fontes, and D. R. Schultz, J. Phys. B **38**, L199 (2005).

[8] M. S. Pindzola, F. Robicheaux, J. P. Colgan, M. C. Witthoeft, and J. A. Ludlow, Phys. Rev. A **70**, 032705 (2004).

[9] M. S. Pindzola, F. Robicheaux, and J. P. Colgan, J. Phys. B **39**, L127 (2006).

[10] B. Peart, D. S. Walton, and K. T. Dolder, J. Phys. B **4**, 88 (1971).

[11] D. J. Yu, S. Rachafi, J. Jureta, and P. Defrance, J. Phys. B **25**, 4593 (1992).

[12] M. Fogle, N. R. Badnell, P. Glans, S. D. Loch, S. Madzunkov, Sh. A. Abdel-Naby, M. S. Pindzola, and R. Schuch, Astron. and Astrophys. **442**, 757 (2005).

[13] S. D. Loch, J. Colgan, M. C. Witthoeft, M. S. Pindzola, C. P. Ballance, D. M. Mitnik, D. C. Griffin, M. G. O'Mullane, N. R. Badnell, and H. P. Summers, accepted in Atomic Data and Nuclear Data Tables (2006).

[14] S. D. Loch, M. S. Pindzola, C. P. Ballance, D. C. Griffin, J. P. Colgan, N. R. Badnell, M. G. O'Mullane, and H. P. Summers, submitted to Atomic Data and Nuclear Data Tables (2006).

[15] D. M. Mitnik, D. C. Griffin, C. P. Ballance, and N. R. Badnell, J. Phys. B **36**, 717 (2003).

[16] D. C. Griffin, N. R. Badnell, and M. S. Pindzola, J. Phys. B **33**, 1013 (2000).

[17] C. P. Ballance, N. R. Badnell, and E. S. Smyth, J. Phys. B **36**, 3707 (2003).

Modeling the Spectral Emission from Tungsten in Tokamaks

S. D. Loch*, M. S. Pindzola*, C. P. Ballance†, D. C. Griffin†, A. D. Whiteford‡, and T. Pütterich◇

* *Department of Physics, Auburn University, Auburn, AL 36849*
† *Department of Physics, Rollins College, Winter Park, FL32789*
‡ *Department of Physics, University of Strathclyde, Glasgow, G4 0NG, UK*
◇ *Max-Planck-Institut für Plasmaphysik, EURATOM Association, D-85748, Germany*

Abstract. We report on recent efforts aimed at improving the atomic data that is used in modeling tungsten spectral emission from tokamak plasmas. Results are shown from recent Dirac R-matrix electron impact excitation calculations of W^{46+} and from configuration-average distorted-wave electron impact ionization calculations. Collisional-radiative calculations are performed using the new data, and the results compared with those obtained using plane-wave Born electron impact excitation data.

Keywords: atomic, collisions, collisional-radiative
PACS: 34

INTRODUCTION

Tungsten is an important element in ITER design, making up most of the wall material at the plasma strike points. As such there has been much interest in the radiative and transport properties of tungsten ions in tokamak plasmas. Experimental studies have included those performed on the ASDEX-upgrade tokamak [1, 2], and measurements made on the JET tokamak [3]. There have also been a range of Electron Beam Ion Trap (EBIT) measurements for tungsten [4, 5, 6]. There is a particular interest in the emission of W^{44+}, W^{45+} and W^{46+}, with spectral lines from these ions being potential temperature and density diagnostics. The W^{46+} ion stage is especially promising as it exists over a broad temperature region, existing from about 1 keV through to about 5 keV.

In order to model the spectra from tungsten one needs large quantities of atomic data, in particular electron impact excitation, ionization and recombination data. Recently electron impact ionization data was calculated for all ion stages of tungsten in the configuration-average distorted-wave approximation [7]. The ADAS project [8] recently calculated plane-wave Born electron impact excitation data for all ion stages of tungsten. Currently the main recombination data used for tungsten modeling is semi-empirical in nature, either coming from the average atom model [9] or from the Burgess-Bethe general program [10].

CP874, *Spectral Line Shapes: 18th International Conference,*
edited by E. Oks and M. Pindzola

In this paper we report on some of the recent efforts aimed at improving the atomic data for tungsten ions. In the next section we describe the collisional-radiative theory which will be used to generate spectrally observable quantities. There then follows a section on atomic data where we present some of the new ionization data, and comment on recent comparisons at ASDEX-upgrade where the data was used. The atomic data section also describes recent Dirac R-matrix calculations for the electron impact excitation data for W^{46+}. In our section on predicted spectra we use the latest atomic data and a collisional-radiative model to produce a range of W^{46+} spectra, and compare with spectra produced using existing plane-wave Born excitation data. Finally, in our summary section we present our conclusions, and describe some areas of future work.

THEORY

Once all of the atomic data is assembled for tungsten one can evaluate the spectral emission for various plasma conditions. We use the quasi-static equilibrium approximation [11, 12] to solve the collisional-radiative equations, using the ADAS suite of codes [8]. In this approximation it is assumed that the ground and metastable levels of each ion stage evolve on a much slower timescale than the excited levels do. The excited levels are considered to be in equilibrium with the ground and metastable populations. Thus, one evaluates the ground and metastable populations via either a transport code calculation, or via an ionization balance calculation. If one considers only the ground states of each ion, and omits the metastables, the ionization balance equations are given by

$$\frac{dN_\sigma}{dt} = -N_e\left[\alpha_{\sigma\to\mu} + S_{\sigma\to\nu}\right]N_\sigma + N_e\alpha_{\nu\to\sigma}N_\nu + N_eS_{\mu\to\sigma}N_\mu, \tag{1}$$

where σ is the ground state of a certain ion stage, ν is the ground state of the next higher ion stage and μ is the ground state of the next lower ion stage. We note that one of these equations exist for each ion stage. To include the effects of metastable levels, one has to modify Eqn. (1) to include the metastable and parent metastable cross coupling coefficients. These are described in more detail by Summers et al. [12]. A further improvement can be made to Eqn. (1) by including the effects of the excited states via the use of generalised collisional-radiative coefficients, rather than just the direct ionization and recombination coefficients. These coefficients include effects such as stepwise ionization via the excited states. However, since the ionization data we will report on is just ground configuration to ground configuration we will use only Eqn. (1) when evaluating our ionization balance results. Extension of this work to include the metastable and excited levels is part of the ongoing work of this project.

One can solve for the equilibrium ground and metastable populations by setting all of the rates of change of populations in Eqn. (1) to zero. An example set of ionization balance results is shown in Fig. (1).

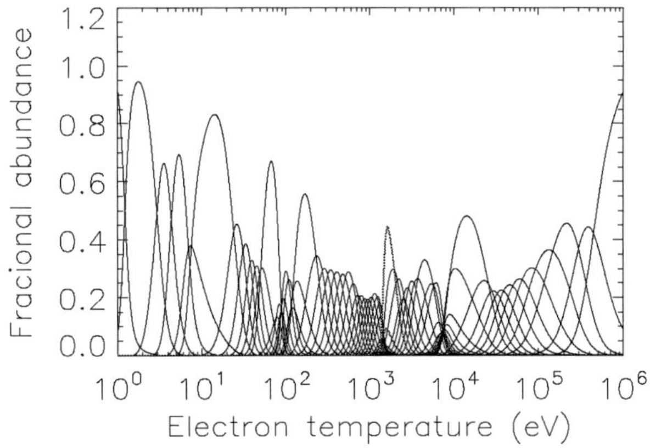

FIGURE 1. Fractional abundance results for all ion stages of tungsten for T_e=5000eV, and $N_e=1\times10^8$ cm^{-3}. Ion stages W^{44+}, W^{45+} and W^{46+} are shown by the lighter lines.

Using the quasi-static approximation one can evaluate the excited populations of each ion stage as a function of the ground and metastable populations. Thus, the excited populations are given by

$$N_j = F^{ex}_{j\sigma} N_e N_\sigma + F^{rec}_{j\nu} N_e N_\nu, \tag{2}$$

where $F^{ex}_{j\sigma}$ and $F^{rec}_{j\nu}$ are the effective contributions to the excited populations due to excitation from the ground and recombination from the next ion stage, and are functions of collisional-radiative matrix elements. The F coefficients are described in more detail by Summers et al. [12]. If metastables were included one needs to modify the above equation to include summations over the ground and metastable contributions. Once the upper level populations have been determined, an emissivity can be evaluated for each spectral line. The ADAS codes archive photon emissivity excitation and recombination coefficients given by

$$PEC^{ex}_{\sigma,j\to k} = A_{j\to k} F^{ex}_{j\sigma} \tag{3}$$

and

$$PEC^{rec}_{\sigma,j\to k} = A_{j\to k} F^{rec}_{j\sigma}. \tag{4}$$

Thus, multiplying these PEC^{ex} and PEC^{rec} coefficients by N_e N_σ and N_e N_ν respectively will return the emissivity of a spectral line. These photon emissivity coefficients can be grouped into spectral features, containing many spectral lines. These coefficients are referred to as feature photon emissivity coefficients and are evaluated from sums of individual PEC coefficients which have had a normalised line profile applied to them, to reflect the spectral line shape. In all of the work presented here the line profiles are assumed to be Doppler profiles.

In this paper we solve for the equilibrium populations. However, we note that transport effects may well be significant for tungsten ions. Thus, our present aim is to investigate the emission characteristics of tungsten, to determine its sensitivity to the quality of the underlying atomic data, and try to identify potential diagnostic spectral lines. Once lines have been identified then the atomic data can be used within transport codes which account properly for transport effects, and comparisons with observed tokamak spectra made.

ATOMIC DATA

Ionization data

Electron impact ionization data was recently calculated for all ion stages of tungsten using the configuration-average distorted-wave method [7]. This data was found to be reasonably accurate for ion stages greater than W^{5+}, when comparisons were made with crossed-beams experimental cross section measurements. Below W^{5+} the CADW results were found to overestimate the ionization cross sections, as usually found from a perturbation theory approach. The good agreement for the higher ion stages can be seen in Fig. (2) where results are shown for W^{9+}. We note that experimental cross section measurements do not exist above W^{10+}, however, we expect the CADW method to produce good results. The data of Loch et al. [7] includes the effects of radiative branching of the autoionizing configurations, which was found to be important above W^{38+}. Above W^{63+} a fully relativistic distorted-wave method was necessary.

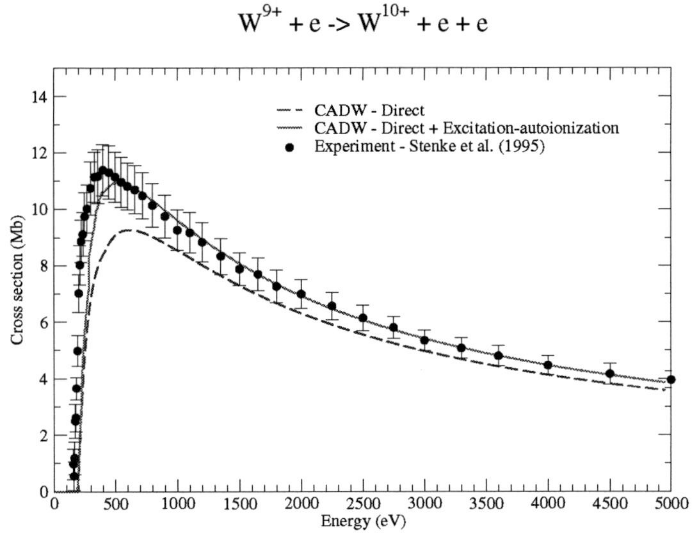

FIGURE 2. The electron impact ionization cross section of W^{9+}. The measurements of Stenke et al. (1995) [17] are also shown. The dashed line shows the direct ionization contribution, and the solid line shows the total direct plus the total excitation-autoionization contribution.

We note that the new ionization data has been used in recent modeling work at ASDEX-upgrade [2]. The new ionization data was found to provide significantly better agreement with experimental measurements on the ASDEX-upgrade tokamak of the measured ion stage abundances, when compared with the results obtained using older semi-empirical ionization data.

Excitation data

Dirac *R*-matrix calculations have recently been completed for W^{46+} [13]. The Dirac *R*-matrix calculation was based on fully relativistic structure and scattering calculations. The calculation contained 115 levels, and is described in more detail in the work of Ballance et al. [13]. The calculation included the effects of radiation damping of the resonant features. It was found that although radiation damping is very important in highly ionized species, its effects were reduced in this case due to the closed-shell ground state of W^{46+}.

This new data presents a significant improvement in data quality compared with the existing plane-wave Born excitation data that is currently used in many modeling codes. The main purpose of this paper is to assess the differences that this new excitation data makes to spectral emission quantities.

PREDICTED SPECTRA

We wished to determine whether the new W^{46+} excitation data would make a significant difference to spectral modeling of W^{46+} emission. Thus, we used the ADAS suite of codes [8] to evaluate feature photon emissivity coefficients for W^{46+}, via Eqn. (2), for the 5-9 $\AA$ wavelength range. Feature photon emissivity coefficients were evaluated for a range of temperatures and densities.

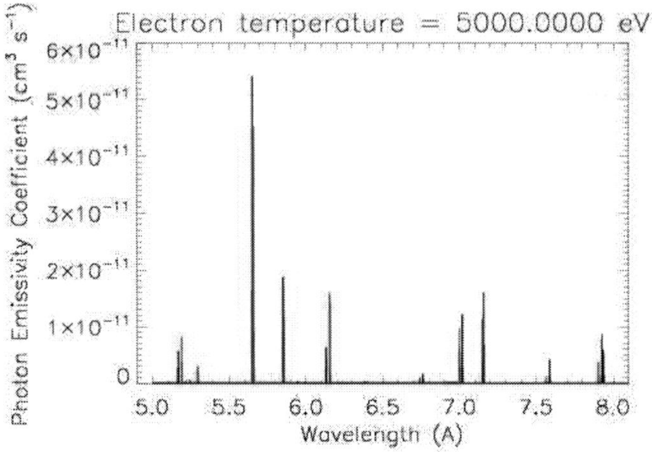

FIGURE 3. The feature photon emissivity coefficient for W^{46+} in the 4.5-8.5$\AA$ range. The black lines show the results using the Dirac R-matrix data, and the lighter lines show the results using the PWB excitation data.

An example of our results is shown in Figs. (3) and (4) for W^{46+} at T_e=5000 eV and N_e=1 × $10^8 cm^{-3}$. We show feature PEC coefficients using both the PWB and the Dirac R-matrix excitation datasets. In general it was found that the PWB data produced good agreement in the wavelength positions of all the strongest spectral lines. The structure used in the PWB data was from the semi-relativistic Cowan code[14], while the Dirac R-matrix data used structure results from the relativistic GRASP code [15, 16]. The PWB data was able to reproduce all of the main features seen using the Dirac R-matrix data within about 30% of the line height. Thus, it is hoped that the more accurate Dirac R-matrix data might allow for an accurate T_e or N_e diagnostic to be developed. One feature of particular interest is the strong line which is observed in ASDEX-upgrade at 7.95 Å. The strength of this line could not be matched by any theoretical results in the work of Pütterich [2]. In our work we see strong emission at this wavelength due to two strong lines. One of these is a magnetic octopole, and is the decay from the first excited level down to the ground. As such it can build up a large population, leading to significant emission. This spectral line is not included at all in the existing PWB datasets. We note that we only predict strong emission of this line at reasonably low densities ($N_e < 1 \times 10^{12} cm^{-3}$), and the line becomes significantly weaker at higher densities. This suggests that the octopole may not be the reason for the strong line observed in the ASDEX experiment. Alternatively it may be that another populating mechanism is important for this ion, such as inner shell ionization from the W^{45+} ion stage. We are currently investigating this possibility.

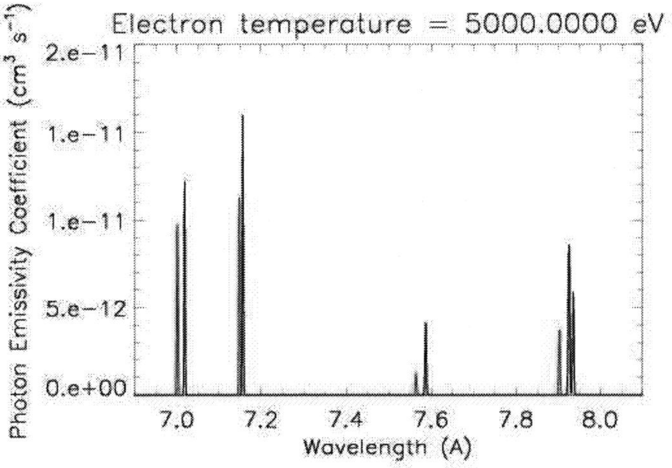

FIGURE 4. The feature photon emissivity coefficient for W^{46+} from 6.9-8.1$\AA$. The black lines show the results using the Dirac R-matrix data, and the lighter lines show the results using the PWB excitation data. Note the presence of the two stong spectral lines near 7.95 $\AA$ in the Dirac R-matrix results.

SUMMARY

We have reported on recent efforts aimed at improving the quality of atomic data used in modeling tungsten emission. Recently published electron impact ionization data has been found to improve the modeling of the ionization balance modeling of tungsten in tokamak plasmas. We also presented results from recent Dirac R-matrix calculations for the electron impact excitation of W^{46+}. This new data was found to produce spectra which agreed with the line position from the PWB datasets currently in use, but differed from the PWB line heights by about 30%. We hope that the more accurate Dirac R-matix excitation data might allow some spectral diagnostics to be developed. We are currently investigating whether alternative populating mechanisms, such as inner shell ionization, might be significant and need to be included in our future collisional-radiative modeling.

ACKNOWLEDGEMENTS

This work was supported by a grant for Scientific Discovery through Advanced Computing by the U.S. Department of Energy. The computational work was carried out at the National Energy Research Scientific Computing Center in Oakland, California.

References

[1] R. Neu, R. Dux, A. Kallenbach, T. Putterich, M. Balden, J. C. Fuchs, A. Herrmann, C. F. Maggi, M. O'Mullane, R. Pugno, I. Radivojevic, V. Rohde, A. C. C. Sips, W. Suttrop, A. Whiteford, and the ASDEX Upgrade team, Nucl. Fusion **45**, 209 (2005).

[2] T. Puetterich, R. Neu, C. Biedermann, R. Radtke, R. and the ASDEX Upgrade Team J. Phys. B **38** 3071 (2005)

[3] M. G. O'Mullane, H. P. Summers, A. D. Whiteford, R. Barnsley, I. H. Coffey, G. Counsell, S. Loch, and Contributors to the EFDA-JET Work Program, Review of Scientific Instruments **74** 2080 (2003).

[4] S. B. Utter, P. Beiersdorfer, E. Träbert, Can. J. Phys. **80** 1503 (2002)

[5] R. Radtke, C. Biedermann, J. L. Schwob, P. Mandelbaum, P. R. Doron, Phys. Rev. A **64** 012720 (2001)

[6] T. Pütterich, R. Neu, C. Biedermann, R. Radtke, R. and the ASDEX Upgrade Team J. Phys. B **38** 3071 (2005)

[7] S. D. Loch, J. A. Ludlow, M. S. Pindzola, A. D. Whiteford, and D. C. Griffin, Phys. Rev. A **72** 052716 (2005)

[8] http://adas.physics.auburn.edu

[9] D. Post, R. Jensen, C. Tarter, W. Grasberger, and W. Lokke, At. Data Nucl. Data Tables **20**, 397 (1977)

[10] N. R. Badnell, M. G. O'Mullane, H. P. Summers, Z. Altun, M. A. Bautista, J. Colgan, T. W. Gorczyca, D. M. Mitnik, M. S. Pindzola, and O. Zatsarinny, Astronomy and Astrophysics **406**, 1151-1165 (2003).

[11] D. R. Bates, A. E. Kingston, and R. W. P. McWhirter, Proc. R. Astron. Soc. London, Ser. A. **267**, 297 (1962)

[12] H. P. Summers, W. J. Dickson, M. G. O'Mullane, N. R. Badnell, A. D. Whiteford, D. H. Brooks, J. Lang, S. D. Loch, and D. C. Griffin, Plasma Phys. Control. Fusion **48**, 263 (2006)

[13] C. P. Ballance and D. C. Griffin, J. Phys. B *in press* (2006)

[14] R. D. Cowan, *The Theory of Atomic Structure and Spectra*, (U. Calif. Press, Berkeley, 1981).

[15] K.G. Dyall, I. P. Grant, C. T. Johnson, F. A. Parpia, E. P. Plummer, Comput. Phys. Commun. **55** 424 (1989)

[16] F. A. Parpia, C. F. Fischer, and I. P. Grant, Comput. Phys. Commun. **94** 249 (1996)

[17] M. Stenke, K. Aichele, D. Hathiramani, G. Hofmann, M. Steidl, R. Volpel, and E. Salzborn, J. Phys. B **28** 2711 (1995)

Use of Quasienergy States for Laser-Aided Diagnostics of Oscillatory Electric Fields in Plasmas

V.P. Gavrilenko[1], E. Oks[2]

[1]*Center for Surface and Vacuum Research, Novatorov Street 40, korp.1, Moscow 119421, Russia*
[2]*Physics Department, 206 Allison Lab., Auburn University, Auburn, AL 36849, USA*

Abstract. Methods for the laser-assisted diagnostics of oscillatory electric fields (OEFs) in plasmas are considered. It is shown that several phenomena can be used for this purpose. First, a tunable laser radiation can scan both the structure of energy levels and spectral line profiles of atoms interacting with the OEFs. Second, the laser radiation can cause a saturation of the atomic transition, including the cases where the saturation parameter depends on characteristics of the OEFs. Third, the laser radiation itself can modify both the structure of energy levels and spectral line profiles in such a way that the modification would depend, in particular, on parameters of the OEFs.

Keywords: Stark effect, plasma spectroscopy, plasma diagnostics, electric fields
PACS: 32.70Jz, 32.60.+i, 32.80.-t, 52.70.Kz

INTRODUCTION

Developing methods for diagnosing oscillatory electric fields (OEFs) $\vec{E}_M(t) = \vec{E}_{0M} \cos\omega_M t$ in plasmas is of a great practical importance: the OEFs control many physical processes in plasmas. The term OEF embraces both the intrinsic eigen-modes of plasmas (plasma turbulence) and extrinsic fields penetrating into plasmas from the outside. To the latter type belongs, e.g., the field of the microwave radiation used for heating in some magnetic fusion machines.

Currently for diagnosing OEFs in plasmas, there are widely employed methods based on the emission spectroscopy [1]. As an example, we note paper [2], where OEFs were revealed and studied in the peripheral plasma of the tokamak T-10 by using profiles of spontaneously emitted spectral lines of deuterium.

Bringing into play a laser radiation can significantly broaden the scope of spectroscopic diagnostics of OEFs in plasmas. The advantage is the possibility of measuring OEFs with a high temporal and spatial resolutions. In the laser-assisted diagnostics of the OEF $\vec{E}_M(t)$, atoms in a plasma interact with both the laser field $\vec{E}_L(t) = \vec{E}_{0L} \cos\omega_L t$ and with the OEF $\vec{E}_M(t)$. We assume that $\omega_M << \omega_L$ and that the frequency ω_L is close to the frequency of the atomic transition b etween the upper and lower energy levels.

CP874, *Spectral Line Shapes: 18th International Conference,*
edited by E. Oks and M. Pindzola

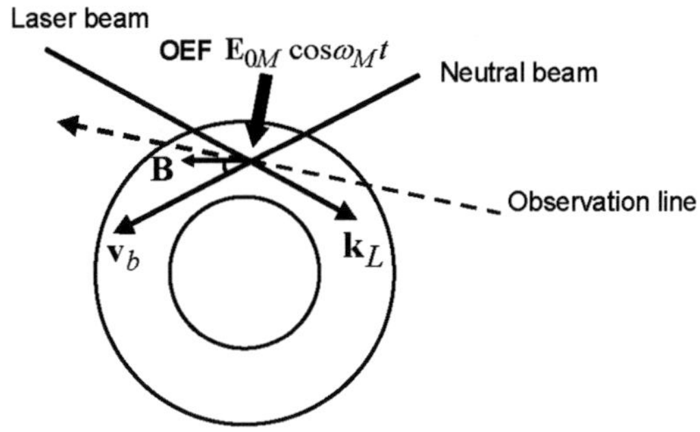

FIGURE 1. Geometry of the observations for laser-assisted diagnostics of oscillatory electric fields in tokamak plasmas.

In magnetic fusion machines, there are various ways for atoms to serve as sensitive "probes" with respect to the OEFs. First, atoms of the main filling gases (hydrogen, deuterium) can be used for diagnosing OEFs in the peripheral plasma. Second, for measuring OEFs in the core plasma one can inject neutral beams (hydrogen, deuterium, helium, litium). Besides, one can also inject pellets (containing, e.g., lithium). As an example, Fig. 1 shows a possible geometry for the laser-assisted diagnostics of OEFs in a tokamak, where atoms of the injected neutral beam play the role of the "probes". In this case, in addition to the OEF, one should also take into account the Lorentz electric field $\vec{E}_b = \vec{V}_b \times \vec{B}/c$, where $\vec{V}_b$ is beam velocity and $\vec{B}$ is the magnetic field.

In this paper we consider physical principles that can serve as the basis for laser-assisted spectroscopic methods for diagnosing OEFs in magnetic fusion plasmas.

ABSORPTION SPECTRA OF ATOMS WITHOUT PERMANENT DIPOLE MOMENTS

The interaction of an atom with the OEF $\vec{E}_M(t)$ can be characterized by the parameter

$$R = \left| d_{\alpha'\alpha''} E_{0M} / [2\hbar(\omega_{\alpha'\alpha''} \pm \omega_M)] \right| \tag{1}$$

where $d_{\alpha'\alpha''}$ is the matrix element of the dipole moment between the closely spaced levels α', α'', and $\omega_{\alpha'\alpha''}$ is the separation between these levels. If $R << 1$ (the case of a weak field $\vec{E}_M(t)$), the field $\vec{E}_M(t) = \vec{E}_{0M} \cos\omega_M t$ can be measured using the fact that the excitation rate for the two-quantum process, involving one quantum $\hbar\omega_L$ of the laser field $\vec{E}_L(t) = \vec{E}_{0L}\cos\omega_L t$ and one quantum $\hbar\omega_M$ of the field $\vec{E}_M(t) = \vec{E}_{0M}\cos\omega_M t$, is proportional to E_{0M}^2. Let us consider, for instance, the excitation scheme shown in Fig. 2, where $\omega_{23} << \omega_{21}$. We assume that in the electric dipole approximation, the single-photon transitions $1 \leftrightarrow 3$ and $2 \leftrightarrow 3$ are allowed, whereas the single-photon transition $1 \leftrightarrow 2$ is forbidden.

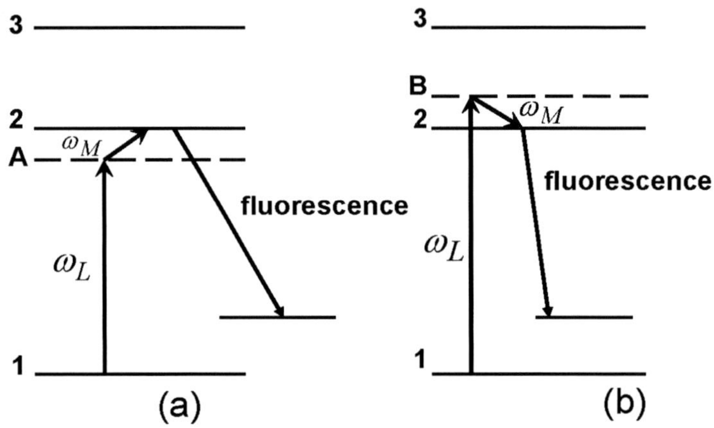

FIGURE 2. Partial scheme of energy levels of an atom. (a) The excitation of the transition $1 \rightarrow 2$ due to the absorption of a laser photon $\hbar\omega_L$ and of an OEF quantum $\hbar\omega_M$. (b) The excitation of the transition $1 \rightarrow 2$ due to the absorption of a laser photon $\hbar\omega_L$ and the emission of an OEF quantum $\hbar\omega_M$.

Using the quantum-mechanical perturbation theory (we assume that $R << 1$ for the levels 2 and 3), one can obtain the excitation rates for the two-quantum transitions shown in Fig. 1:

$$W_{1\to 2}^{(+)} = \{E_{0L}^2 E_{0M}^2 d_{13}^2 d_{32}^2 / [8\hbar^2(\omega_{32} + \omega_M)^2]\} L(\omega_{21} - \omega_L - \omega_M, \gamma),$$
$$W_{1\to 2}^{(-)} = \{E_{0L}^2 E_{0M}^2 d_{13}^2 d_{32}^2 / [8\hbar^2(\omega_{32} - \omega_M)^2]\} L(\omega_{21} - \omega_L + \omega_M, \gamma). \quad (2)$$

In Eqs. (2), $L(\omega,\gamma) \equiv (\gamma/2)/(\omega^2 + \gamma^2/4)$, γ is the width (FWHM) of the line. The value of $W_{1\to 2}^{(+)}$ corresponds to the process shown in Fig. 1(a), whereas the value of $W_{1\to 2}^{(-)}$ corresponds to the process shown in Fig. 1(b). The excitation rate for the direct single-photon transition $1 \rightarrow 3$ induced by the laser field is as follows:

$$W_{1\to3} = [d_{13}^2 E_{0L}^2 /(2\hbar^2)] L(\omega_{21} - \omega_L, \gamma) . \tag{3}$$

Figure 3 shows the absorption spectrum consisting of the lines $W_{1\to2}^{(\pm)}$ and $W_{1\to3}$ for the transitions $1 \to (2,3)$. One can see that both the ratios $W_{1\to2}^{(+)} / W_{1\to3}$ and $W_{1\to2}^{(-)} / W_{1\to3}$ are proportional to E_{0M}^2. Therefore, the ratios $W_{1\to2}^{(+)} / W_{1\to3}$ and $W_{1\to2}^{(-)} / W_{1\to3}$ can be used to measure E_{0M} by scanning the laser frequency around the frequencies ω_{21} and ω_{31} and observing the fluorescence signals at transitions originated from the levels 2 and 3.

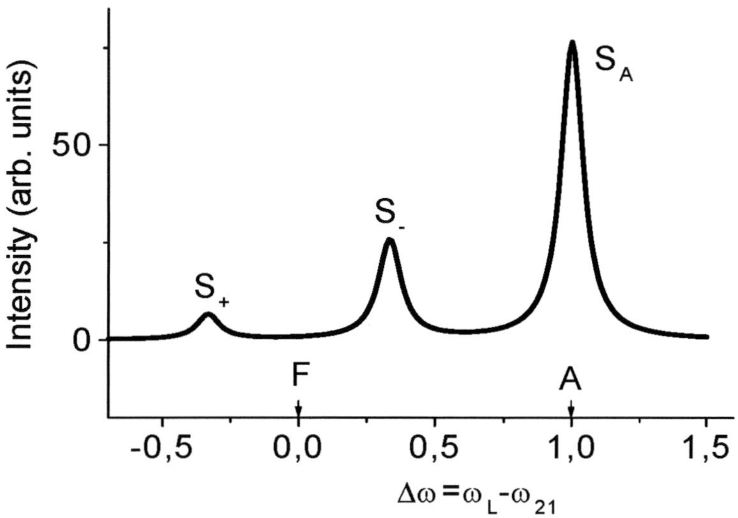

FIGURE 3. Absorption spectrum for the transitions $1 \to (2,3)$ shown in Fig. 2. $S_A \equiv W_{1\to3}$ is the intensity of the allowed line $1 \to 3$; $S_+ \equiv W_{1\to2}^{(+)}$ is the intensity of the line corresponding to the excitation scheme (a) ; $S_- \equiv W_{1\to2}^{(-)}$ is the intensity of the line corresponding to the excitation scheme (b) (cf. Fig. 2). The separation between the lines S_- and S_+ is equal to $2\omega_M$.

In the case where the parameter R in Eq. (1) is of the order or greater than unity, the formalism of quasienergy states (QES) is more appropriate for treating the interaction of the atom with the OEF $\vec{E}_M(t)$. According to the Floquet theory (see, e.g., Ref. [3]), for the atom interacting with the OEF $\vec{E}_M(t) = \vec{E}_{0M} \cos \omega_M t$, among the wave functions (WFs) of the atom, there exists a basis of the WFs of QES

$$\Phi_n(\vec{r},t) = \exp(-i\varepsilon_n t/\hbar) \sum_k \exp(-ik\omega_M t) \phi_{nk}(\vec{r}) \tag{4}$$

where ε_n is the quasienergy of the level n, and the functions $\phi_{nk}(\vec{r})$ are time independent. We note that the term "quasienergy states" was introduced in Refs. [4,5]. The quasienergies ε_n and the functions $\phi_{nk}(\vec{r})$ can be obtained either numerically, or analytically by solving the Schrödinger equation. Generally they depend on the parameters of the field $\vec{E}_M(t) = \vec{E}_{0M} \cos\omega_M t$. Taking into account Eq. (4), one can see that the excitation of the transition $b \to a$ can occur when the following resonance condition is satisfied

$$\omega_L \approx (\varepsilon_a - \varepsilon_b)/\hbar + k\omega_M, \qquad k = 0, \pm 1, \pm 2, \ldots$$

where ε_a and ε_b are the quasienergies of the levels a and b, respectively. Using Eq. (4), we find that for determining the excitation rate for the transition $b \to a$, we can use the formula (3), in which the dipole matrix element d_{13} should be replaced by the effective dipole matrix element $D_{ba}^{(k)} = \langle \varphi_b \mid d \mid \phi_{ak} \rangle$, where φ_b is the WF of the lower level b. (We assume here that strong the OEF $\vec{E}_M(t)$ perturbs only the upper state a.) As a result, the excitation rate for the transition $b \to a$ can be expressed as

$$W_{b\to a} = [|< \varphi_b \mid d \mid \phi_{ak} >|^2 \, E_{0L}^2 /(2\hbar^2)] L((\varepsilon_a - \varepsilon_b)/\hbar - \omega_L + k\omega_M, \gamma).$$

Let us assume that the levels b and a coincide, respectively, with the levels 1 and 2 shown in Fig. 2. In the case of a weak OEF $\vec{E}_M(t)$, according to Eq. (2) the ratio $W_{1\to 2}^{(+)}/W_{1\to 2}^{(-)}$ does not depend on E_{0M}. However, for the strong OEF $\vec{E}_M(t)$ this ratio is a function of the parameters of the OEF. Figure 4 shows, as an example, the ratio $W_{1\to 2}^{(+)}/W_{1\to 2}^{(-)}$ versus the amplitude of the OEF $\vec{E}_M(t)$, calculated for two two-quantum transitions in helium atoms: the transitions $2^1P \to 4^1F$ and $2^3P \to 4^3F$. For helium atoms, the levels $2^{2S+1}P$, $4^{2S+1}F$, and $4^{2S+1}D$ $(S = 0; 1)$ play the role of the levels 1, 2, and 3 shown in Fig. 2. Using the dependence of $W_{1\to 2}^{(+)}/W_{1\to 2}^{(-)}$ on the parameters of the OEF, one can measure these parameters in a plasma by observing the laser-induced fluorescence (LIF) signal at the transition from the level 2 to some lower level q. The advantage of this method is that both the fluorescent lines originate from the same level 2 - therefore, a possible collisional transfer of the population from the level 2 to nearby levels does not interfere with the measurements of the OEF.

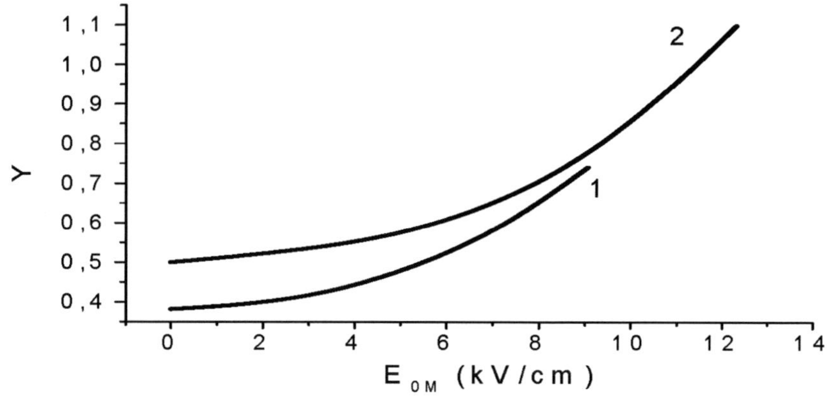

FIGURE 4. The ratio $Y = W^{(+)}_{1\to 2} / W^{(-)}_{1\to 2}$ versus the amplitude E_{0M} of the OEF for two two-quantum transitions in helium atoms: $2^1P \to 4^1F$ (curve 1) and $2^3P \to 4^3F$ (curve 2). The OEF frequency in this example ω_M was 2.4×10^{11} s^{-1}.

ABSORPTION SPECTRA OF ATOMS POSESSING PERMANENT DIPOLE MOMENTS

Let us consider laser-induced transitions in atoms possessing permanent dipole moments ($d_{aa} \equiv <\varphi_a | d | \varphi_a > \neq 0$). Hydrogen atoms are one of the examples of such atoms. In the case where $d_{aa} \neq 0$, the WF of a QES of the atom, which is in the state a and interacts with the OEF $\vec{E}_M(t) = \vec{E}_{0M} \cos\omega_M t$, has the form [6]

$$\Psi_a(\vec{r},t) = \exp[-id_{aa}E_{0M} \sin\omega_M t /(\hbar\omega_M)]\varphi_a(\vec{r}). \tag{5}$$

Let the frequency ω_L of the laser radiation be close to the frequency of the transition $b \leftrightarrow a$. Performing the Fourier series expansion in Eq. (5) and taking into account the formula $\exp(-ix\sin\omega_M t) = \sum_{n=-\infty}^{+\infty} J_n(x)\exp(-in\omega_M t)$, where $J_n(x)$ are the Bessel functions, we find that the effective matrix element of the dipole moment between the levels a and b has the form: $D^{(k)}_{ab} = J_k(\Delta\beta_{ab})d_{ab}$, where $\Delta\beta_{ab} \equiv (d_{aa} - d_{bb})E_{0M} /(\hbar\omega_M)$. Using the above expression for $D^{(k)}_{ab}$, we obtain the increase of the population ΔN_a of the upper level a

$$\Delta N_a \equiv N_a - N_{a0} = 2^{-1} G_k (N_{b0} - N_{a0}) / [1 + (\omega_{ab} + k\omega_M - \omega_L - \sigma_k)^2 \tau_{12}^2 + G_k],$$
$$G_k \equiv 4 W_{ab}^2 J_k^2 (\Delta\beta_{ab}) \tau_{12} \Gamma^{-1}, \quad \sigma_k = 2\omega_M^{-1} W_{ab}^2 g_k (\Delta\beta_{ba}), \tag{6}$$
$$g_k(\nu) = \sum_{r=1}^{\infty} [J_{k-r}^2(\nu) - J_{k+r}^2(\nu)]/r, \quad W_{ab} \equiv d_{ab} E_{0L} / (2\hbar), \qquad k = 0, \pm 1, \pm 2, \ldots$$

Here τ_{12} and Γ^{-1} are transverse and longitudinal relaxation times, respectively. The quantity τ_{12} is inversely proportional to the impact width of the spectral line $a \to b$; the quantity Γ controls the relaxation rate of the levels a and b.

In the case of a weak laser field the expression for ΔN_a in (6) simplifies to:

$$\Delta N_a = \frac{d_{ab}^2 E_{0L}^2}{2\hbar^2} \frac{J_k^2(\Delta\beta_{ab}) \tau_{12} \Gamma^{-1} (N_{b0} - N_{a0})}{1 + (\omega_{ab} + k\omega_M - \omega_L)^2 \tau_{12}^2}, \qquad k = 0, \pm 1, \pm 2, \ldots \tag{7}$$

In fact, Eq. (7) determines Blokhinzew's spectrum [6], where the satellite at the frequency $\omega_L = \omega_{ab} + k\omega_M$ has the intensity proportional to $J_k^2(\Delta\beta_{ab})$.

From Eq. (6) it follows that that in the general case the spectrum of the atomic absorption of the laser radiation consists of a set of satellites at the frequencies

$$\omega_L = \omega_{ab} + k\omega_M - \sigma_k, \qquad k = 0, \pm 1, \pm 2, \ldots, \tag{8}$$

the half-width of the k-th satellite being $\Delta\omega_{k,1/2} = 2(1 + G_k)^{1/2} / \tau_{12}$. We point out that the term σ_k in the denominator of the expression for ΔN_a (see Eq. (6)) is related to the dynamic Stark shift of the quasienergy levels. It depends on both the laser field, and the OEF $\vec{E}_M(t)$. We note that the expression for σ_k was first obtained in Refs. [7,8]. Figure 5 shows, as an example, the absorption spectrum of the laser radiation by atoms possessing the permanent dipole moments in the states a and b. It consists of the unshifted line S_0 and of the satellite lines $S_{\pm k}$ (k=1,2,..). The intensity of these lines, according to Eq. (6), is the function of the amplitude E_{0M} of the OEF. For Fig. 5, the matrix elements of the dipole moment were assumed to be as follows: $d_{ab} = 2.15 e a_0$, $d_{aa} - d_{bb} = 3.5 e a_0$ where a_0 is the Bohr radius. Profiles shown in Fig. 5 were calculated at $E_{0L} = 15$ kV/cm, $E_{0M} = 7$ kV/cm, $\omega_M = 2.4 \times 10^{11}$ s^{-1}. For the profile (a), the separation between the satellites S_{+k} and S_{-k} is equal to $2k\omega_M$. One can see from Fig. 5 that the allowance for the dynamic Stark shifts σ_k can be important for calculations of the absorption spectra of atoms.

Equations (6) can be used for measuring the amplitude of the microwave field E_{0M} in plasmas. For this purpose one should record the wavelength-integrated intensity of LIF $I_f \propto \Delta N_2$ during the transition from the upper level a to one of the lower levels versus the laser field intensity $I_L \propto E_{0L}^2$: $I_f^{-1} \propto \left(1 + I_{k,satur} I_L^{-1}\right)$, $I_{k,satur} I_L^{-1} \equiv G_k^{-1}(E_{0M})$. Here k is the number of microwave quanta involved in the resonance (see Eq. (8)). The amplitude E_{0M} can be measured in two ways. In the first way, one could measure the ratio of slopes of experimental dependences $I_f^{-1}(I_L^{-1})$ at two different values of k ($k = k'$, and $k = k''$), and use the dependence of the ratio $G_{k'} / G_{k''}$ on E_{0M}. In the second way, one could use the dependence of the dynamic Stark shift σ_k on the parameters of the OEF $\vec{E}_M(t)$. By scanning the laser frequency around the frequency ω_{ab} and by recording peaks of the LIF intensity, one could tune to multi-quantum resonances (8) corresponding to two different indices $k = k_1$, and $k = k_2$. The laser frequency should be kept constant. In this situation one would have:

$$\omega_{L,1} = \omega_{ab} + k_1 \omega_M - \sigma_{k_1}, \qquad \omega_{L,2} = \omega_{ab} + k_2 \omega_M - \sigma_{k_2}. \tag{9}$$

Since ω_M is known, from Eqs. (9) one would determine the dynamic Stark shifts σ_{k_1} and σ_{k_2}. Then, using the fact that the ratio $\sigma_{k_2} / \sigma_{k_1}$ depends on E_{0M} but does not depend on E_{0L}, one could finally find the OEF amplitude E_{0M}.

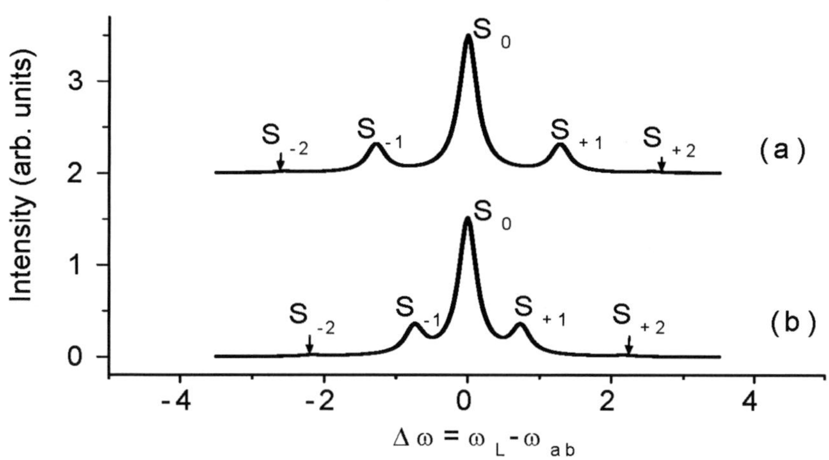

FIGURE 5. Absorption spectra of the laser radiation in the presence of the field $E_M(t) = E_{0M} \cos\omega_M t$ by atoms possessing permanent dipole moments in their states a and b. Profile (a) was calculated without the allowance for the dynamic Stark shifts (DSS) σ_k of the satellites, whereas profile (b) was calculated with the allowance for the DSS σ_k of the satellites.

Now let us consider a situation where a hydrogen (or deuterium) beam travels through a magnetic fusion plasma containing OEF $\vec{E}_M(t)$. In this case, hydrogen atoms are subjected to the combined electric field $\vec{\varepsilon}(t) = \vec{E}_b + \vec{E}_M(t)$, where $\vec{E}_b = \vec{V}_b \times \vec{B}/c$ is the Lorentz electric field. The field $\vec{E}_b(t)$ causes the static Stark splitting of the hydrogen energy level of the principal quantum number n. By a proper choice of the beam velocity, it is possible to realize the situation where this Stark splitting (in the frequency scale) coincides with the frequency ω_M of the OEF $\vec{E}_M(t)$, thus resulting in a resonance. The resonance condition has the form:

$$\omega_M = \frac{3\hbar}{2me} nE_b . \tag{10}$$

Under the condition (10), the probability, that the hydrogen atom is in a Stark sublevel of the energy level n, oscillates at the Rabi frequency, which is proportional to the OEF amplitude. This causes an additional splitting of those hydrogen spectral lines, for which the energy level n serves as the upper level or as the lower level. This additional splitting is linear with respect to the OEF amplitude.

As an example, Fig. 6 shows the spectrum of the hydrogen line L_α under the resonance (10) for the upper level $n = 2$. The additional splitting, being linear with respect to the OEF amplitude, can be used as sensitive tool for measuring the OEF amplitude in magnetic fusion plasmas. For tuning into the resonance (10), one should vary the beam energy. The additional splitting would manifest in the spectrum of the fluorescence excited by the laser tuned to the atomic transition, for which the level n (where n satisfies Eq. (10)) serves as the upper level or as the lower level.

For an estimate, let us assume the OEF frequency $\omega_M = 8.7 \times 10^{11}$ s^{-1}, the magnetic field $B = 1$T, the hydrogen beam velocity $\vec{V}_b$ being orthogonal to the magnetic field $\vec{B}$. For fulfilling the resonance (10) for the energy level $n = 3$ (which is the upper level of the H_α line), we find that the energy of the hydrogen beam should be 30 keV.

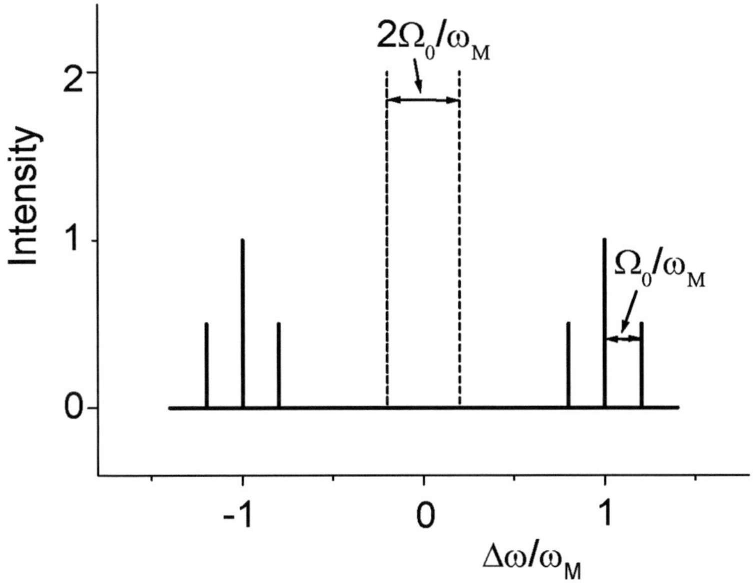

FIGURE 6. Spectrum of the hydrogen line L_α in two orthogonal electric fields: a Lorentz field $\vec{E}_b$ and an oscillatory field $\vec{E}_M(t) = \vec{E}_{0M} \cos\omega_M t$ - under the resonance (10) for the level $n = 2$. The observation is perpendicular to both $\vec{E}_b$ and $\vec{E}_M(t)$. The spectrum was calculated for two orientations of the linear polarization analyzer: parallel to $\vec{E}_b$ (solid line) and parallel to $\vec{E}_M(t)$ (dashed line). The Rabi frequency controlling the splitting is $\Omega_0 = \dfrac{3\hbar}{2me} E_{0M}$.

CONCLUSION

In this paper we presented the main physical principles that can serve as the basis for laser-assisted spectroscopic methods for diagnosing OEFs in plasmas. As atomic "probes", one can use either atoms possessing permanent dipole moments (e.g., hydrogen or deuterium atoms) or atoms that do not possess permanent dipole moments (e.g., helium or lithium atoms).

We showed that several phenomena can be used for laser-assisted diagnostics of OEFs in magnetic fusion plasmas. First, a tunable laser radiation can scan spectral line profiles and/or the structure of energy levels of atoms interacting with OEFs. In this case, parameters of the OEFs can be deduced from the spectrum of fluorescence, resulting from the atomic transition from an excited state to a lower state. For this case we assumed that the laser intensity is relatively weak, so that it does not modify the structure of the atomic energy levels.

Second, the laser radiation can cause the saturation of the atomic transition. In this case for some atomic transitions the saturation parameter depends both on the laser intensity and on the parameters of the OEF. The latter can be used for diagnosing the

OEF. For this case one needs a sufficiently intense laser radiation that can cause the saturastion.

Third, a combined effect of the OEF and of a strong laser field can modify both the structure of energy levels and spectral line profiles. In this case, the character of the modification is controlled by the quantity, in which parameters of both the OEF and the laser field are represented in an entangled way.

ACKNOWLEDGMENTS

The research described in this publication was made possible in part by Award No RU-P1-2594-MO-04 of the U.S. Civilian Research & Development Foundation for the Independent States of the Former Soviet Union (CRDF).

REFERENCES

1. E. Oks, *Plasma Spectroscopy. The Influence of Microwave and Laser Fields*, Berlin: Springer, 1995.
2. V.P. Gavrilenko, E. Oks, V.A. Rantsev-Kartinov, *JETP Letters* **44**, 207 (1985).
3. N.B. Delone and V.P. Krainov, *Atoms in Strong Light Fields*, Berlin: Springer, 1985.
4. Ja.B. Zel'dovich, *Sov. Phys. – JETP* **24,** 1006 (1967).
5. V.I. Ritus, *Sov. Phys. – JETP* **24,** 1041 (1967) .
6. D.I. Blochinzew, *Phys. Z. Sow. Union* **4,** 501 (1933).
7. V.P. Gavrilenko, E. Oks, *Phys. Rev. Lett.* **74,** 3796 (1995).
8. V.P. Gavrilenko, E. Oks, *J. Phys. B: At. Mol. Opt. Phys.* **28,** 1433 (1995).

Improved Interpretation of Stark Broadening Data from Detached Divertor Plasmas

S. Lisgo[1,2], N. Brooks[3], E. Oks[4], M. Groth[5], A.W. Leonard[3], D. Volodko[4], and the DIII-D Team

[1]*University of Toronto Institute for Aerospace Studies, Toronto, M3H 5T6, Canada*
[2]*UKAEA Fusion Association, Culham Science Centre, Abingdon, Oxon OX14 3DB, UK*
[3]*General Atomics, San Diego, California 92186-5608, USA*
[4]*Physics Department, Auburn University, Auburn, AL 36849-5311, USA*
[5]*Lawrence Livermore National Laboratory, Livermore, California 94550, USA*

Abstract. Stark broadening analysis of high-n Balmer series transitions (n→2, where n = 6-10) in deuterium is benchmarked against electron density measurements from a Thomson scattering system in the divertor region of the DIII-D tokamak. The time-averaged Thomson data show $n_e = (6.0\pm1.1)\times10^{20}$ m^{-3} in the low T_e region (~1 eV) of a detached divertor plasma. Analysis of the measured, Stark-broadened, Balmer-series line widths using the *conventional* theory gives $n_e = (4.5\pm0.9)\times10^{20}$ m^{-3}. A more accurate analysis technique, referred to here as the *highly-advanced* theory (HAT) [E. Oks, *Stark Broadening of Hydrogen and Hydrogenlike Spectral Lines in Plasmas: The Physical Insight*, Alpha Science International, Oxford, UK, 2006], produces $n_e = (5.0 \pm 0.5)\times10^{20}$ m^{-3}. In addition, a temperature of (1.5±0.5) eV is determined (the conventional theory does not allow deducing the temperature). Also the HAT reproduces the shape of the dependence of the measured width of the high-n Balmer lines on the principal quantum number n much better than the conventional theory. The improved diagnostic results favor the employment of the HAT when analyzing Stark broadening data from tokamak divertors.

Keywords: Stark broadening, tokamak, DIII-D, Thomson scattering
PACS: 39.30.+w, 52.55.Fa

INTRODUCTION

The lifetime of plasma facing components is a critical issue for next-generation fusion devices, including the proposed International Thermonuclear Experimental Reactor, ITER [1]. The physics solution for the steady-state heat flux to the divertor target plates, where the plasma-material surface interactions are concentrated, involves operating with a detached plasma, which is characterized by low plasma temperatures and a reduced ion flux to the target [2]. Unfortunately, a quantitative understanding of detachment has proved elusive, hampering efforts to accurately predict detached divertor performance in next-step machines.

The Alcator C-Mod tokamak utilizes a high magnetic field that allows operation at higher core densities than other devices; these high core densities naturally give rise, with the limited auxiliary heating power available on C-Mod, to cold, detached plasmas in the divertor region. Detailed

CP874, *Spectral Line Shapes: 18th International Conference*,
edited by E. Oks and M. Pindzola

spectroscopic studies of detachment were carried out in an effort to characterize the associated plasma conditions [3-5], with electron temperatures of $T_e \approx 0.5$ eV in the divertor volume obtained by fitting the distribution of high-n Balmer emissions to the Saha equation (where "n" is the principal quantum number of the upper level), and electron densities of $n_e = (0.5\text{–}2.0)\pm0.3\times10^{21}$ m^{-3} deduced from Stark broadening of the same lines using conventional theory [6,7]. The full spectroscopic and Langmuir probe data sets were compared with output from the OSM-EIRENE "plasma reconstruction" code, but agreement was only marginally within experimental uncertainties [8]. Accurate measurement and simulation are critical for detached plasma analysis because volume recombination processes can be significant below ~1 eV and are highly nonlinear in both n_e and T_e [9].

In order to evaluate the accuracy of the Stark broadening analysis methods, the study has been repeated on the DIII-D tokamak [10], where the divertor is fitted with a Thomson scattering system [11] that can provide an independent measurement of n_e.

EXPERIMENTAL SETUP

DIII-D is a medium size diverted tokamak with major radius R = 1.66 m and minor radius a = 0.67 m. The discharge being investigated has a lower single null in the poloidal magnetic field, line averaged electron density $\overline{n}_e = 8.1\times10^{19}$ m^{-3} during the flat-top, an axial toroidal field of B_T = 2 Tesla, plasma current of I_p = 1.1 MA, and 6.6 MW of neutral beam heating, with the $B\times\nabla B$ drift toward the lower divertor. Time traces of the principal plasma parameters are shown in Fig. 1. Enhanced confinement, or H-mode, is indicated by the regular D_α fluctuations that are characteristic of H-mode related Edge Localized Modes, or ELMs [12]. These ELMs are associated with periodic bursts of particles and energy out of the confined plasma, which are then transported to the walls of the vaccum vessel. The average time between ELMs, τ_{ELM}, for this discharge is ~5 ms.

A schematic cross-section of the lower divertor is presented in Fig. 2, along with the diagnostic lines-of-sight that are of interest here.

The line shapes of the high-n Balmer series lines (n→2, with n ranging from 6 to 10) are measured in the wavelength interval 397 < λ < 410 nm along a vertical path through the divertor at R=1.49 m. The Czerny-Turner spectrometer has an instrumental broadening of ~0.2 nm and a dispersion of 0.0226 nm/pixel. The divertor plasma density is high enough in this discharge, and the temperature low enough, that the high-n Balmer lines exhibit Stark broadening that is significantly greater than the instrumental broadening. Unfortunately, the spectrometer integration time is 10 ms, or approximately two ELM periods, and so the measured FWHMs of these lines are always ELM-averaged quantities, complicating the analysis; see Section 3.

The divertor Thomson scattering (DTS) system views the electron-scattered light from a Nd:YAG laser at eight points along its vertical path, measuring n_e and T_e (for T_e > ~1 eV). Each laser pulse lasts only 10 ns, a duration much shorter than τ_{ELM}; the laser pulse repetition rate is 50 Hz. The laser is at the same radial location as the high-resolution spectrometer view, but displaced toroidally, allowing a direct comparison of $n_{e,Stark}$ and $n_{e,DTS}$ at identical points in the poloidal plane.

The tomographic inversion of a tangentially viewing 2D CCD camera image gives the poloidal distribution of D_γ (5→2) in the divertor [13,14]; see Fig. 3, where the inversion was performed for a 1 ms integration over the ELM peak. Assuming D_γ to be representative of the higher-n Balmer lines, the emission along the spectrometer line-of-sight is clearly localized to the private-flux region (PFR) below the x-point. As a result, only channels 0–3 of the DTS system need to be considered in the present analysis. These high levels of D_γ emission indicate that the divertor is partially detached on both the inner and outer target plates.

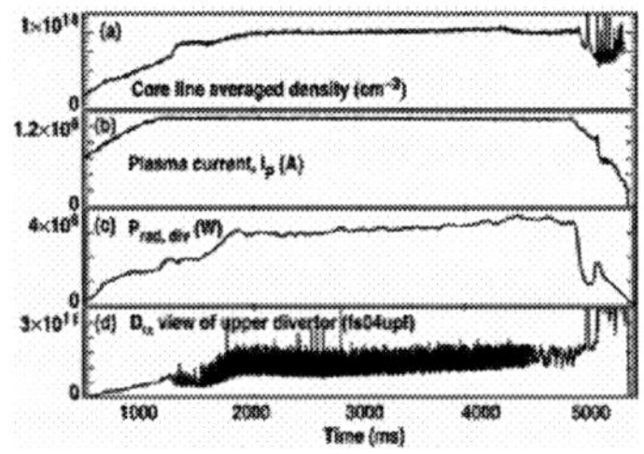

FIGURE 1. Relevant time traces for discharge 123410. (a) Average plasma density in the confined plasma (core), (b) plasma current, (c) radiated power in the divertor region, (d) fast D_α trace (20 kHz sampling) in the upper divertor (a lower divertor view was not available). The high-frequency D_α fluctuations indicate rapid ELMs.

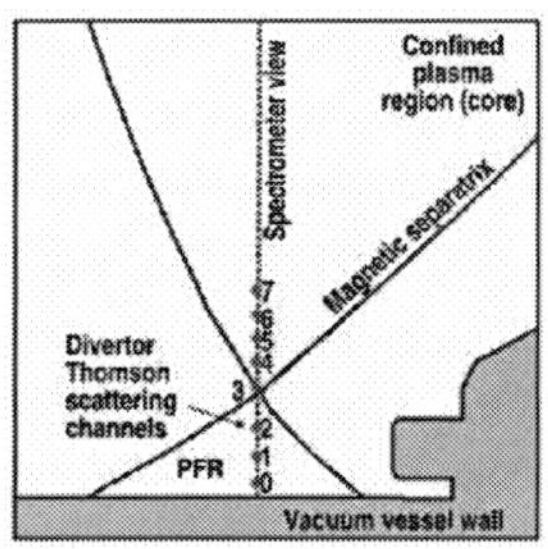

FIGURE 2. Schematic of a poloidal cross-section of the DIII-D lower divertor region. The high-resolution spectrometer view (dashed line) and divertor Thomson channels (dots, marked 0 to 7) are shown. The "separatrix" is the last closed magnetic flux surface, and marks the interface between the confined plasma and the boundary plasma that interacts with the walls of the containment vessel. The private flux region (PFR) is below the x-point and between the legs of the separatrix.

ANALYSIS OF DIVERTOR THOMSON DATA

As stated in the introduction, the objective is to compare estimates of n_e from Stark broadening analysis with n_e data from DTS, in an effort to evaluate the accuracy of competing Stark broadening models. Unfortunately, the divertor plasma fluctuations related to ELMs are substantial, as seen in Fig. 1d, which means that the measured Stark broadened density is a function of the density variation over the ELM. Therefore, a comparison with $n_{e,DTS}$ requires the calculation of a weighted average ($\tau_{DTS} << \tau_{ELM}$) over the ELM:

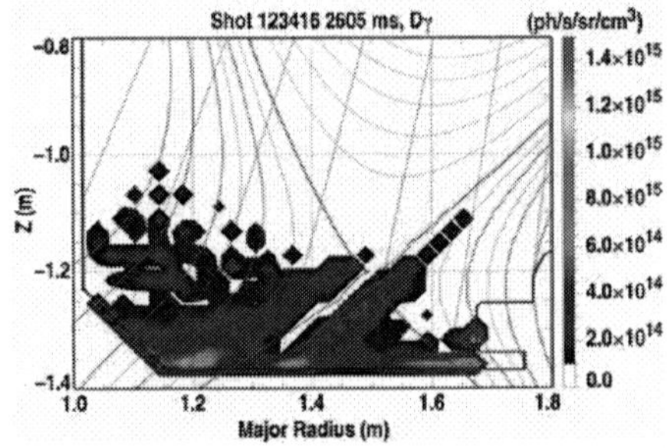

FIGURE 3. Tomographic inversion of 2D CCD tangentially viewing camera image of the divertor, in D_γ light. The emission is concentrated in the PFR, so that only this region needs to be considered in the DTS analysis. Reproduced from [15], courtesy of M. Groth.

$$\langle n_{e.div} \rangle = \int_{Z_0}^{Z_x} \int_{0}^{t_{ELM}} \varepsilon(Z,t) n_{e,div} \, dt \, dZ, \tag{1}$$

where Z is the vertical coordinate along the line-of-sight, t is time, and ε is the normalized emissivity of the high-n Balmer line that is used as the weight function. The spatial variation of the divertor plasma along the spectrometer view, as indicated by the D_γ inversion in Fig. 3, is (coarsely) resolved by the four DTS channels between the x-point and the divertor floor. The DIII-D discharge being investigated was successfully repeated 22 times, giving excellent samples of the temporal variations of n_e and T_e over the ELM cycle.

The DTS data for channels 0 to 3 are plotted in Fig. 4 as a function of the time elapsed between the laser pulse and the start of the previous ELM, τ_{ELM}, which is determined from a library of ELM profile analysis tools [16]. The DTS measurements are filtered so that data were plotted only for typical ELMs, i.e. where the D_α peak is less than 1 ms after the start of the ELM and the time to the start of the next ELM is 4.5–6 ms.

The polynomial fits to the DTS data in Fig. 4 are interpolated along Z in order to produce $n_e(Z,t)$ and $T_e(Z,t)$ profiles along the spectrometer line-of-sight, with linear interpolation results shown in Fig. 5. The $\varepsilon(n_e(Z,t),T_e(Z,t))$ values in Equation (1) are then set to the emissivity for a given high-n Balmer transition using the ADAS database [17]; see Fig. 5c. Given the detached divertor plasma conditions in these discharges, it is assumed that the high-n emissions are dominated by volume recombination processes and that $n_{D+} = n_e$ locally.

Equation 1 is not the best method of calculating $\langle n_{e,div} \rangle$. Ideally, a Stark broadening model should be used to predict the line shape at each point along the spectrometer view, with the resultant superposition of line shapes compared to the measured profile (assuming instrumental and Zeeman broadening are properly taken into account). This exercise is relegated to future work.

The above exercise gives $\langle n_{e,div} \rangle = 6.0 \pm 0.5 \times 10^{20}$ m^{-3} when only considering uncertainties related to the interpolation process. If the stated uncertainties in the individual DTS data measurements are also included, then the absolute error increases to $\langle n_{e,div} \rangle = 6.0 \pm 1.1 \times 10^{20}$ m^{-3}. The average density depends weakly on the particular high-n line that is used as the weight function.

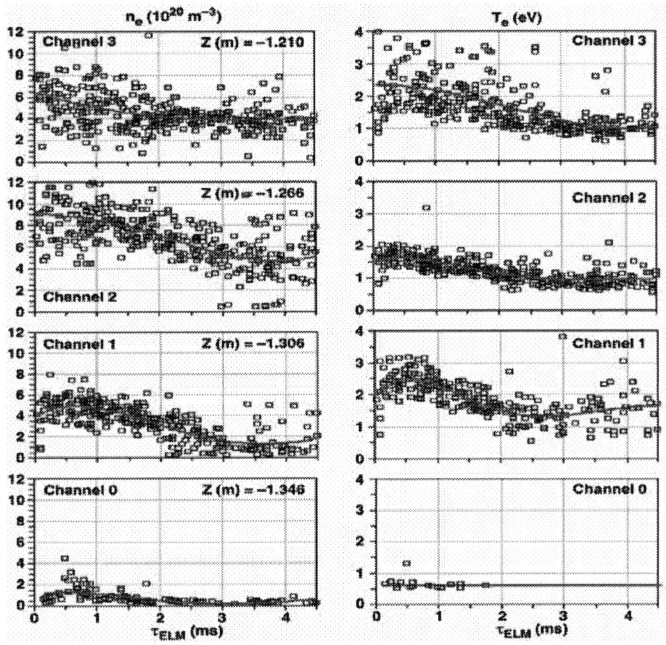

FIGURE 4. DTS n_e (left column) and T_e data plotted as a function of time since the start of the most recent ELM, relative to the associated laser pulse. The lines are polynomial fits to the data. The energy and particles expelled from the core during the ELM are apparent, although much of the energy has been dissipated by the dense plasma in the detached divertor. The low n_e indicated for channel 0 give spurious T_e data.

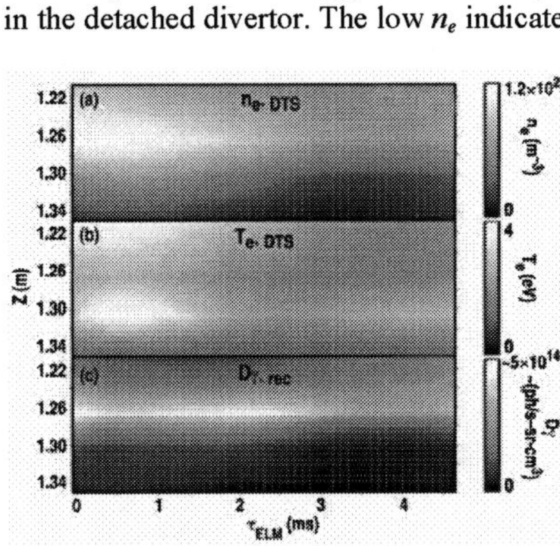

FIGURE 5. Divertor Thomson scattering data interpolated temporally over an ELM cycle and spatially between the lower divertor floor and x-point. The Balmer emission proves the weight function when averaging $n_{e,DTS}$ along the spectrometer line-of-sight. The D_γ signal from volume recombination is shown here, and there is little variation in the distribution for the higher-n lines.

Figure 6 shows the shapes of the measured deuterium Balmer lines $n \rightarrow 2$, with n = 6-10. Acceptable signal-to-noise was obtained by integrating over the period during which the x-point was held stationary and plasma conditions were constant.

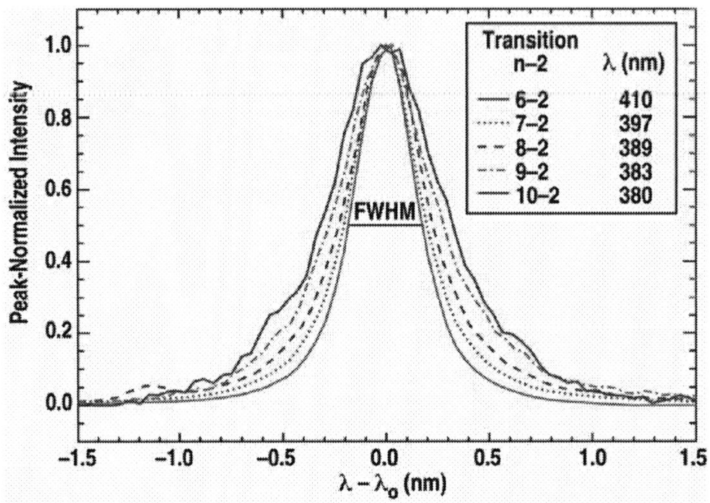

FIGURE 6. Profiles of high-n deuterium Balmer lines. The signals were integrated over the portion of the discharge during which conditions were held constant.

The experimental FWHM values from Fig. 6 are plotted versus n in Fig. 7. Also shown are fits from the "conventional" [7] and "highly-advanced" [18,19] Stark broadening theories.

The highly-advanced theory (HAT) [18,19] is based primarily on a generalization of the formalism of dressed atomic states (DAS) in plasmas. DAS is the formalism initially designed to describe the interaction of a monochromatic (or quasi-monochromatic) field – e.g., laser or maser radiation – with gases. Later it was applied for the interaction of a laser or maser radiation with plasmas [20]. The generalization of DAS consists in using atomic states dressed by a broad-band field of plasma electrons and ions [18,19] (rather than by a monochromatic field).

The employment of the generalized DAS by the HAT [18,19] enables an analytic description of the coupling between electron and ion microfields mediated by the radiating atom. This indirect coupling increases nonlinearly with the increase of the principal quantum number n and with the decrease of the temperature. Therefore it becomes important for high-n lines even at moderate electron densities.

Accurate analytical results for the ion-dynamical broadening are the second outcome of employing the generalized DAS in the HAT. The ion-dynamical broadening is obtained in a non-binary description of the ion microfield and without using the impact approximation. For the edge plasmas of tokamaks, the ion-dynamical corrections could be relatively significant for Balmer lines originating from levels with $n < 8$ at the electron densities $n_e \sim 10^{20}$ m^{-3} or for Balmer lines originating from levels with $n < 6$ at the electron densities $n_e \sim 10^{21}$ m^{-3}.

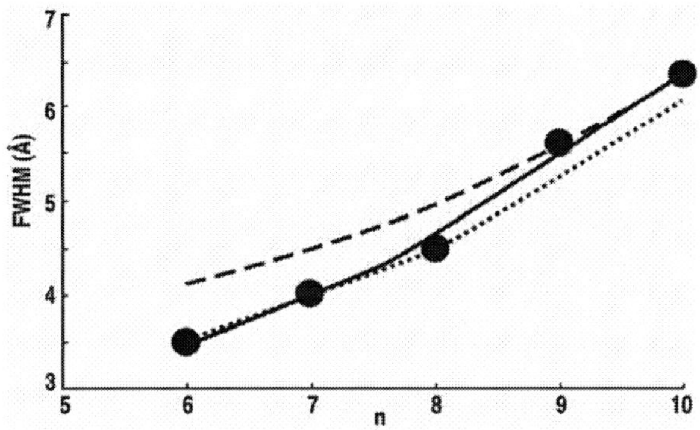

FIGURE 7. FWHM versus the principal quantum number, n, of deuterium high-n Balmer lines, as measured in the DIII-D divertor. Fits for Stark broadening theories are shown: highly-advanced theory with $n_e = 5.0\times10^{20}$ m^{-3} and $T = 1.5$ eV (solid line), and conventional theory with $n_e = 3.9\times10^{20}$ m^{-3} (dashed line) and $n_e = 5.1\times10^{20}$ m^{-3} (dotted line).

For the above reasons the application of the HAT to the analysis of experimental high-n lines enhances the accuracy of the density determination and, additionally, yields the *temperature*, which the conventional theory doesn't provide at all.

From Fig. 7 it is seen that the HAT more accurately reproduces the shape of the experimental FWHM dependence than the conventional theory. The shape of the FWHM dependency in the HAT is generally more complicated than that in the conventional theory because of the competition of the two additional effects included in the HAT, but neglected in the conventional theory: indirect coupling and ion-dynamical broadening. Broadening due to the indirect coupling increases with n, while that due to the ion-dynamical broadening decreases with n.

Plasma parameters were obtained from the best fit to the shapes of the experimental Balmer lines presented in Fig. 6. The best fit for the conventional theory yields $n_{e,Stark} = (4.5\pm0.9)\times10^{20}$ m^{-3} and no information on the temperature. The most probable values of the densities, deduced from individual line profiles fitted by the conventional theory, vary by about 13%. The best fit for the HAT yields $n_{e,Stark} = (5.0\pm0.5)\times10^{20}$ m^{-3} and the temperature $T = 1.5\pm0.5$ eV. The most probable values of the densities, deduced from individual line profiles fitted by the HAT, vary only by about 5%.

The plasma parameters deduced from the Stark broadening of the experimental high-n Balmer lines can be now compared with $\langle n_{e,div}\rangle = (6.0\pm1.1)\times10^{20}$ m^{-3} and $T_e = 1$–2 eV determined from the experimental DTS ELM data. It is seen that the latter agrees with the temperature deduced via the HAT (while the conventional theory was unable to deduce the temperature from the high-n lines) and also agrees with the electron density deduced via the HAT slightly better than with the electron density inferred via the conventional theory. Last but not least, the HAT reproduces the shape of the dependence of the measured width of the high-n Balmer lines on the principal quantum number n much better than the conventional theory*/.

In the experimental profiles of the $9\rightarrow2$ and $10\rightarrow2$ lines (D_9 and D_{10}) one can discern slight "shoulders" – especially in the D_{10} profile. On the one hand, this could be a result of a spatial non-uniformity of the plasma along the line of sight, with the shoulders being emitted from a higher density region and the bulk of the line from a lower density region. On the other hand, for our experimental conditions, Balmer lines D_n with $n > 8$ could have some shoulders, though less pronounced than in our experimental profiles, even if the plasma were spatially uniform. In the discussion below a physical explanation is offered for slight shoulders in the case of a spatially uniform plasma.

The positions of Stark components of hydrogen/deuterium lines are controlled by (proportional to) the following combination of quantum numbers

$$X_{nn'} = nq - n'q', \quad (2)$$

where n and n' are the principal quantum numbers of the upper and lower energy levels, respectively; $q = n_1 - n_2$ and $q' = n'_1 - n'_2$ are the electric quantum numbers of the upper and lower Stark sublevels, respectively; n_1, n_2, n'_1, n'_2 are the standard parabolic quantum numbers defined in textbooks on quantum mechanics (see, e.g. [23], sect. 37). For the most intense Stark component of a given high-n line, in accordance to Ref. 20, page 121, one has $X_{nn'}^{(1)} = n$ for the lines containing no central ("unshifted") components ($n + n'$ is even) or $X_{nn'}^{(1)} = 2n$ for all other high-n lines. Specifically, for the Balmer lines ($n' = 2$) corresponding to $n > 4$, one has:

$$X_{n2}^{(1)} = n \text{ (for even } n); \qquad X_{n2}^{(1)} = 2n \text{ (for odd } n). \quad (3)$$

For the second most intense Stark component of the Balmer lines of $n > 4$, one has:

$$X_{n2}^{(2)} = n(n-1) - 2. \quad (4)$$

The ratio $X_{n2}^{(2)}/X_{n2}^{(1)}$ increases with n:

$$X_{n2}^{(2)}/X_{n2}^{(1)} = (n-1-2/n) \text{ (for even } n); \quad X_{n2}^{(2)}/X_{n2}^{(1)} = (n-1-2/n)/2 \text{ (for odd } n). \quad (5)$$

This means that as n increases, the second most intense Stark component moves farther out into the wings relative to the position of the most intense Stark component.

Now let us look at the ratio $I_{n2}^{(2)}/I_{n2}^{(1)}$ of the intensity $I_{n2}^{(2)}$ of the second most intense Stark component to the intensity $I_{n2}^{(1)}$ of the most intense Stark component. For $n > 4$, both within the sequence of odd n and within the sequence of even n, as n increases, this ratio $I_{n2}^{(2)}/I_{n2}^{(1)}$ first decreases, then reaches a minimum, and then increases. The ratio $I_{n2}^{(2)}/I_{n2}^{(1)}$ reached its minimum at $n = 7$ for odd n or at $n = 8$ for even n.

The increase of the ratio $I_{n2}^{(2)}/I_{n2}^{(1)}$ starting from $n = 9$ combined with the fact that the second most intense Stark component moves further to the wings relative to the position of the most intense Stark component as n increases is the underlying physical reason for possible shoulders in the Balmer lines of $n > 8$. However, these shoulders will show up only if the electron density is sufficiently low.

Indeed, the above reasons relate to quasistatically-broadened profiles of Balmer lines. The dynamical broadening (by most of the electrons and some of the ions) tends to smooth out any structure resulting from the quasistatic broadening, since the former scales essentially linearly with the electron density n_e, whereas the latter scales as $n_e^{2/3}$. Therefore, at sufficiently large electron densities, the dynamical broadening will smooth out any structure in quasistatic profiles, including the possible shoulders.

The fact that Stark profiles can have structures, such as peaks, dips and/or a second maximum/shoulder (in each wing of the line), is not new. The structures can be found both within the purely quasistatic calculations of 1959 [21] and within the calculations by the conventional theory of 1970 [6]. The latter fact had not been explicitly claimed in [6], but in paper [5] of 1995 - while discussing the calculations from [6]. It should be emphasized that the calculations performed in [5] by the conventional theory were conducted in a more approximate way than the calculations from [6]. Indeed, instead of profiles containing structures, yielded within the conventional theory by rigorous calculations in [6], the authors of [5] substituted such profiles by Lorentzians of the same halfwidth – as noted in [5]. Obviously, the Lorentzians do not have any structures. Therefore, the calculations from [5] could not reveal any structures at all.

Let us now discuss later calculations performed by various authors for modeling experimental profiles of high-n Balmer lines observed from divertors of other tokamaks. In papers [3] of 1998 and [4] of 1999, from fitting experimental profiles of high-n Balmer lines observed from the divertor of Alcator C-Mod, the authors deduced the electron density $n_e = (1.2 - 1.8)\times10^{21}$ m^{-3}. The modeling of the profiles was done employing the approximate version of the conventional theory, in the same way as in [5]. Of course, such modeling cannot produce shoulders in the line profiles, as explained above. Besides, the higher electron density in the divertor of Alcator C-Mod divertor compared with DIII-D caused a significant overlap between the wings of adjacent high-n Balmer lines, making the profile fitting to individual high-n lines like D_9 and D_{10} more problematic than in the case of DIII-D. As for the experimental profiles, they did not show any shoulders. Most probably this was due to the fact that the electron density in the experiments [3, 4] was about three times higher than in the DIII-D experiment, so that the dynamical broadening would have smoothed out any possible shoulders.

In paper [22] of 2002, from fitting experimental profiles of high-n Balmer lines observed from the divertor of JET, the authors deduced the electron density $n_e = 8\times10^{19}$ m^{-3}. The modeling of the profiles was done employing the frequency fluctuation model (FFM). The FFM is much more advanced than the conventional theory. In this simulation model, closely lying Stark components of a spectral line are artificially bunched together and considered as one "radiative channel". So, the spectral line is represented by these radiative channels, the number of which is much smaller than the number of the Stark components, what simplifies the computations. The ion dynamics is then taken into account approximately by a mixing of the radiative channels via some stochastic Markovian process. Though only approximate in its description of ion dynamics and the structure of the Stark components and lacking entirely the coupling of electron and ion microfields contained in HAT, the FFM is usually much more accurate than the conventional theory. The profiles calculated by the FFM in [22] did not seem to show shoulders, perhaps due to the approximate treatment of the structure of the Stark components.

Coming back to the shoulders observed in the lines D_9 and D_{10} in our experiment: they could result either from a spatial non-uniformity of the plasma along the line of sight or - in the case of a quasi-uniform plasma - from the specific properties of the Balmer lines of $n > 8$ outlined above. This problem requires a further study.

DISCUSSION AND CONCLUSIONS

High resolution measurements of Stark broadened, high-n Balmer series lines were made for a rapidly ELMing plasma discharge. The ELM-averaged values of n_e deduced from Stark broadening analysis of the Balmer line profiles by the conventional theory and by the highly-advanced theory were compared with a column-averaged n_e derived from divertor Thomson scattering data. The n_e inferred via the highly-advanced theory is in better agreement with the DTS value than that inferred via the conventional theory. The temperature deduced from the highly-advanced theory gives good agreement with the DTS analysis, whereas the conventional theory of Stark broadening provides no information about temperature. Finally, the highly-advanced theory gave a significantly better match to the distribution of FWHM with n. Overall, the highly-advanced theory provided the best representation of the experimental data.

REFERENCES

1. ITER Expert Groups, *Nuclear Fusion* **39** (12), 2137-2627 (1999).
2. P. C. Stangeby, *Nuclear Fusion* **33** (11), 2503-2504 (1993).
3. J. L. Terry, B. Lipschultz, et al., *Physics of Plasmas* **5** (5), 1579 (1998).
4. B. Lipschultz, J. L. Terry, et al., *Physics of Plasmas* **6** (5), 1907 (1999).
5. B. L. Welch, H. R. Griem, et al., *Physics of Plasmas* **2** (11), 4246-4251 (1995).
6. R. Bengston, J. D. Tannich and P. Kepple, *Physical Review A* **1**, 532 (1970).
7. H. R. Griem, *Spectral Line Broadening by Plasmas*, Academic Press, New York, 1974.
8. S. Lisgo, et al., *J. Nuclear Materials* **337** (1-3), 139-145 (2005).
9. S. Lisgo, "Interpretive Modeling of the Alcator C-Mod Divertor," Ph.D. Thesis, University of Toronto, 2003. http://starfire.utias.utoronto.ca/divimp/publications/
10. J. L. Luxon, T. C. Simonen, R. D. Stambaugh, *Fusion Science and Technology* **48** *(2)*, 807-827 (2005).
11. S. L. Allen, D. N. Hill, T. N. Carlstrom, et al., *J. Nuclear Materials* **241**, 595-601 (1997).
12. W. Fundamenski, R. A. Pitts, et al., *Nuclear Fusion* **45** (8), 950-975 (2005).
13. M. Fenstermacher, W. H. Meyer, et al., *Review Scientific Instrumments* **68** (1), 974-977 (1997).
14. M. Groth, M. E. Fenstermacher, et al., *J. Nuclear Materials* **313**, 1071-1075 (2003).
15. M. Groth, *DIII-D Science Meeting talk,* March 2006. groth@fusion.gat.com
16. G. D. Porter, T. A. Casper, J. M. Moller, *Physics of Plasmas* **8** (12), 5140-5150 (2001).
17. H. P. Summers, et al., *Plasma Physics and Controlled Fusion* **44**, B323-B338 (2002).
18. E. Oks, *Stark Broadening of Hydrogen and Hydrogenlike Spectral Lines in Plasmas: The Physical Insight,* Alpha Science International, Oxford, UK, 2006.
19. E. Oks, "Stark Widths of Hydrogen Lines in Plasmas: a Highly-Advanced Non-Simulative Semiclassical Theory and Tables," in *this Volume.*
20. E. Oks, *Plasma Spectroscopy: The Influence of Microwave and Laser Fields*, Springer, Berlin, 1005.
21. A.B. Underhill and J.M. Waddell, *National Bureau of Standards Circular No. 603*, Washington DC, 1959.
22. M. Koubiti, H. Capes, L. Godbert-Mouret, Y. Marandet, A. Meigs, S. Loch, R. Stamm, and H. Summers, "Diagnostics of Detached Plasmas Using High-n Lines and Continuum Spectral of D and He", in *Spectral Line Shapes v. 12*, edit. C.A. Back, AIP Conf. Proc. 645, AIP, New York, 2002, p. 67.
23. L. D. Landau and E. M. Lifshitz, *Quantum Mechanics*, Pergamon, Oxford, 1965.

Spectral Line Shapes as a Diagnostic Tool in Magnetic Fusion

R. Stamm[1], H. Capes[2], A. Demura[3], L. Godbert-Mouret[1], M. Koubiti[1]
Y. Marandet[1], M. Mattioli[4], J. Rosato[1], F. Rosmej[1] and K. B. Fournier[5]

*[1]Physique des Interactions Ioniques et Moléculaires
Université de Provence, Marseille, France
[2]Département de Recherches sur la Fusion Contrôlée,
CEA, Cadarache, France
[3]HEPTI, RRC "Kurchatov Institute", Moscow, Russia
[4]Consorzio RFX, Associazzone Euratom-Enea sulla Fusione, Padova, Italy
[5]Lawrence Livermore National Laboratory, P.O. Box 804, Livermore, CA 94551, USA*

Abstract. Spectral line shapes and intensities are used for obtaining information on the various regions of magnetic fusion devices. Emission from low principal quantum numbers of hydrogen isotopes is analyzed for understanding the complex recycling mechanism. Lines emitted from high principal quantum numbers of hydrogen and helium are dominated by Stark effect, allowing an electronic density diagnostic in the divertor. Intensities of lines emitted by impurities are fitted for a better knowledge of ion transport in the confined plasma.

Keywords: Zeeman effect, Stark broadening, Doppler broadening, edge plasmas, turbulence, transport
PACS: 32.70.Jz, 52.25.Gj, 52.25.Os, 52.55.Fa, 52.70.-m

INTRODUCTION

Similarly to spectral observations and modelings in astrophysics, researches in line broadening for magnetic fusion devices are inspired by a vision. One dreams that the line shapes will easily convey all the information on the plasma containing the emitter, revealing density and temperature of the charged and neutral particles, the bulk motions of the plasma, and the small scale turbulent fluctuations. Such hope is however only partially fulfilled, since an accurate determination of the plasma parameters generally requires a careful and complex analysis, combining detailed data in atomic physics, a knowledge of the plasma statistical properties, and in some cases a description of hydrodynamic turbulence.

Spectral line widths and shifts have been widely used for determining the plasma parameters and dynamics since the early days of research in magnetic fusion [1]. While these first observations were made in the UV and XUV range, and concerned impurities in the core plasmas, more recent studies were mainly devoted to the emission of hydrogen isotopes in the edge plasma. These studies enabled a better understanding of the edge plasma composition and behavior, and helped to elucidate the complex recycling processes of the hydrogen isotopes [2]. Part of the studies were

CP874, *Spectral Line Shapes: 18th International Conference,*
edited by E. Oks and M. Pindzola

focused on the measure of the electron density by using the Stark broadening of high principal quantum number transitions of the Balmer series of deuterium [3]. Other informations like the normal and reversed ion flows in the DIII-D divertor [4] have been obtained thanks to line shapes observations of carbon ions.
In this paper, we will review some recent analyses of spectral line shapes and intensities observed in tokamaks like TORE-SUPRA, TEXTOR, JET or ALCATOR C-MOD. We will show that, depending on the plasma parameters and on the line observed, Doppler, Stark broadening and Zeeman effect may be observed, sometimes superimposed. Our toolbox for modeling such spectra is composed of population kinetic models, line shape codes, including ultra-fast Lyman line profile calculations designed for a study of line radiation transport. In a first section, we will discuss the problem of neutrals in the edge and divertor plasmas, and illustrate with the examples of hydrogen isotopes and neutral helium the dominant broadening mechanisms in the various plasmas encountered. The specific problem of line shapes for opacity calculation will be discussed for the Lyman alpha line. For divertor plasmas, and for high density plasmas such as expected in ITER, Stark broadening will be discussed in the second section. The third section will be devoted to the emission of impurity lines, generally observed in the confined core plasma. A fit of the measured spectral lines usually requires the use of accurate atomic data and an impurity transport model. Several examples will be presented, with measured data obtained on the JET tokamak. Finally, we will discuss the effect on the line shapes of the ubiquitous turbulent fluctuations of the hydrodynamic variables [5], and briefly present a statistical formalism able to take their effect into account on the observed line shapes.

NEUTRAL EMISSION IN EDGE AND DIVERTOR PLASMAS

Hydrogen Isotopes in an Ionizing Plasma

Line emission of hydrogen isotopes is a unique source of information for understanding the composition and dynamics of the edge plasma. On all tokamaks, one observes a strong emission of the optically thin Balmer alpha line. For most presently operating tokamaks, this line is broadened by Doppler effect and exhibits a structure sensitive to the magnetic field. A multi Gaussian fit of this line revealed the presence of at least three populations having different temperatures, with the coldest one being as low as 0.2 eV [6]. This cold population actually gives the sharpest structure of such lines, which in a first approximation appear as being separated in a Lorentz triplet (or doublet in an observation parallel to the magnetic field), broadened by Doppler effect in the conditions of most current tokamaks. It has soon been realized that these different populations of neutral deuterium are generated by the various recycling mechanisms [2], each one leaving a specific signature on the D_{α} line shape. In all cases, a cold population with a temperature of the order of the eV or less results from the dissociation of deuterium molecules released by the wall. Warmer populations with a temperature of about 10 eV or more are produced by the reflection of a neutral following the impact on the wall of an incident ion, or by charge exchange with an energetic ion. Some care should be taken for the modeling of this line shape, first for the case of the charge exchange population, for which a neutral transport calculation should be carried out [7]. Moreover, for the central part of the line resulting mainly from the cold atom population, a detailed structure of the line retaining fine structure and Zeeman effect should be used, especially for low magnitude magnetic field cases. Indeed, a calculation of D_α for a a magnetic field of 1 T, a density $N_e=10^{12}$ cm^{-3} and an electronic temperature $T_e=1$ eV, reveals that the complex structure of the line has to be taken into account for an accurate diagnostic.

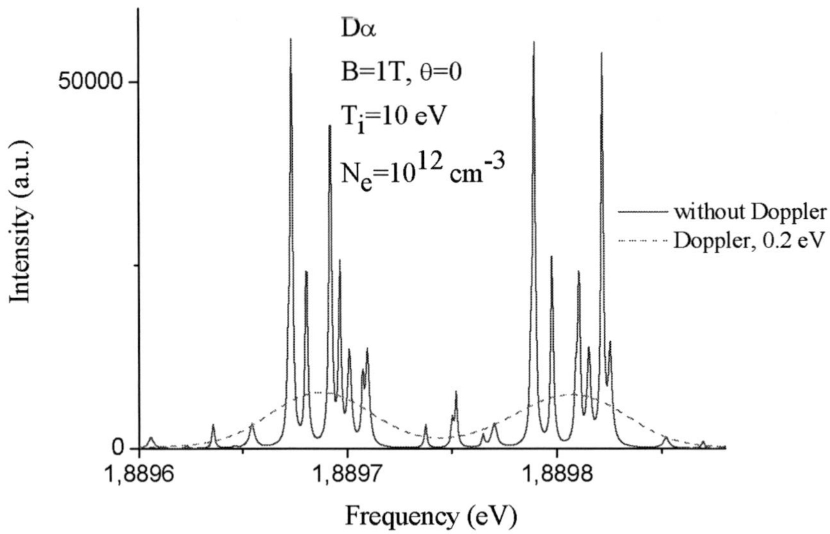

FIGURE 1. Profile of the deuterium Balmer alpha line for edge plasma conditions and an observation parallel to the magnetic field. Zeeman, Stark and fine structure effects are retained.

On Fig. 1, the numerous resulting components are drawn with a full line for the case without Doppler effect, with each component broadened by Stark effect treated in the impact approximation both for electrons and ions. If a convolution with a Gaussian Doppler profile at 0.2 eV is performed (dotted line), it can be seen that most of the structure is lost, but a closer look reveals that the resulting profile remains asymmetric.

Line Shapes for Opacity Calculations

For the conditions of the ITER divertor, the mean free path of a Lyman alpha photon may be shorter than 2 mm, and the plasma is thus optically thick for this radiation. Models show that this can strongly modify the ionization-recombination balance. Recently, a photon transport study was developed for the Monte-Carlo neutral transport code EIRENE [8,9], and showed the need for incorporating detailed line shapes in the modeling. This has been the motivation for providing an ultra fast, but accurate line shape model for the optically thick lines. For the Lyman alpha line, and for most of the plasma conditions encountered in edge plasmas, the main broadening mechanisms are Doppler, Zeeman and Stark effect. Since the final convolution with the simulated velocity distribution function is performed by the transport code, we present on Fig. 2 the Lyman alpha line shape without Doppler effect, but retaining Zeeman and Stark effect. Here the calculation has been performed analytically, by using an impact approximation for both ions and electrons [10,11].

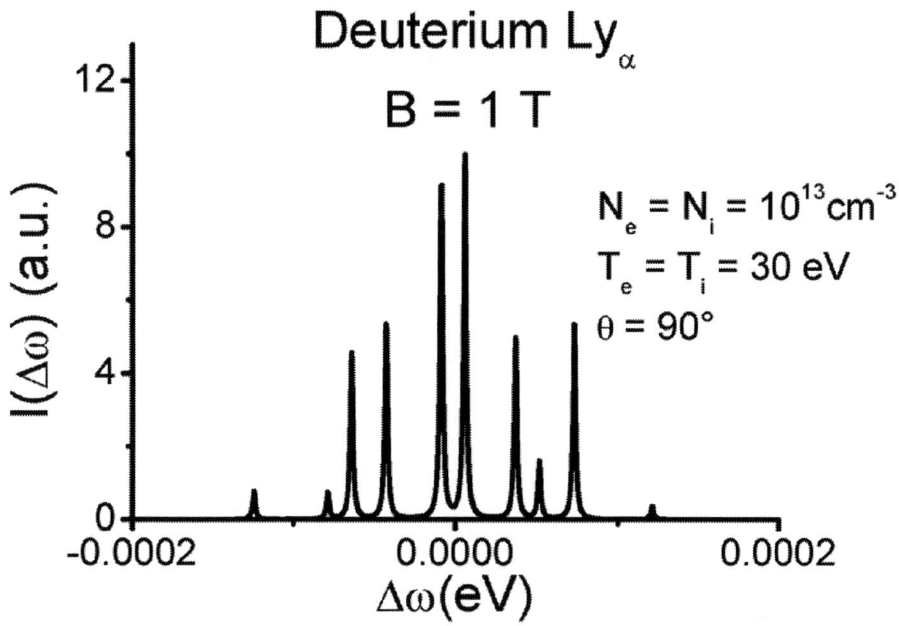

FIGURE 2. Profile of Lyman alpha line for edge plasma conditions and an observation perpendicular to the magnetic field (θ=90°). Zeeman, Stark and fine structure effects are retained.

Using this modeling, it can be seen on Fig. 2 that Lyman alpha consists in ten components for this observation perpendicular to the magnetic field, and that the asymmetry resulting from the fine structure of the line is clearly seen. For the conditions of this calculation, with $N_e=10^{13}$ cm^{-3} and equal electronic and ionic temperature, $T_e=T_i=30$ eV, the validity of the impact approximation for the ions is well insured. Using the EIRENE transport code, calculations of the ionization balance for the conditions of the ITER divertor are in progress by using these detailed profiles. For higher densities and/or lower temperatures, the Stark broadening model for ions should take into account the effect of many body dynamic interactions with the emitter. These conditions will be discussed in the next section.

ROLE OF STARK EFFECT

Recombining Divertor Plasmas

In the conditions of most presently operating tokamaks, Stark effect mainly affects hydrogen isotopes or helium lines originating from high principal quantum numbers. Such lines are often seen in divertor plasma conditions, where three body recombinations are followed by radiative cascades allowing the observation of transitions with upper principal quantum number as high as n=15-20. On Fig. 3, we have plotted a part of the spectra of He I 1s2p ^{3}P-1snd ^{3}D diffuse line series, observed in the divertor of the JET tokamak.

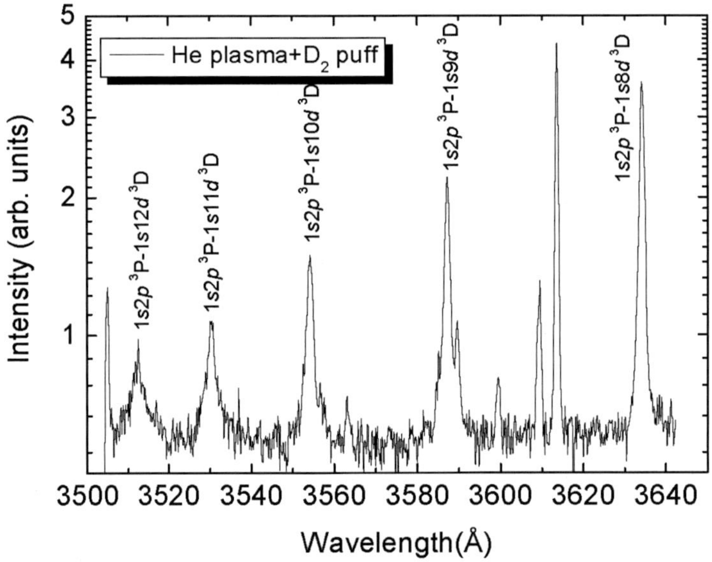

FIGURE 3. Spectral lines of He I 1s2p ^{3}P-1snd ^{3}D diffuse series emitted in the divertor of the JET tokamak.

Lines with upper principal quantum number n=8 to n=12 are seen on Fig. 3, and are found to be dominated by Stark effect for typical divertor conditions ($N_e=5\times10^{12}$-5×10^{14} cm^{-3}, T_e=0.2-5 eV). For the 1s2p ^{3}P-1s8d ^{3}D line at 3635 Å, estimates of the collision time $\tau\sim d/v_{th}$ (with d the average distance between charged particles, and v_{th} their thermal velocity) for ions and electrons, compared to the line time of interest t_l (inverse of the line width), suggest that a standard Stark approach may be used, with the electrons described by an impact operator ($\tau_e << t_l$), and the ions by a static approximation ($\tau_i >> t_l$). Using theoretical Stark line profiles, it will thus be possible to perform a density diagnostic by fitting the shape of the line. The line shape resulting from our model is plotted on Fig. 4 for a density of N_e=5x10^{13} cm^{-3}, and for three temperatures T_e=1, 2 and 5.8 x10^4 K. As the temperature is increased, thus reducing the impact width of the Stark components, it can be seen that the line exhibits satellite components near to the main line. The component appearing on both sides of the main peak of the line correspond to the forbidden transitions 2 ^{3}P - 8 ^{3}P, ^{3}F, ^{3}G which have been made observable by the mixing of the levels induced by the electric microfield.

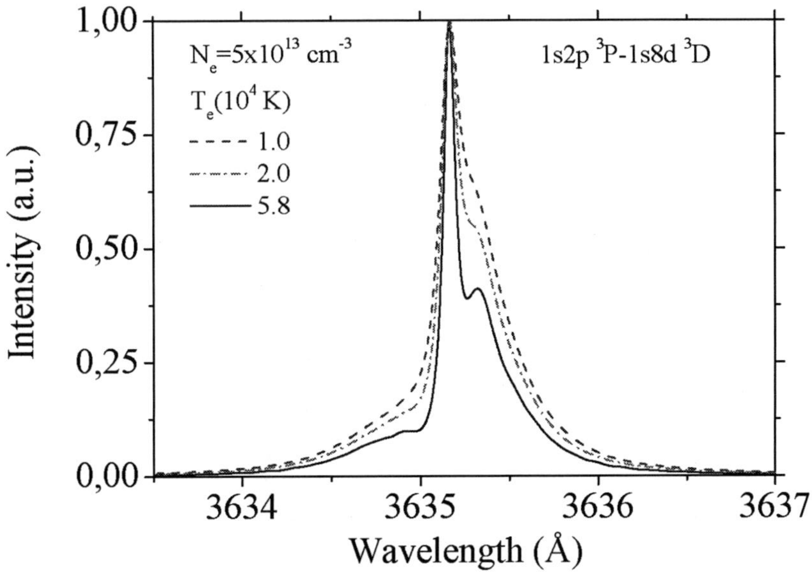

FIGURE 4. Model for the spectral line of He I 1s2p ^{3}P-1s8d ^{3}D, for N_e=5x10^{13} cm^{-3}, and T_e=1,2 and 5.8 x10^4 K.

Note that divertor plasmas can be strongly inhomogeneous, and that in such cases a comparison to experimental spectra will require a specific analysis.

Tokamaks with Dense Plasmas

A few magnetic confinement fusion devices are already operated with densities of the order of 10^{14} cm^{-3} or higher. For such conditions, found for instance in the Alcator C-Mod tokamak, Stark broadening of lines emitted in the divertor plasma not only affects the high n lines, but has to be taken into account even for the low lying lines. The study of such conditions has recently gained a new interest with the decision of building the ITER tokamak, which is designed for high density steady state discharges. One should stress that some care must be taken for modeling the line shapes in such plasmas. First of all, the low lying lines will be affected simultaneously by Zeeman and Stark effects, requiring a specific approach [12,13,14]. Secondly, the standard Stark broadening approach will rarely be valid. Indeed, for low n lines, ion dynamics affects the line shapes, even for the highest densities found in divertors. On the other hand, the impact approximation for the electrons becomes more and more questionable as lines with large principal quantum numbers are considered (n>10-12). Although it remains to perform a detailed analysis of the error made by using the binary impact approximation for cases where multiple electronic collisions act on the emitter, other models may be used to address this issue. Indeed, this problem is similar to the ion dynamics problem, since it involves many body dynamic interactions between charged particles and the emitter. Approaches such as simulations coupled to the integration of the Schrödinger equation, model microfield methods (MMM [15], FFM [16]), or analytical models [17] should be validated and compared before using them for extensive applications to divertor conditions.

SPECTRAL ANALYSIS OF IMPURITIES

Intrinsic and/or purposely seeded impurities are always present in magnetic confinement fusion (MCF) devices. Spectrometers observing plasma emission from the visible wavelength range up to the X-ray range are installed on these devices. Most spectroscopy is usually dedicated to the determination of both the plasma chemical composition (i. e., the identification of the impurity elements and the evaluation of their concentration) and the impurity transport coefficients. For this purpose only a few strong lines are considered and are simulated using impurity ion transport codes, that describe ionisation, recombination and radial transport of the ions of a given impurity species. They require as input data the radial profiles of the electron temperature $T_e(r)$ and of the electron density $N_e(r)$, which are needed to evaluate ionisation and recombination rate coefficients (respectively, S_{ion} and α_{rec}) and two transport coefficients, a diffusion coefficient D(r) and a convection velocity V(r).

Until recently, simulations of full spectra were less frequent. These include, e. g., full simulation in the XUV range of n=2 to n=2 L-shell Ne, Ar and Fe spectra [18, 19], of n=3 to n=2 L-shell Ne and Ar spectra [20, 21], and of n=3 to n=3 M-shell Fe and Ni spectra [22, 23]. The analysis of spectra emitted by hot and well-diagnosed magnetic fusion plasmas is important since it touches on atomic physics issues to be used in the analysis of high spectral resolution astrophysical observations, i.e. on the determination of the electron temperature, electron density and elemental abundances. In most cases atomic data are supposed to be valid and the simulations give D(r) and V(r), like, e.g., in Refs [20, 21]. But with high quality and large amount of spectroscopic data, it is possible to go further and to verify also S_{ion} and α_{rec}. To the best of our knowledge, the first of such comparisons has been reported by May et al. [24], analysing in the FTU tokamak space-resolved distributions of laser blow-off (LBO) injections of Fe and Ge ions and of intrinsic Ni ions. Using data from Ref. [25] for Fe and Ni and from Ref. [26] for Ge, the F-like to B-like charge states could not be simulated unless the relevant α_{rec}/S_{ion} ratios were increased. In another paper on

M-shell Ni spectra in the 144-180 Å region [23] both D(r) and V(r) were known from similar discharges and with best fits on the strongest lines of the involved Ni X-XV ions, the best sets of data for both ionisation and recombination have been obtained.

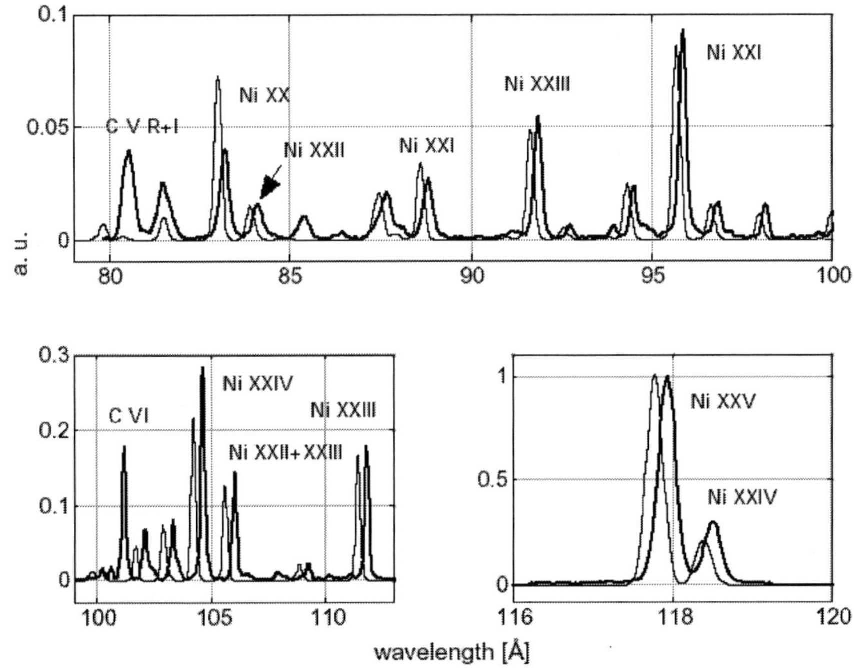

FIGURE 5. Experimental (thick line) and simulated (thin line) normalised Ni spectra from a limiter JET discharge heated by a 6 MW ICRH pulse with the most intense Ni and C line identified. For clearness, three partial spectral ranges are considered and the thin spectrum has been slightly shifted towards the blue.

The ionisation and recombination rates have to be regularly updated following the publication of new theoretical calculations and new experimental data. Astrophysicists usually consider only elements up to Cu (Z=29) or Zn (Z=30). An article that updates recommended data from recent references is published every five or six years allowing determination of charge state distributions (CSD). At present, the most recent article in this context is the 1998 paper from Mazzotta *et al.* [25], which has been extended in Ref. [26] from Cu to Ge (Z=32). Assessments and updates of these atomic data have been started both by us and at the Columbia University under the leadership of D. Savin [27]. Concerning both radiative (RR) and dielectronic (DR) recombinations it was concluded that, for highly ionised ions down to the recombining Na-like isosequence, the calculations from the DR+RR coordinated program under the leadership of N. Badnell of the University of Strathclyde should be used [28, 29] as reference calculations. Where measurements exist, these theoretical data agree with recent laboratory work to within typically 35%

or better at the temperatures relevant for collisionally ionized plasmas. For the less ionized sequences (up to recombining Mg-like ions), the Columbia University group claims [27] that there are no new reliable recombination data, and that the Mazzotta's

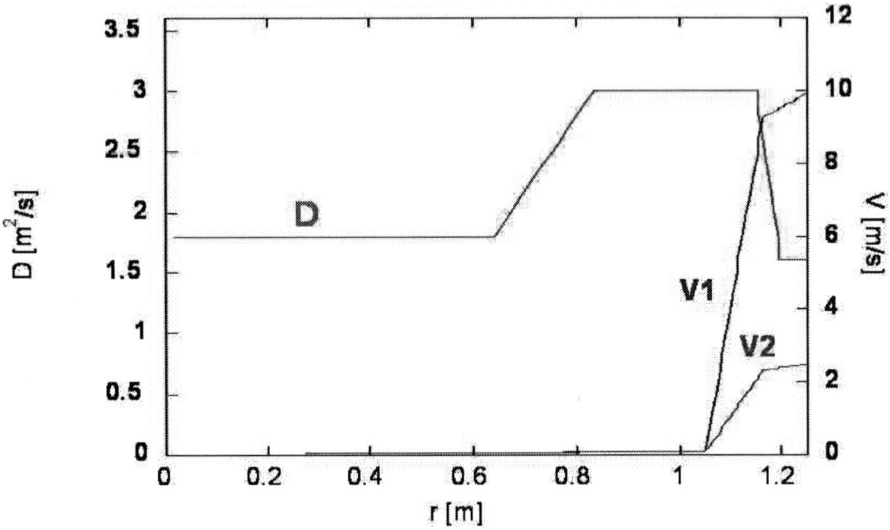

FIGURE 6. D(r) and V(r) radial profiles for the Mo-Ge LBO. To simulate closely the outflow of as many as possible ionisation stages, the simulation requires a time-dependent V(r). Critical for it is the simulation of the soft X-ray brightness. Shown are the limit profiles: V1 up to 100 ms and V2 between 300 and 600 ms after the injection. Inward velocities are positive.

rates should be kept for calculations of charge state distribution (CDS), in spite of the almost general use of the Burgess-Merts formula, which is known to underestimate the DR rates at low T_e.

A corresponding review of the available ionization data is not yet available and, at the present, Mazzotta's ionization rates should be kept as the reference ones to be used in CSD evaluations, the only exception being Ni, since the assessed data of Ref. [30] were not considered. For Kr and Mo (elements without astrophysical interest) our review is presented in Ref. [31].

Here we present the analysis of two JET tokamak Ni and Mo spectra from pre-divertor limiter discharges. The intrinsic Ni spectrum in the 80-120 Å spectral range was obtained during a steady state phase of a 5 MA discharge heated by 6 MW ion cyclotron resonance heating (ICRH) pulse ($T_e(0)$~5 keV, $N_e(0)$~2 10^{13} cm^{-3}). The Mo spectrum in the 35-50 Å spectral range was obtained with a LBO injection from a composite Mo-Ge target into a 3 MA discharge heated by a 8 MW ICRH pulse ($T_e(0)$~6 keV, $N_e(0)$~3.5 10^{13} cm^{-3}).

After evaluation of radial CSD, post-process subroutines are used to produce simulated spectra, using line photon emission coefficients (PECs) from collisional-radiative models (CRM). Like for Refs [18-23], the PECs have been obtained using the HULLAC code.

We perform best fits of the simulated spectra to the peaks of the strongest lines (i.e., to the main contributors to the detected spectra). If sp_{exp} (λ) and sp_{sim} (λ) are, respectively, the normalized experimental and simulated spectra, C_{mult} is a multiplicative constant, whose value is obtained by minimization of $\Sigma_i (C_{mult}\, sp_{sim}(\lambda_i) - sp_{exp}(\lambda_i))^2$ over the selected lines. If a verification of the atomic data is possible, the 'best' spectra are the ones with the lowest estimated sample deviation defined as the square root of the sum to be minimized.

Comparisons of experimental and simulated Ni spectra are given in Fig. 5, where, for clearness, the simulated thin line spectrum has been slightly shifted towards the blue. For this simulation both transport coefficients are the same as for the Ar seeded discharge of Ref. [21], implying, as found in several JET spectroscopic data analysis, the presence of a peripheral transport barrier, obtained reducing D and/or increasing V inward. The lack of other experimental spectroscopic data and/or of time resolved spectra hinders any verification of the used atomic data. Intensity ratios of L-shell lines can be density-sensitive and applicable to the diagnostic of astrophysical sources. In Ref. [19] experimental Fe line ratios are presented and compared with the predictions of the HULLAC code. Corresponding not-discussed Ni line ratios of the spectrum of Fig. 5 appear to be in reasonable agreement with the predictions.

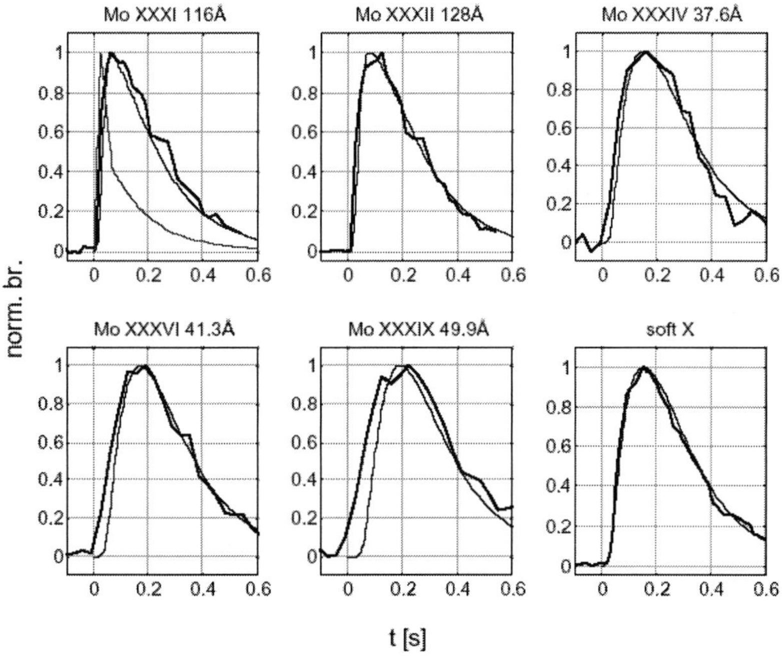

FIGURE 7. Experimental (thick line) and simulated (thin line) normalised brightnesses. The additional thin line (superposed to the Mo XXXI line) shows the Mo XIII line brightness, following the incoming neutral flux.

The transport coefficients D(r) and V(r) of the Mo-Ge LBO (shown in Fig. 6) have been obtained by simulations of Mo and Ge line brightnesses. They show the presence of both a peripheral barrier at the plasma edge and of an improved confinement core region.
As a general remark, it should be stressed that the values found for D by this fit are much larger than those predicted by theories based on collisional diffusion (so called neoclassical diffusion). The reason for the larger observed values is that the diffusion is dominated by turbulence both for impurities and the whole plasma. The question of turbulence and its possible alteration of line shapes will be introduced in the last section.
In Fig. 7 six normalized brightnesses are shown. The boundary condition (source function) of the incoming neutral flux at the last mesh has been chosen to follow the lowest ionisation stages observed [31] (Mo XXIV-XXV and Ge XXI).
The relative amplitudes of the lines from different ionisation stages depend on the transport parameters and on the atomic rates. The 35-50Å experimental spectrum, obtained 200 ms after the LBO injection, is shown (solid) in Fig. 8. The not-shown simulated spectrum shows B-like and Be-like ion lines clearly stronger than found experimentally. It is then necessary to modify either the recombination or the ionisation rates of the involved iso-sequences to reduce the core ionisation degree. One of the satisfactory simulations is shown (thin line, slightly shifted towards the blue for clearness) in Fig. 8. It has been obtained by multiplying, for recombining F-like to Be-ions, the α_{rec} rates by the following multiplicative factors f = 1.0, 1.1, 1.6, 1.8, 1.25, 1.25. The simulation was practically equivalent (i.e., with practically the same estimated sample deviation) by dividing by f the S_{ion} rates for ionising Ne-like to B-like ions. An idea of the uncertainties associated with the choices of the multiplicative factors f is given by the fact that the sample deviations are similar when decreasing and increasing the f-values by about 0.1-0.15 for two consecutive ionisation stages.

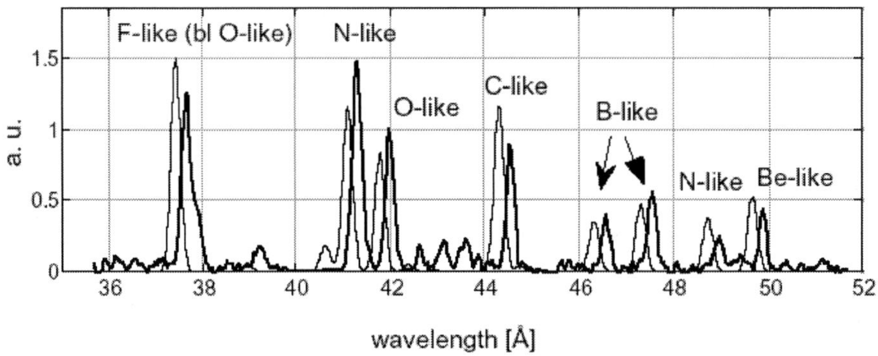

FIGURE 8. Experimental (thick line) and simulated (thin line) spectra, 200 ms after the LBO injection. For clearness, the thin spectrum has been slightly shifted towards the blue. The recombination rates of the six shown iso-sequences have been increased as explained in the text.

It is known that, both in astrophysics and in MCF devices, the CSD is reduced by recombining charge exchange (CX) collisions. In the latter devices, the considered CX collisions are with neutral hydrogen isotopes with poorly known radial profiles. Moreover, excepted for light elements, only scaled cross sections versus scaled energies are available.

The CRMs generally consider each ion independently from others ignoring ionisation and recombination into and out of the ground and excited energy levels. Processes included are radiative decays and electron collisions (proton collisions are sometimes included only between the lowly lying states).

In the Marseilles laboratory, CRMs have been considered adding CX reactions populating the excited states [32]. In fact, a self-consistent analysis of the particle transport and charge exchange (including also charge exchange from excited states) with a self-consistent multilevel collisional-radiative approach has shown that charge exchange phenomena, particle transport and atomic data are very closely related, as proved for the argon spectra in Ref. [33].

EFFECT OF HYDRODYNAMIC TURBULENCE

Edge plasmas of tokamaks are affected by low frequency turbulence. This turbulence is produced by small scale instabilities, induced by density, temperature and velocity gradients. Here we focus on drift wave turbulence, whose effects on the fluctuation of hydrodynamic variables has been studied by fluid models [34]. Such models predict fluctuations rates for the velocity, density and temperature reaching several tens of percent. For tokamak edge plasmas their characteristic fluctuation time is of the order of a few 10 μs, and their typical spatial length is about 10 Larmor radius (~cm). This characteristic turbulent time is usually much larger than the typical time scale of the atomic processes, such as for instance the line shape time of interest. On the other hand, the acquisition time τ_m for a line shape measure is generally much larger than this turbulent fluctuation time. A line shape measurement thus performs a spatial and time average over fluid cells having each a specific set of values for the density, velocity and temperature

$$I_{mes}(\omega) = \frac{1}{\tau_m} \int_0^{\tau_m} \frac{1}{L} \int_0^L I_0(\omega, z, t)\, dz\, dt, \tag{1}$$

where I_0 is the local profile at location z and time t, and L is the length of the line of sight. It has been shown [5] that the measured profile may be expressed as an integral over the fluid variables of the local profile, weighed by the probability density function (PDF) of the fluctuating hydrodynamic variables. The PDF of only one hydrodynamic variable has been studied theoretically [35,36], and has been measured in a few devices [37,38]. Since this quantity is of great interest for characterizing the edge plasma turbulence, the link made with the line shape may lead to a valuable diagnostic of turbulence by spectroscopy. Recent advances in this direction for Doppler line shapes [7] and Stark profiles [39] are reported in this volume.

CONCLUSION

Spectral lines have been extensively used since the beginning of spectroscopic diagnostics in magnetic fusion. The observation of line intensities, widths and shifts brings information on the plasma parameters and dynamics. A reliable diagnostic requires the use of realistic atomic data, atomic kinetics, and line shape models. The complex recycling process of hydrogen isotopes is clarified by the observation of neutral deuterium Balmer alpha lines broadened by Doppler and Zeeman effects. An accurate model for this diagnostic requires the full calculation of Zeeman effect by retaining the fine structure. A similar model is necessary for the calculation of opaque Lyman transitions used in Monte Carlo codes for predicting the divertor conditions. Stark effect affects line shapes of hydrogen and helium lines emitted from levels of high principal quantum numbers. These lines are used for an electronic density diagnostic in the edge regions. For the condition of ITER, Stark effect will affect most of the hydrogen lines due to higher charged particles densities. Models designed to perform a density diagnostic will thus often require the treatment of many body dynamic effects. As a general remark, an improved treatment of the spectral line shape should consider simultaneously atomic kinetics, radiation transport and diffusion processes that would affect the observations performed from non-uniform, non-equilibrium and non-stationary tokamak plasmas with complex composition.
Spectroscopy in the core plasma concerns principally the impurities. The fit of their spectra simultaneously improves the knowledge of atomic data, and the complex physics of transport. Remarkably, such fits indicate that the impurity transport is turbulent, this phenomena being especially marked while approaching the edge region. Since it is the whole plasma which is turbulent, this has been the starting point for an analysis of a possible role of hydrodynamic turbulence on the line shapes.

ACKNOWLEDGMENTS

This work is part of a collaboration (LRC DSM 99-14) between the Laboratory of Physique des Interactions Ioniques et Moléculaires and the association Euratom-CEA in Cadarache. The work of K.B. Fournier has been performed under the auspices of the US Department of Energy by the University of California Lawrence Livermore National Laboratory under Contract No. W-7405-Eng-48.

REFERENCES

1. S. Suckewer and E. Hinnov, Phys. Rev. Lett. 41, 756 (1978)
2. J. D. Hey, C.C. Chu, P. Mertens, S. Brezinsek, 17th Int. Conf. Spectral Line Shapes, Ed. E. Dalimier, Frontier group 2004, p.207
3. B. Welch *et al.*, 13th Int. Conf. Spectral Line Shapes, Ed. M. Zoppi, AIP 386, p.113
4. R. C. Isler *et al.*, Phys. Plasmas 6, 541 (1999)
5. Y. Marandet *et al.*, Europhysics Letters 69, 531(2005)

6. Ph. Mertens *et al.*, Plasma Phys. Control. Fusion 43, A349 (2001)
7. Y. Marandet *et al.*, this volume
8. D. Reiter, S. Wiesen and M. Born, Plasma Phys. Control. Fusion 44, 1723 (2002)
9. V. Kotov *et al.*, to appear in Contributions to Plasma Physics
10. H. R. Griem, A. C. Kolb and K. Y. Shen, Phys. Rev. 116, 4 (1959)
11. J. Rosato *et al.*, to appear in Journal of Nuclear Materials.
12. S. Günter and A. Könies, JQSRT 62, 425 (1999)
13. L. Godbert-Mouret *et al.*, JQSRT 71, 365 (2001)
14. M. Adams *et al.*, Phys. Rev. E 66, 066413 (2002)
15. C. Stehlé, Eur. Phys. J. D. 11, 491 (2000)
16. B. Talin *et al.*, Phys. Rev. A 51, 4917 (1995)
17. D. B. Boercker, C. A. Iglesias, and J. W. Dufty, Phys. Rev. A 36, 2254 (1987)
18. M. Mattioli M *et al.*, Plasma Phys. Control. Fusion 44, 33 (2002)
19. K. Fournier *et al.*, Astrophys. J. 561, 1144 (2001)
20. M. Mattioli *et al.*, Phys. Rev. E 60, 4760 (2000)
21. M. Mattioli *et al.*, J. Phys. B 34, 127 (2001)
22. K. Fournier *et al.*, Astrophys. J. 550, L117 (2001)
23. M. Mattioli *et al.*, J. Phys. B 37, 13 (2004)
24. M. May *et al.*, Phys. Rev. E 64, 036406 (2001)
25. P. Mazzotta *et al.*, Astron. & Astrophys. Suppl. 133, 403 (1998)
26. G. Mazzitelli and M. Mattioli, At. Data Nucl. Data Tables 82, 313 (2002)
27. P. Bryans *et al.*, to be published in Astrophys. J. Suppl.
28. N. Badnell N *et al.*, Astron.& Astrophys. 406, 1151 (2003) and following articles in A&A
29. N. Badnell, submitted to Astrophys. J. Suppl.
30. M. Pindzola *et al.*, Physica Scripta T37, 35 (1991)
31. M. Mattioli *et al.*, to appear in J. Phys. B
32. F.B. Rosmej, R. Stamm, V. Lisitsa, Europhysics Letters 73, 342 (2006)
33. F.B. Rosmej *et al.*, Plasma Phys. and Control. Fusion 41, 191 (1999)
34. W. Horton, Rev. Mod. Phys. 71, 735 (1999)
35. Y. Marandet and J. Dufty, Contrib. Plasma Phys., 46 (7-9), 197 (2006)
36. J. Krommes, Phys. Reports 360, 1 (2002)
37. J. Graves *et al.*, Plasma Phys. Control. Fusion 47, L1(2005)
38. A. Ferreira, Plasma Phys. Control. Fusion 46, 669(2004)
39. J. Rosato *et al.*, this volume

Coupling of Helium and Impurities to the H/D Background in Magnetically Confined Plasmas

F.B. Rosmej[1], R. Stamm[1], V.S. Lisitsa[2]

[1] *PIIM, Université de Provence, UMR 6633, Centre de St. Jérôme, 13397 Marseille Cedex 20, France*
[2]*Russian Research Center Kurchatov, 123182 Moscow, Russia*

Abstract. Alpha particles and impurity ions are coupled to the hydrogen/deuterium background plasma via charge exchange processes. The corresponding change of the charge state distribution of the "radiating test particle" is identified to be of similar nature like the particle transport. Traditional analysis in terms of diffusion and convection coefficients (D, V) is therefore difficult to interpret as the neutral density distribution is not well known, charge exchange processes are treated as a free parameters and excited state contributions are diverging. A self-consistent excited states coupling approach (SCESCA) of impurity and plasma background atomic kinetics is proposed which removes these difficulties. SCESCA provides an ideal basis for a pure diffusion analysis in space which is requested to test global effects of turbulent transport theories. Simulations for argon and helium are carried out which are contrasted to the data.

Keywords: particle transport, diffusion, radiation emission, charge exchange, diagnostic.
PACS: 32.70.Fw, 52.25.Vy, 52.25.Ya

INTRODUCTION

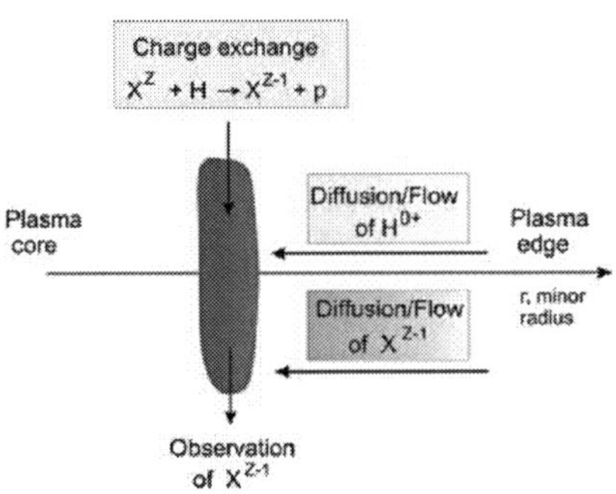

Figure 1. Superposition of transport effects of particles X and charge exchange with the background plasma H/D.

The confinement of the fusion plasma is one of the most important issues and intensive efforts have therefore been devoted to the understanding of the particle transport. However, the physical processes that underlie plasma transport in torodially confined plasmas are not so well understood. The plasma transport induced by Coulomb collisions (so called classical or neo-classical transport) is often much less than what is actually observed [1, 2] and thus the transport is called anomalous.

Methods which determine the particle transport independent of theoretical plasma models are therefore of fundamental importance in the magnetic fusion research. Spectroscopic methods have turned out to be very effective. One of the most powerful spectroscopic methods is based on the space and time resolved observation of line

CP874, *Spectral Line Shapes: 18th International Conference,*
edited by E. Oks and M. Pindzola

emission from impurity ions [1-3]. The emission is simulated from an atomic physics model with given temperature and density profiles [1, 4] and contrasted with the experimental data. Diffusion and convective velocity parameters, D and V, respectively are then determined by matching the experimental observations [1-3].

However, under real experimental conditions of fusion plasmas, the impurity ions do interact with the plasma background H/D via charge exchange (Cx) which in turn leads to a change of the radial charge state distribution (CSD) of the impurity ions. This effect has a large impact for the analysis of the particle transport because diffusion in space (particle transport) and diffusion in charge states (charge exchange) are of similar nature in the framework of the traditional particle transport analysis via diffusion coefficients D and convective velocities V [5-7]. The superposition of charge exchange and particle transport can be illuminated as follows (see also Figure 1):

$$n_e n^{Z+1}\left(\langle DR\rangle + \langle RR\rangle + \left[\langle Cx\rangle + diffusion/convection\right]\right) = n_e n^{Z}\langle I\rangle \qquad (1)$$

Eq. (1) demonstrates that not only charge exchange and particle transport are strongly related to each other, but dielectronic/radiative recombination (DR, RR) and ionization (I) rates too [5-8].

IMPURITY RADIATION AND CHARGE EXCHANGE

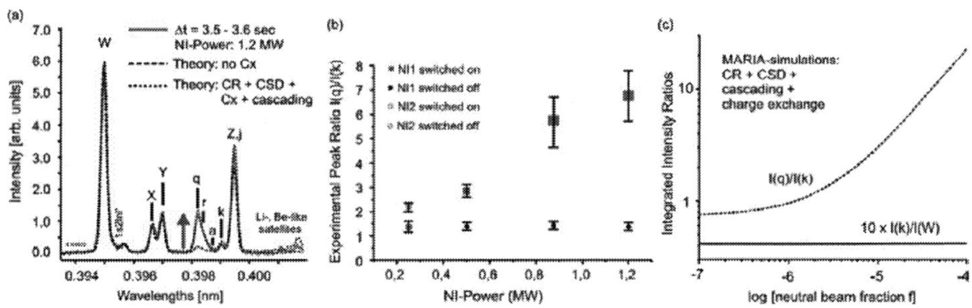

Figure 2. a) Argon impurity x-ray spectra observed in TEXTOR during neutral beam injection "___" and SCESCA simulations ".......", b) observed q-satellite intensity, c) simulated intensity ratios

Figure 2a shows the argon x-ray emission during neutral beam injection (NI) observed at the TEXTOR tokamak, more details are described elsewhere [7]. Figure 2b shows a strong rise of the q-satellite intensity with increasing power of the NI. This observation suggests charge exchange phenomena to be the cause. Figures 2a, c show simulations in the framework of SCESCA where charge exchange is calculated from the coupled H/D populations itself [6, 9]. Figure 2c shows a strong dependence of the q- and k-satellite ratio from the neutral beam fraction $f = n_{H/D}/n_e$. After switching off NI at t = 3.7 sec the neutral background is still enhanced to a level of about $f = 6\cdot 10^{-6}$ (Δt=3.9–4.0 sec) whereas during NI $f = 1.7\cdot 10^{-5}$ (Δt=3.5–3.6 sec). Excellent agreement with the observations (dotted curve in Figure 2a) are seen. Simulations not taking into account charge exchange (dashed curve in Figure 2a) considerably underestimate the intensities for the q-satellite and also for the forbidden lines (x,y,z). Charge exchange driven signatures in forbidden lines have first demonstrated by multi-level collisional radiative (CR) cascade simulations [6] and have been recently repeated and confirmed

[10]. However, the claimed surprise [10] due to large Z-line intensities in simulations [7] are irrelevant as the plasma parameters in [10] and [7] are quite different. Moreover, unlike claimed in [10], the present simulations and those in [7] are in excellent agreement with the observations.

HELIUM EMISSION

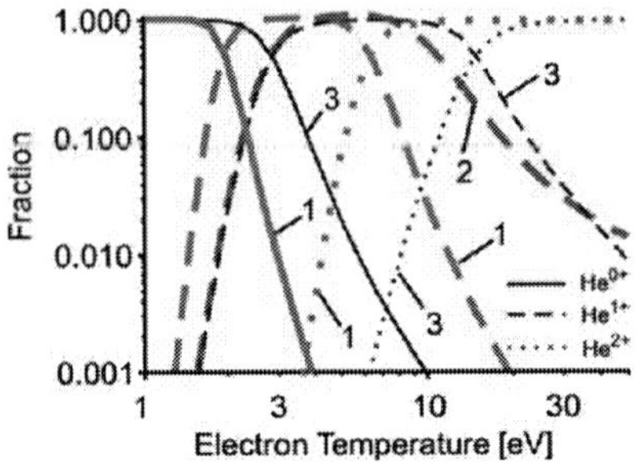

Figure 3. SOPHIA simulations of the CSD of helium, 1) SCESCA, 2) diffusion, 3) Cx ***not*** self-consistent

Fig. 3 shows the CSD of He: '1' includes Cx with H/D in the framework of SCESCA [9], '3' includes Cx in the traditional approach according eq. (1), '2' shows diffusion where Cx is ***not*** taken into account at all. As dashed curves '2' and '3' are very similar , the superposition of Cx and diffusion effects (Fig. 1, eq. (1)) can be clearly seen. We note that simulations '3' are diverging due to the excited states contributions : $\sigma_{Cx} \sim n^4$. The SCESCA removes this divergence via the thermalization of the CR-coupled highly excited states of helium and H/D [9].

CONCLUSION

The self-consistent excited states coupling approach SCESCA of impurities and alpha particles to the H/D background removes the free parameters for the charge exchange from the theory and converges all excited states contributions on a unique footing. This approach provides an ideal basis for independent spectroscopic analysis of turbulent particle transport phenomena.

ACKNOWLEDGMENTS

This work is part of a collaboration (LRC DSM 99-14) between PIIM and the Département de Recherches sur la Fusion Contrôlée, CEA Cadarache. Partial financial support from ARCUS-Program (n° P003-5a) is also acknowledged.

REFERENCES

1. W. Engelhardt, in "Course on diagnostics for fusion reactor conditions", vol. 1, p. 1, Varenna, Italy, 1982, EUR 8351-IEN.
2. D. Pasini, M. Mattioli, A.W. Edwards, R. Gianella, R.D. Gill, N.C. Hawkes, G. Magyar, B. Saoutic, Z. Wang, D. Zasche, Nuclear Fusion **30**, 2049 (1990).
3. R.A. Hulse, Nuclear Technology/Fusion **3**, 259 (1983).
4. C. de Michelis, M. Mattioli, Nuclear Fusion **21**, 677 (1981).
5. V.A. Abramov, V.S. Lisitsa, A.Yu. Pigarov, JETP Lett. **42**, 356 (1985).
6. F.B. Rosmej, V.S. Lisitsa, Phys. Lett. A, **244**, 401 (1998).
7. F.B. Rosmej, V.S. Lisitsa, D. Reiter, M. Bitter, O. Herzog, G. Bertschinger, H.-J. Kunze, Plasma Physics and Controlled Fusion **41**, 191 (1999).
8. V.A. Shurygin, Plasma Physics Reports **30**, 443 (2004).
9. F.B. Rosmej, R. Stamm, V.S. Lisitsa, Europhysics Letters **73**, 342 (2006).
10. P. Beiersdorfer, M. Bitter, M. Marion, R.E. Olson, PRA **72**, 032725 (2005).

Shear Stabilization of Turbulence in Tokamaks: An Application for Line Shape Analysis of Impurity Spectra Excited by Interaction with a Hydrogen Neutral Beam

William L. Rowan, I. O. Bespamyatnov, and R. V. Bravenec

Fusion Research Center, The University of Texas at Austin, Austin, TX 78712

Abstract. The edge of a tokamak plasma can be characterized by regions of sharp gradients which indicate good plasma confinement by the magnetic fields. Unfortunately, these gradients and the regions of high confinement do not extend very far into the plasma. Here we will report our experiments which are intended to understand why the region of good confinement does not extend further into the plasma. Plasma parameters which are critical to interpretation of the experiments are derived from line shape analysis of impurity spectra. To support this interpretation, an accurate model for the line shape is needed and will be presented here. The model takes into accounts effects of magnetic field, details of the excited state of the beam, motion of the emitting ion, and blends of interfering spectra. The methods for extraction of the useful spectral parameters, intensity, spectral width spectral shift, and uncertainties in each, will be described.

INTRODUCTION

This paper describes an application of spectral line shapes which elucidates the physics of plasma confinement in magnetic fields in the presence of turbulence. The spectral line shape analysis yields the intensity, width and shift of the spectral line which are then used to infer the relative density of the emitting impurity, its temperature and velocity. These quantities are measured for one spectral line but in the presence of line blends and taking into account distortions from simple line shapes due to fine structure, Zeeman splitting, finite decay time, and atomic cross sections. The plasma parameters derived from the spectral analysis contribute to estimates of plasma flow and radial electric field in the plasma. These quantities have a significant impact on the stabilization of plasma turbulence. When incorporated into detailed plasma simulations which embody state-of-the-art theories of transport in turbulent plasmas, these quantities play a major role in understanding plasma confinement in the regions of interest for these experiments.

Confinement in tokamaks

In some instances, the edge of a tokamak plasma is characterized by sharp gradients which indicate good plasma confinement by the magnetic fields. In figure 1, each of the curves represents the electron temperature at the periphery of the plasma for a different type of confinement. The region of particular interest is the gradient region near 0.89 m.

CP874, *Spectral Line Shapes: 18th International Conference,*
edited by E. Oks and M. Pindzola

The blue curve represents a plasma of low confinement (L-Mode). In this case, it is an Ohmically heated plasma. With application of RF heating, the gradient steepens (red curve) as the plasma makes a transition to a new confinement mode, H-mode. Compared to L-mode, H-mode is also marked by a reduction in plasma turbulence.

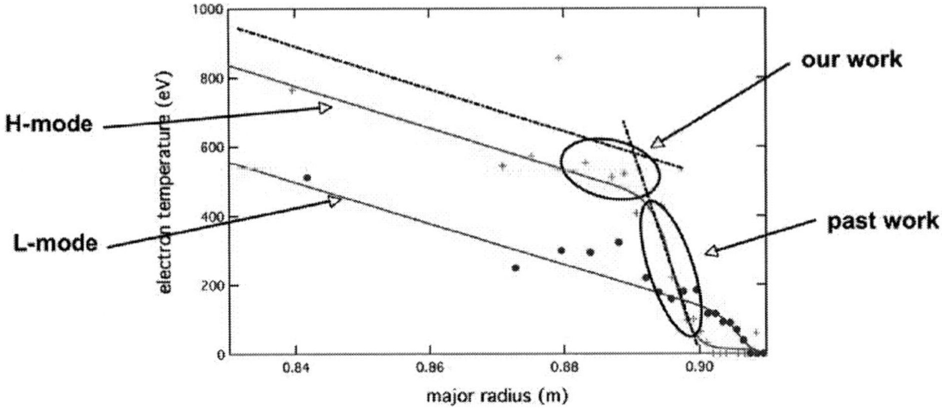

FIGURE 1. Examples of measured electron temperature profiles for L-Mode (blue) and H-mode (red) for a portion of the Alcator C-Mod cross section (magnetic axis is approximately at 0.67 m). The dashed line illustrates the implications of extending the region of good confinement just one centimeter further into the plasma.

Unfortunately, the H-mode gradient and the improved confinement that it witnesses does not extend very far into the plasma. In this application, we explore the physics which prevents this. Our reason for desiring that the region of high confinement extend further into the plasma is shown quite clearly by the imaginary profile sketched in Fig. 1 as a dashed line. If the region of high confinement extended further into the plasma then there would be higher reactivity in the plasma core, a fusion goal. There are good plasma physics reasons for studying this as well: Turbulence is important in many areas beyond plasma physics, and it plays an important role in plasmas in areas other than fusion related ones.

Past work in understanding the stabilization of turbulence in H-mode, has concentrated on understanding the physics of the gradient region (see Fig. 1). We take a different point of view in our recent work and ask an equally important question: Why is it that the region of good confinement does not extend further into the plasma? In other words, what is special about the region that causes transport to change from high confinement to low confinement.

An advantage of exploring this region of reduced gradients is that we have an opportunity to deploy the most detailed physics models for fluctuation induced transport in tokamak plasmas. These models contain the most advanced description of plasma behavior in the presence of plasma turbulence and are believed to cover all relevant spatial and temporal scales. A limitation of the models is that they do not necessarily properly represent plasmas in strong spatial gradients. Therefore, we investigate microstability at the top of the H-mode pedestal, pushing as close to the steep-gradient region as possible. In this way, we hope to gain insight into the relevant instabilities being stabilized in the steep-gradient region. These models are applied to the problem via simulation of turbulence due to measured macroscopic plasma profiles. Here, we employ the GS2 code.[1]

Approach and Role of Spectral Line Shapes

The approach that we have chosen depends heavily on measurement and analysis of spectral line shapes and on accurate prediction of the turbulent modes. The physics that we know seems to suggests this scenario. Plasma transport is driven by plasma turbulence. The source energy for the turbulence is in the plasma gradients - a necessary consequence of magnetic confinement. Suppression of the turbulence has been observed and is correlated with improved confinement and sharper gradients: H-Mode, for example. One turbulence suppression mechanism has been identified: flow-shear stabilization (discussed later). We have criteria for predicting the suppression, but the mechanism is only studied for a limited range of parameters. From what we do know, flow-shear stabilization can be characterized by T_i, v, and E_r. All of these are accessible via spectral line-shape analysis and form the goals of the empirical work described here. In addition, we have a new-found ability to predict unstable modes from plasma profiles which was discussed above.

WHAT WE NEED TO KNOW ABOUT TURBULENCE AND SHEAR SUPPRESSION

This critical issue in plasma confinement is discussed in detail elsewhere.[2,3] Here, a qualitative description is provided which captures the important physics. Turbulence in plasmas is manifested as broadband fluctuations $\Delta\omega \sim \omega \leq 500kHz$ in density, temperature, and other plasma parameters. It grows from noise level fluctuations in the plasma and draws its energy from the density and temperature gradients which are an unavoidable consequence of magnetic confinement. The noise level fluctuations are due to numerous non-linearly coupled and saturated unstable modes. The primary interest here is in turbulence driven by ion temperature gradients which is widely known and feared simply as ITG turbulence.

Flow-Shear Suppression of Turbulence

Microscopic turbulence generates macroscopic transport because the turbulence is correlated over some spatial scale, e.g., d_l in panel a of Figure 2. It is the correlation that allows the turbulence to take the plasma particles on a random walk down to the edge of the plasma. Physically, a random walk with step size d_l is summarized by a diffusive transport $\propto (d_l)^2$. At least, that is what happens when the plasma flow field is homogenous. The new idea of flow fields is introduced in panels b and c of Fig. 2.

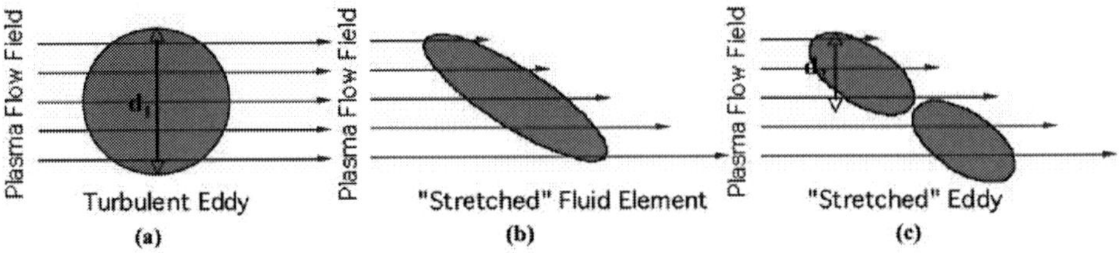

FIGURE 2. A qualitative explanation of flow shear stabilization. Turbulent eddies induce random walk of particles and energy on the spatial scale of the eddy. In the presence of an inhomogeneous plasma flow field, the spatial scale of the random walk decreases and the transport goes with it.

As indicated in panel (b) of Fig. 2, different parts of the fluid are advected at different speeds by an inhomogeneous flow field. The fluid element that encompassed the turbulent eddy in panel (a) is stretched by the flow. The eddy or the region over which the turbulence is correlated cannot continue to occupy the entire fluid element. The eddy is coherent only over a distance like d_1 along the stretched fluid element. Thus, it appears to break into multiple eddies with characteristic dimension d_1, and the spatial scale of the transport transverse to the flow field is reduced to d_2. Physically, the transport scale has become smaller and the diffusive transport $\propto (d_2)^2$. This is an altogether satisfying outcome. It is this flow shear suppression that we are trying to measure

Evaluation of Flow-Shear Stabilization from Line Shape

It is the contention here that line shape analysis is the fundamental tool for the evaluation of flow shear stabilization. Turbulence prediction requires a complete description of the plasma including $T_e(r)$, $T_i(r)$, $n_e(r)$, $n_i(r)$, $n_z(r)$. Of these fields, two, $T_i(r)$ and $n_z(r)$ follow from line shape analysis. The first prediction of turbulence theory that is needed is the growth rates of the turbulence modes. The only one of interest to us is the maximum growth rate γ This mode is suppressed by an inhomogeneous plasma flow field which is simply expressed as

$$v = \frac{E \times B}{B^2}$$

where E is an electric field with a component transverse to B, the confining toroidal magnetic field in the tokamak.[4] This is a direct consequence of an analysis of the motion of a particle in a strong magnetic field. The criterion for shear suppression[2] is then $\omega_{E\times B} > \gamma$
where

$$\omega_{E\times B} = \frac{(RB_\theta)^2}{B}\left(\frac{\partial}{\partial\psi}\right)\frac{E_r}{RB_\theta}$$

The source of E_r in plasmas remains a major unsolved problem. It can be measured, of course, and we extract it here from our measurements and quite reliable conservation equations. One component of the momentum balance equation in a plasma can be rearranged to yield

$$E_r = \frac{kT_\sigma}{e_\sigma}\frac{1}{n_\sigma}\frac{\partial n_\sigma}{\partial r} + \frac{1}{e_\sigma}\frac{\partial kT_\sigma}{\partial r} - v_{\sigma\theta}B_z + v_{\sigma z}B_\theta$$

The quantities expected from line shape analysis and contributing to E_r are $T_i(r)$, $n_z(r)$, , n_σ, $v_{\sigma\tau}$, $v_{\sigma z}$

The remainder of this paper discusses the complications of line shape analysis for spectral lines emitted by plasmas in Alcator C-Mod. Before that, the results of the line shape analysis will be presented since the problems encountered are quite complicated and the following graphs offer assurance that these can be overcome.

The plasma parameters for a particular C-Mod discharge are shown in Fig. 3.

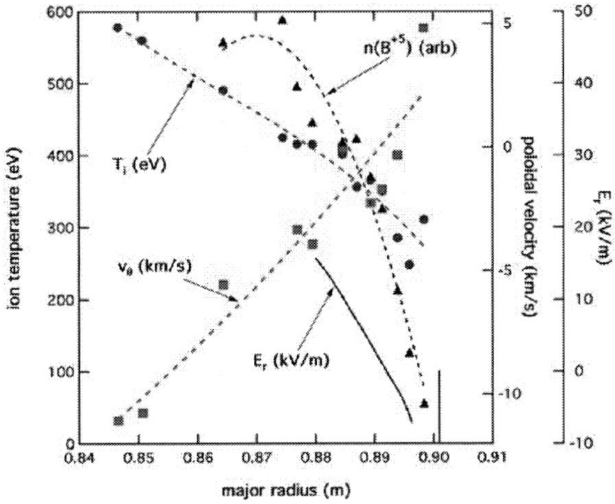

FIGURE 3. Ion temperature (T_i), relative impurity density ($n(B^{+5})$), poloidal velocity or velocity transverse to the confining toroidal magnetic field (v_θ), and an inferred radial electric field (E_r) in an Alcator C-Mod discharge. These quantities are the result of line shape analysis.

THE SPECTRUM

The spectrum that we will interpret is excited by charge exchange. A beam of hydrogen neutrals is injected into the plasma as shown in Fig. 4. The beam neutrals charge exchange with fully stripped B^{+5} to form excited hydrogen-like B^{+4}.

$$H_{Beam} + B^{+5} \rightarrow H^{+} + B^{+4}(j) \tag{1}$$

Relaxation via radiative decay produces the emission whose spectrum we observe

$$B^{+4}(j) \rightarrow B^{+4}(k) + h\nu \tag{2}$$

The equations (1) and (2) require inclusion of more detailed physics for quantitative analysis of the spectrum. Of course, it is the detailed physics that can make this an interesting line shape problem. Before describing the features of the spectral interpretation, we will describe the observed spectrum

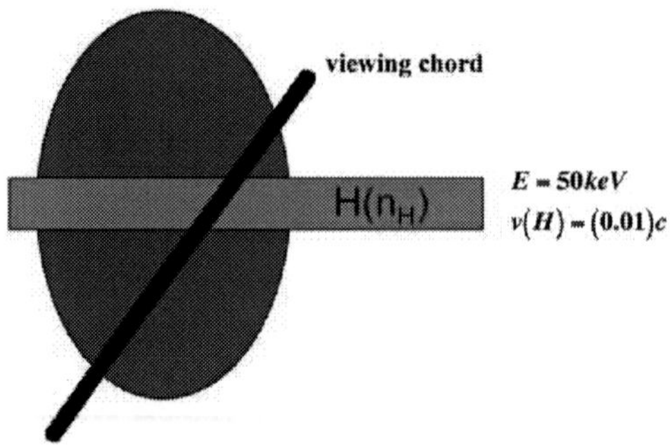

Figure 4. A beam of hydrogen neutrals (red) is injected in the plasma (blue). The interaction of beam and plasma produces emission that is observed along the viewing chord.

A typical spectrum is shown in Fig. 5. Its features are a B^{+1} line[5] at 4940.38, (2s3d $^1D^2 \rightarrow$ 2s4f $^1F^3$), a continuum composed mainly of bremsstrahlung, and two B^{+4} lines, one at 4944.67, n=7→6 and a second at 4950.77, n=11→8. The 4944.67, n=7→6, B^{+4} line contains three superposed components due to electron ion collisions, thermal charge exchange, and active charge. The active charge exchange component is identified by comparing spectra with and without excitation by the neutral beam. Thermal charge exchange and electron impact components are inferred from the presence of the required reactants and confirmed via prediction.

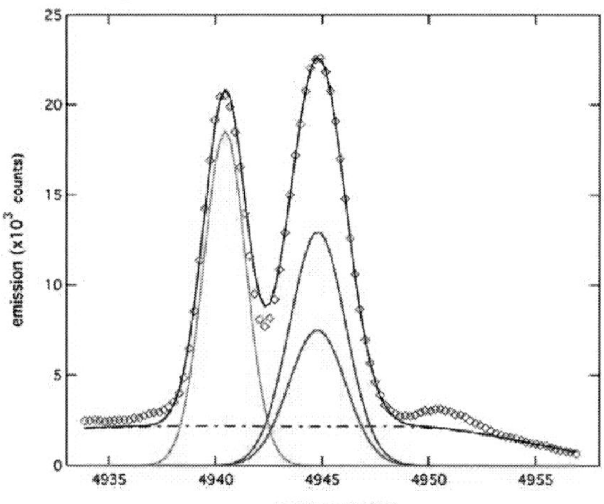

FIGURE 5. An emission spectrum observed from Alcator C-Mod. The total spectral blend (diamonds) is composed of 4940.38, (2s3d $^1D^2 \rightarrow$ 2s4f $^1F^3$) B^{+1} (green), 4944.67, n=7→6, B^{+4}, 4950.77, n=11→8, B^{+4}, and a continuum background. The 4944.67, n=7→6, B^{+4} is composed of components due to active charge exchange (red) and thermal charge exchange and electron ion collisions (blue).

The spectral components are separated in an empirical analysis which employs non-linear least squares fitting to a model. The intensity and shape of the thermal charge and electron impact components of the B^{+4} line is measured with the beam turned off although this might possibly be better done via prediction.[6] The intensity of the active charge exchange portion of the spectrum is due to charge exchange cross sections with ground state beam neutrals <u>and</u> the excited states. C-Mod has typical particle density of

$1 \times 10^{20} m^{-3}$ and that high density results in significant excitation. For C-Mod, the confining magnetic fields range from 3 T to 8 T. The Zeeman splitting for high n transitions is significant. When modified by Doppler broadening of the components, this produces something like a broadening effect and can even introduce a shift in the centroid of the spectral shape. Further distortions in the line shape are introduced by charge exchange cross section effects. Each of these will be discussed in turn.

Spectral Intensity

The beam is not monoenergetic. For reasons associated with beam production, all positive hydrogen ion beams are composed of three or four components each with a different energy. The beam used in these experiments is composed of four components (E, E/2, E/3, and E/18) where E is the main energy of the acceleration. Eqn 1 then becomes

$$\sum_{i=1}^{4} H_{Beam}(E_i) + B^{+5} \rightarrow H^+ + B^{+4}(j) \tag{3}$$

Each of the beam components is excited by collisions with plasma particles. Each of the excited states contributes to the density of B^{+4} with a different cross section. Thus the concentration of each component must be measured. An example of the measurement is in Figure 6. The spectrum shown is taken at approximately 50° to the beam trajectory and shows the various components in a typical beam. Each component is represented by a Doppler shifted component of H_α. Using known cross sections[7], each component density can be inferred.[8]

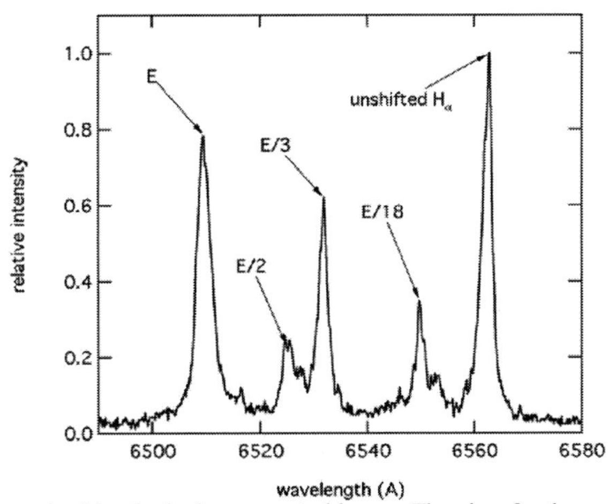

FIGURE 6. Spectrum emitted by the hydrogen neutral beam. The view for the spectral measurement is has a component anti-parallel to the beam trajectory. Each energy component is represented by a Doppler shifted H_α. The unshifted component is emitted by the background target gas.

Excitation of each of the components is then reliably derived from a collisional-radiative model for the beam. In the case of C-Mod, this is particularly important because the plasma density is relatively high at $1 \times 10^{20} m^{-3}$. Compared with other applications, the collisional radiative model for the beam is unique in that Lorentz ionization limits the number of available excited states.[9] In C-Mod, this means that the Lorentz ionization limit is approximately n=7. An example for the n=2 and n=3 states is shown in Fig. 7.

The emission rate coefficients corresponding to excitation of B+4 by each beam component are in some cases available. In this case, all of the cross sections were not available and were supplemented by scaling from known cross sections.[10] The result of this is shown in Fig. 8. Clearly, there is a strong energy dependence which differs among the components. This illustrates that qualitative as well as quantitative results are dependent on having the correct cross sections.

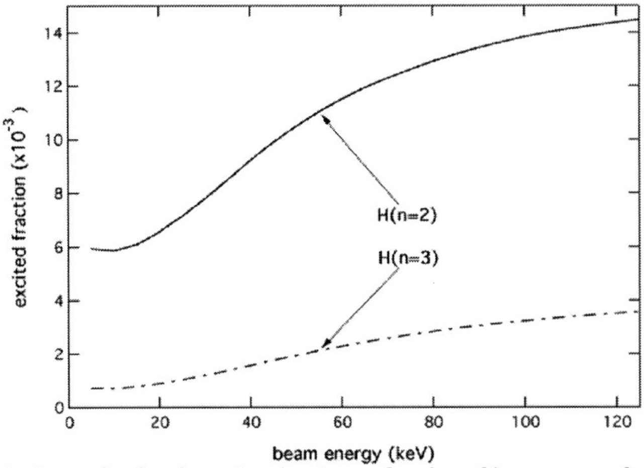

FIGURE 7. Excited state fraction for n=2 and n=3 as a function of beam energy for a typical C-Mod plasma.

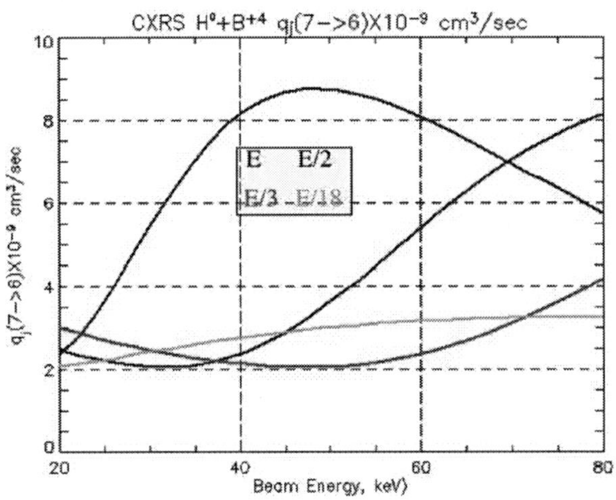

Figure 8. Emission coefficients for the n=7 to n=6 transition in B^{+4}. The coefficient for each energy component is shown.

Spectral Width and Shift

The spectral line array consists of 31 fine structure components. Of course, the components have been known to arbitrarily high accuracy for many years.[11,12] The distribution is shown in Figure 9.

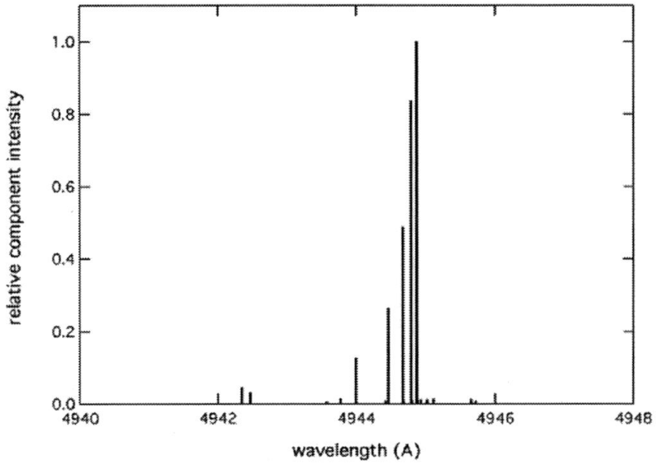

FIGURE 9. The distribution of wavelengths and intensities of the j-resolved components of the spectral line. The intensities follow from full mixing.

The relative intensity of each component is usually assumed to be determined by full mixing of the n=7 levels. This particular point has not been closely studied and it is possible that this is a distribution driven by the charge exchange process. Because the Doppler width for the components tends to merge them, the individual lines cannot be resolved and must be considered as a single spectral structure. The complexity of the underlying structure introduces some interesting and non-intuitive effects which are summarized in Figure 10. The centroid of the spectrum shifts to shorter wavelength as the temperature of the emitting ion increases. The Zeeman effect reduces the effect of width on shift because the intensity of the outlying transitions is reduced.

These problems do not affect analysis so long as the full set of fine structure components are used for the line fitting model. They are important because of the advantages that would be offered by a simplified analysis technique which would lead to faster analysis. Tokamak discharges last for only a couple of seconds and are generated at 10 to 15 minute intervals. Efficient operation of these expensive devices require reasonably good analysis immediately after a shot and in time to impact planning for the next shot. In this case, many spectra must be analyzed between shots: 45 spatial channels and 100 spectra per channel for a 1 s shot. Simple ideas such as representation of the line as a single Gaussian to acquire T_i or a centroiding technique for line shift to acquire rotational velocity will not adequately describe the line and would need to be supplemented by something like the data of Figure 10 recast as a correction factor. One slightly more complex method[13] that has achieved some currency is a line model consisting of three gaussian components which were intended to reflect one piece of physics -- Zeeman splitting into three components. Since the Zeeman effect is smaller than the range of the fine structure components, this might seem to be inappropriate. In fact though, it does work to some degree. The loss with this type of fit is that the line centroid shifts with increasing temperature and accurate shifts cannot be derived from this type of analysis. Corrections are still required. It might be better to devise a better "simplified" model. There is then still a need for a simplified method for fast analysis. Of course, it would need to be firmly based on the results of the physics based line mode.

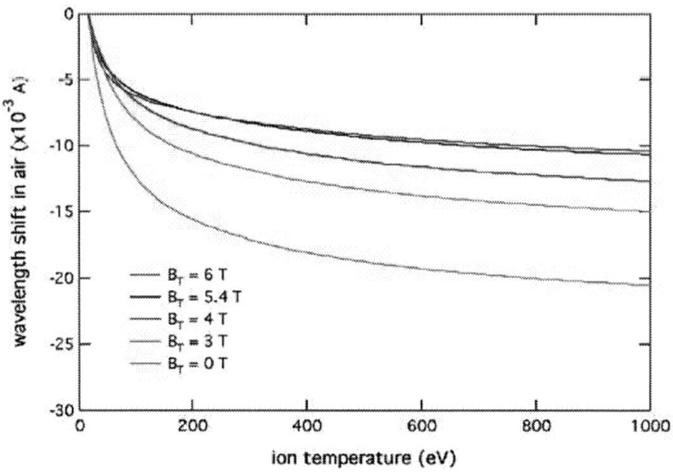

FIGURE 10. The centroid of the spectral structure increases with increasing ion temperature and decreasing magnetic field. This is due to the asymmetric distribution of fine structure components.

Cross Section Effect

The "cross section effect"[14,15] can most easily be described by contrast with the Doppler broadened line shape. For a Doppler broadened line shape, the particle velocity distribution is directly reflected in an emission distribution. If the particle velocity distribution along the line of sight is

$$f = \frac{n}{\sqrt{\pi}}\sqrt{\frac{m}{2kT}}\exp\left(-\frac{mv^2}{2kT}\right)$$

then a simple transformation between velocity and the Doppler shifted wavelength yields the Doppler broadened and Doppler shifted line shape.

$$I\ d(\Delta\lambda) = I_0 n\left(\frac{c}{\lambda_0}\sqrt{\frac{m}{2\pi kT}}\right)\exp\left(-\frac{c^2(\Delta\lambda)^2 m}{2\lambda_0^2 kT}\right) d(\Delta\lambda)$$

In the CXRS case, the line shape is distorted by the energy dependence of the cross section for views with a component along the beam. We are attempting to measure the velocity distribution of B^{+5}.

$$H_{Beam} + B^{+5} \rightarrow H^{+} + B^{+4}(j), \quad q(E_{col})$$

The emission is from B^{+4} which is formed according to a collision cross section $q(E_{col})$ as shown in Figure 11.

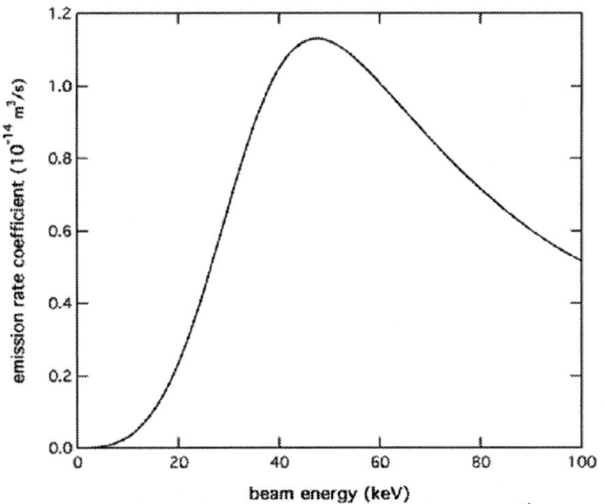

FIGURE 11. A emission rate coefficient for the n=7 to n=6 transition in B^{+4}.

This allows the line profile to be distorted by the energy dependence of the cross section for views with a component along the beam trajectory.
The distribution is then replaced by this from which the line profile is derived by the familiar transformation between Doppler shifted wavelength and velocity

$$f(v_z)dv_z = dv_z \int_{-\infty}^{\infty} dv_x \int_{-\infty}^{\infty} dv_y \frac{1}{\sqrt{\pi} v_{th}} \exp\left[-\left(\frac{v_x - r_x}{v_{th}}\right)^2\right] \frac{1}{\sqrt{\pi} v_{th}} \exp\left[-\left(\frac{v_y - r_y}{v_{th}}\right)^2\right] \frac{1}{\sqrt{\pi} v_{th}} \exp\left[-\left(\frac{v_z - r_z}{v_{th}}\right)^2\right] q(E_{col})$$

A Comparison of the cross section effect for a view along the beam (0 degrees) and one at an angle to the beam (40 degrees) is shown in Figure 12. This effect would be eliminated for a view perpendicular to the beam, but for the finite lifetime effect.

The basic motion of a magnetically confined ion its gyration in the plane transverse to the confining magnetic field. Even though precautions are taken to assure that the viewing chord is perpendicular, the ion may still charge exchange while it is traveling along the beam trajectory. It will spontaneously decay after it has executed some fraction of its orbit and thus emit while traveling parallel to the view chord. The cross section is mixed into the view in spite of the precaution of measuring perpendicular to the beam. An example of this also is shown in Figure 12.

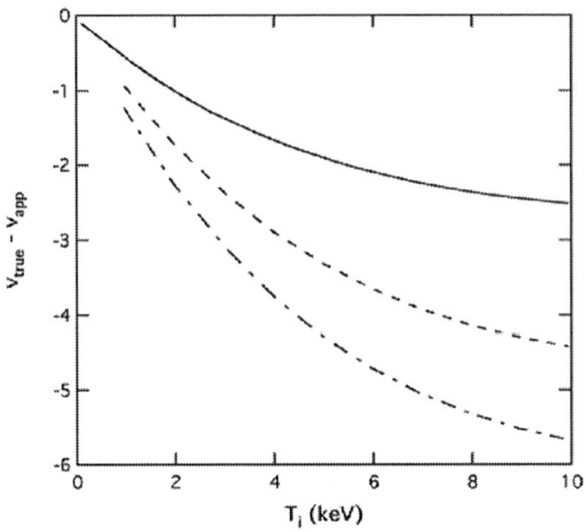

FIGURE 12. The difference between true rotation velocity and that inferred without correction for cross section effects and finite life time effects. The dash-dot line is the difference for a view chord aligned along the beam trajectory. The dashed line is the difference if the angle between the view chord and the beam trajectory is 40 degrees. The solid line is the difference if the angle between view chord and the beam trajectory is 90 degrees and the finite lifetime effect is omitted.

CONCLUSION

In this paper, a line shape analysis is described which leads to a much improved understanding of turbulence in magnetically confined plasmas. Though only a fraction of this problem in plasma physics, the analysis is complex and requires construction of a detailed model for the line and pains-taking extraction of useful data from the measured spectra via the model.

This application also illustrates the need for improved techniques. For example, simplified models would be useful to make analysis available more quickly. Based on the example discussed here, this requires that the full complexity of the line formation be understood. The penalty for not following this procedure is that important information may be not be available. Also, there remain interesting topics for study in the physics as for example uncertainty in the intensity distribution among fine structure components due to the charge exchange process.

ACKNOWLEDGEMENTS

This work was supported DOE grants DE-FG02-96ER54373 and DE-FG02-97ER54415. We thank the C-Mod team for the example data used in figure 1.

REFERENCES

1 Jenko, F., Dorland, W., Kotschenreuther M., and Rogers, B. N., *Phys. Plasmas* **7**, 1904 (2000).
2 Burrell, K. H., *Phys. Plasmas* **4**, 1499 (1997).
3 Terry,P. W., *Rev. Mod. Phys.* **72**, 109 (2000).
4 Chen, Francis F., *Introduction to Plasma Physics and Controlled Fusion*, New York and London: Plenum Press, 1984, chapter 2.
5 Olme, A., *Physica Scripta* **1**, 256 (1970).
6 von Hellermann, M. G., Bertschinger, G.,Biel, W. , Giroud, C., Jaspers, R. , Jupen, C. , Marchuk, O., O'Mullane, M. , Summers, H. P. , Whiteford, A. and Zastrow, K-D. , *Physica Scripta* **T120**, 19 (2005).
7 Williams, I. D., Geddes, J., and Gilbody, H. B., *J. Phys. B: At. Mol. Phys.* 15, 1377 (1982).
8 Rowan, William L.,Sampsell, M. B., Granetz, R. S., *Review of Scientific Instruments* **75**, 3487 (2004).
9 Boley, C. D., Janev, R. K., and Post, D. E., *Phys. Rev. Lett.* **52**, 534 (1984).
10 Bespamyatnov, I. O., Rowan, W. L., Granetz, R. S., and Beals, D. F., accepted for publication in *Rev. Sci. Instrum.*
11 Garcia, J. D. and Mack, J. E., *JOSA* **55**, 654 (1965).
12 Erickson, G. W., *J. Phys Chem Ref Data* **6** 831 (1977).
13 Blom, A. and Jupen, C., *Plasma Phys. Control. Fusion* **44**, 1229 (2002).
14 von Hellermann, M., Bregert, P., Frieling, J., Konigt, R., Mandls, W., Maast, A. and Summers, H. P., *Plasma Phys. Control. Fusion* **37**, 71 (1995)
15 Bell, R. E. and Synakowski, E. J., "New Understanding of Poloidal Rotation Measurements in a Tokamak Plasma," , in Atomic Processes in Plasmas: Twelfth Topical Conf, edited by R. C. Mancini and R. A. Phaneuf, Woodbury, New York: American Institute of Physics, 2000.

Line Shapes in a Turbulent Divertor Plasma

J. Rosato[1], R. Stamm[1], Y. Marandet[1], L. Godbert-Mouret[1], M. Koubiti[1], F. Rosmej[1], and H. Capes[2]

[1]*PIIM, UMR 6633 CNRS-Université de Provence, Centre de Saint-Jérôme, Case 232, F-13397 Marseille Cedex 20, France*
[2]*DRFC, Association Euratom-CEA, Centre de Cadarache, F-13108 Saint-Paul lez Durance Cedex, France*

Abstract. In this work we investigate line shapes in a divertor plasma and we study the effects due to the hydrodynamic turbulence.

Keywords: Line shapes, Turbulence, Divertor plasmas.
PACS: 32.70.Jz, 52.35.Ra

INTRODUCTION

The divertor plasmas of tokamaks are usually affected by low frequency turbulence [1,2]. Such kind of turbulence involves fluctuations of density and temperature at the hydrodynamic scale. The radiation of the hydrogen isotopes is related to the fluctuations since the emission strongly depends on the fluid parameters of the plasma. The line profiles are expected to be altered by the turbulence. Following a study made in a recent work on the wings of the D_α profile [3], we successively investigate here the core of D_{10} and D_α lines. These lines are commonly used as a diagnostic tool, respectively for the plasma density and for the neutral velocity distribution function. We first present the model which allows in a general case to describe line shapes with hydrodynamic turbulence, and we apply this model on the mentioned line shapes.

FORMALISM

Let us consider a plasma in which fluid quantities X = (N,**u**,T,...) are fluctuating. The measured line shape I(ω) results from a spatial and a temporal integration of the local line shape $I_0(\omega,X(z,t))$, respectively along the line of sight and during the spectrometer acquisition time. Assuming this time to be much larger than the time scale of turbulence, the formalism developed in [3] allows to describe the fluid fields X as a set of stochastic variables. The measured line shape is thus written as a statistical average

$$I(\omega) = \int W(X) I_0(\omega, X) dX \,, \tag{1}$$

where W(X) is the joined probability density function (PDF) of the fluid fields integrated on the line of sight.

CP874, *Spectral Line Shapes: 18th International Conference,*
edited by E. Oks and M. Pindzola

EFFECTS OF DENSITY FLUCTUATIONS ON D_{10} LINE SHAPE

We investigate in this section the profile of D_{10} line in presence of density fluctuations. This line, observed in detached plasmas, is mainly broadened by Stark effect. Assuming the density to be of the order of 10^{13} cm^{-3} and the ion and electron temperatures to be of the order of 1 eV, the local line shape $I_0(\omega,N)$ can be described by the standard model [4]

$$I_0(\omega,N)=\frac{-1}{\pi}\rho(N)\operatorname{Im}\int P(\vec{E},N)\sum_{ii'ff'}d_{if}d^*_{i'f'}\left\langle\left\langle i'f'\right|(\omega-L_0+\vec{d}.\vec{E}+i\Phi(N))^{-1}\left|if\right\rangle\right\rangle. \quad (2)$$

The density appears in the ion microfield PDF $P(\vec{E},N)$, in the collision operator $\Phi(N)$, and in the population of the emitting sublevels $\rho(N)$. The latter has been here calculated with the collisional-radiative code SOPHIA [5]. The PDF of density W(N) has been assumed to be a Gamma distribution, in accordance with experimental measurements in the scrape of layer of the TCV tokamak [6]. We have compared the profile of D_{10} with $\Delta N/\langle N\rangle = 50$ % to the same profile without fluctuations. On Fig. 1 are plotted the normalized line shapes with $\langle N\rangle = 10^{13}$ cm^{-3} and $T_e = 1$ eV. The density fluctuations lead to a weak broadening of the line shape, by less than 20 %, validating the previous density diagnostics using Stark broadening.

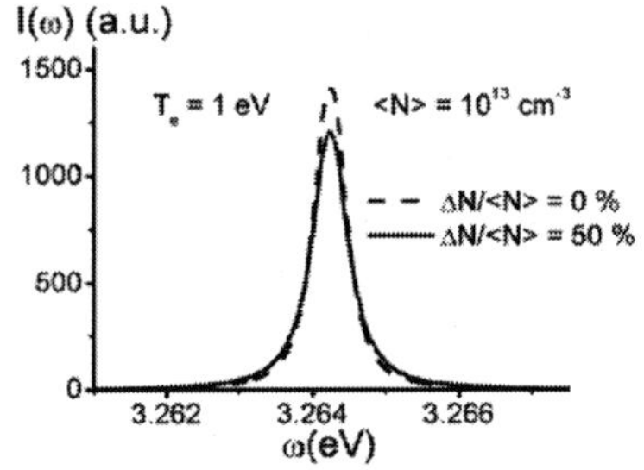

FIGURE 1. Profile of D_{10} line with 50 % of density fluctuations compared to the same profile without fluctuations, with $\langle N\rangle = 10^{13}$ cm^{-3} and $T_e = 1$ eV.

ION TEMPERATURE FLUCTUATIONS AND D_α LINE SHAPE

We now investigate the effect of the fluctuations of the ion temperature T on the D_α line shape. This profile is mainly broadened by Doppler effect combined to Zeeman splitting. We focus here on the population of neutrals created by charge exchange processes, and we consider one component of the Zeeman splitting. Assuming the electron and the ion temperature to be equal, the local line shape is thus given by $I_0(\Delta\omega,T) = \rho_3(T)f(c\Delta\omega/\omega_0,T)$, where ω_0 is the central pulsation, $\Delta\omega$ is the pulsation detuning, f is the VDF of the ions, which is locally Maxwellian, c is the speed of light, and $\rho_3(T)$ is the population of the emitting sublevels of the (n = 3)-level. The latter, also calculated by the SOPHIA code, has a maximum at $T_{max} \approx 1.5$ eV. We have compared the profile of D_α with 50 % of temperature fluctuations to the same profile without fluctuations, for $\langle T\rangle = 20$ eV $> T_{max}$ and for $\langle T\rangle = 1$ eV $< T_{max}$, by using a

Gamma PDF for temperature fluctuations. The results are displayed on Fig. 2. The fluctuations of temperature significantly narrow the line shape when <T> > T_{max}, whereas the profile is broadened when <T> < T_{max}. These opposite effects result from the range of temperature acting in the integral in Eq. 1. The maximum of ρ(T) induces a shift of the values toward lower (resp. larger) temperatures if T_{max} < <T> (resp. T_{max} > <T>). In the first case the measured line thus appears narrowed, and in the second case the line shape is broadened.

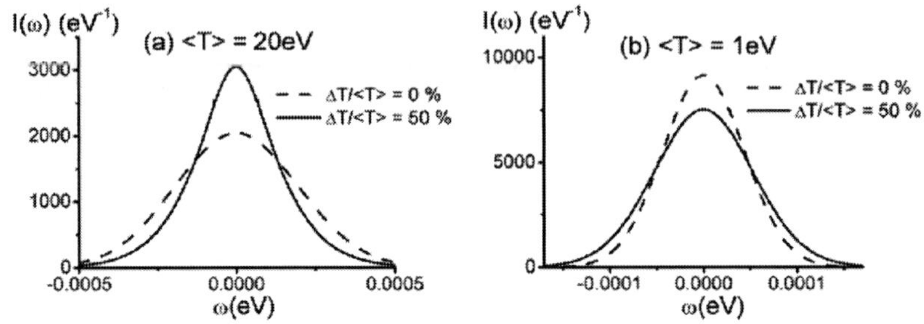

FIGURE 2. Profile of D_α line with 50 % of ion temperature fluctuations compared to the same profile without fluctuations, (a) for <T> = 20 eV, (b) for <T> = 1 eV.

SUMMARY

We have investigated the D_{10} line and the charge exchange contribution to the D_α line in a divertor turbulent plasma. The D_{10} line, broadened by Stark effect, presents a weak additional broadening with fluctuations of density. This additional broadening remaining small even for large fluctuation rates, we validate the previous density diagnostics of detached plasmas using Stark broadening. The charge exchange part of the D_α line, broadened by Doppler effect, can be significantly altered by fluctuations of ion temperature. The measured line shape is narrowed for an average temperature larger than 1.5 eV, and is broadened in the opposite case. Our results show that an accurate diagnostic of the edge plasmas requires to take into account the turbulence in line profile models, and encourage us to pursue studies on line shapes in turbulent divertor plasmas.

ACKNOWLEDGMENTS

This work has been performed for the Euratom-CEA collaboration LRC DSM 99-14.

REFERENCES

1. I. Garcia-Cortés *et al.*, Plasma Phys. Control. Fusion **38**, 2051-2062 (1996).
2. R. A. Moyer *et al.*, J. Nucl. Mater. **241-243**, 633-638 (1997).
3. Y. Marandet *et al.*, *Europhys. Lett.* **69**, 531-537 (2005).
4. H. R. Griem, A. C. Kolb and K. Y. Shen, Phys. Rev. **116**, 4-16 (1959).
5. F. Rosmej, SOPHIA-code, in press.
6. J. P. Graves *et al.*, Plasma Phys. Control. Fusion **47**, L1-L9 (2005).

Doppler Line Shapes, Turbulence and Neutral Transport in Tokamak Edge Plasmas

Y. Marandet*, P. Börner†, S. Brezinsek†, J.W. Dufty**, Ph. Mertens†, D. Reiter†, G. Sergienko† and U. Samm†

*PIIM, CNRS-Université de Provence, Marseille, France
†Institut für Plasmaphysik, Forschungszentrum Jülich, Association EURATOM-FZJ, Trilateral Euregio Cluster, Germany www.fz-juelich.de/ipp
**Department of Physics, University of Florida, P.O. Box 118440, Gainesville, Florida 32611-8440, USA

Abstract. First investigations of the effect of ion temperature fluctuations on Doppler spectral line shapes emitted in tokamak edge plasmas are reported on. It is shown that hot charge exchange neutrals have to be taken into account in the analysis.

Keywords: Plasma Turbulence, Probability Distribution Function, Neutral transport, tokamak
PACS: 32.70.Jz, 52.35.Ra

INTRODUCTION

We focus on the practical implementation of the model presented in Ref. [1], where it was shown that the effect of fluid fields fluctuations on the line profile is entirely characterized by the shape of their Probability Density Function (PDF). A simplified model for the ion temperature fluctuations in edge plasmas, and the resulting Doppler profiles, are discussed [2]. The contribution of charge exchange neutrals coming from the inner plasma, is evaluated using the Monte Carlo neutral transport code EIRENE [3]. First qualitative comparison with experimental spectra is performed.

PDF OF ION TEMPERATURE FLUCTUATIONS

Ion temperature T_i obeys a fluid equation in edge plasmas, for which the Braginskii closure [4] holds. Application of the drift ordering ($\widetilde{l}/L_\nabla \ll 1$, where $\widetilde{l}$ and L_∇ resp. characterize turbulence and mean field length scales), which is relevant for turbulent fluctuations, leads to a simplified equation for temperature fluctuations $\widetilde{T}_i$, namely

$$\left[\partial_t + \widetilde{\mathbf{v}}_E \cdot \nabla - \kappa_{i\perp} \nabla^2\right] \widetilde{T}_i = \widetilde{\mathbf{v}}_E \cdot \nabla \overline{T}_i, \tag{1}$$

where $\widetilde{\mathbf{v}}_E$ is the fluctuating electric drift velocity, $\kappa_{i\perp}$ the ion thermal diffusivity perpendicular to the magnetic field, and $\nabla \overline{T}_i$ the average temperature gradient, which provides a source term for fluctuations. Here mean flows have been neglected, as well as energy exchanges with other species [2]. As a first approximation, one might consider that $\widetilde{\mathbf{v}}_E(\mathbf{r},t)$ is determined independently from $\widetilde{T}_i$, i.e. treat $\widetilde{T}_i$ as a passive scalar [2]. The PDF of $\widetilde{T}_i$

CP874, *Spectral Line Shapes: 18th International Conference*,
edited by E. Oks and M. Pindzola

can then be calculated following a Lagrangian formulation developed for neutral fluids [5]. This calculation assumes that $\widetilde{\mathbf{v}}_E(\mathbf{r},t)$ is described by a Gaussian functional PDF whose correlation function is known. The resulting PDF has a Gaussian core but tails decaying as an exponential. It is plotted on Fig. 1a, for typical values of $g\xi$ (mean gradient times velocity correlation length) and the Peclet number Pe $= 100$, which measures the relative importance of collisional diffusion and convection. A simple saddle point calculation then shows that the corresponding Doppler profile has an exponential decay too, a result confirmed by the numerical calculations plotted on Fig. 1b. Comparison to a Gaussian Doppler profile, which would be obtained in the turbulence free case, show that deviations essentially occur in the line wings. The main interest of this model is to show that non-gaussian temperature PDF should be expected in edge plasmas, since even this simplified model generates non-gaussianity. Note that stronger effects on the profiles are obtained in the cases considered in Ref. [6].

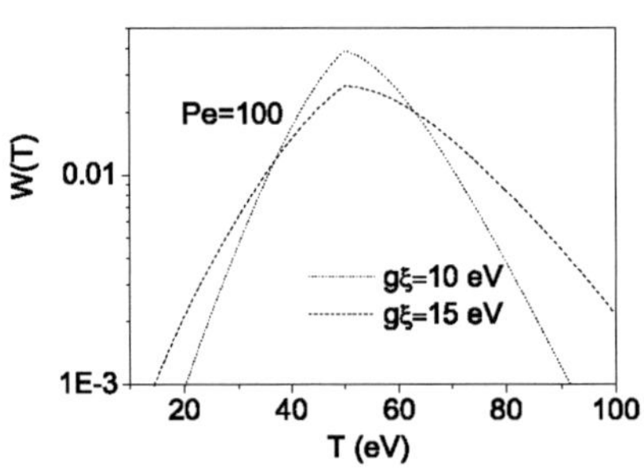

a)

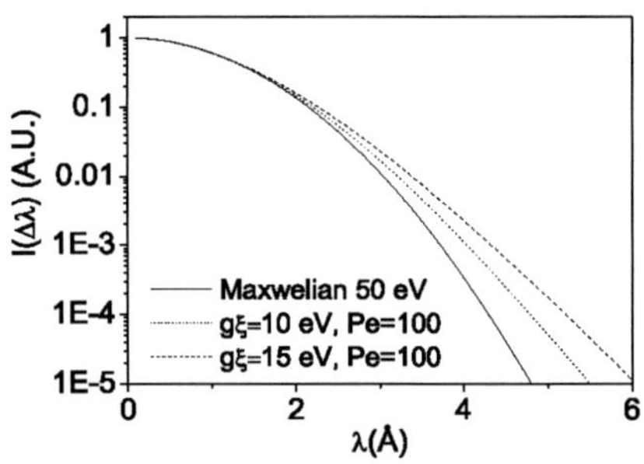

b)

FIGURE 1. a) Plot of the temperature PDF for $T_0 = 50$ eV, Pe=10^2, $g\xi = 10$ eV, (dotted line), and $g\xi = 15$ eV (dashed line) . b) Resulting Doppler profile on a logarithmic scale (dotted and dashed lines). The solid line is the Gaussian Doppler profile for 50 eV.

HOT CHARGE EXCHANGE NEUTRALS AND TENTATIVE COMPARISON TO EXPERIMENTS

In order to account for neutrals created by charge exchange (CX) in the inner regions of the plasma where the temperature is higher, which have long mean free paths, we use the EIRENE neutral transport code [3]. The latter is run in a 3D cylindrical geometry relevant to that of the TEXTOR tokamak [7], but does not take turbulence into account. The plasma profiles $n(r)$, $T_{e,i}(r)$ are obtained and/or extrapolated from measurements. The code simulates the neutral velocity distribution, integrated along the viewing chords. First results are plotted on Fig. 2a, where it is seen that the neutral VDF f(v) presents an exponential like hot tail corresponding to non-local CX neutrals (here v is expressed in terms of the corresponding Doppler shift). The three curves correspond to the spectra at different radial locations, and show a larger contribution of hot neutrals as one moves deeper into the plasma, which is physically sound. The first experimental spectra obtained in the TEXTOR tokamak with a high resolution (resolving power R$= 10^5$) Littrow spectrometer coupled to a high dynamic range (18-bits) 2D CCD camera, exhibit the same qualitative behavior, see Fig. (2)b. Work is ongoing to perform more quantitative comparisons. The point here is to show that this non local CX contribution will

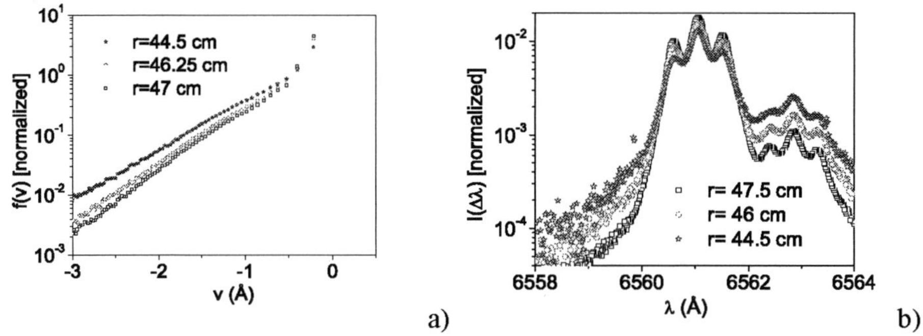

FIGURE 2. a) Simulated VDF for 3 different chords providing radial resolution. b) High resolution $D\alpha/H\alpha$ spectra measured in TEXTOR.

have to be taken into account when performing line wings analysis. It has to be clearly distinguished from turbulence effect, since the latter might produce similar behaviors.

CONCLUSION AND PERSPECTIVES

We have shown that turbulence would affect Doppler line wings, on the basis of a very simplified but physically relevant turbulence model. Line wing analysis should however be carried out carefully because of the hot charge exchange neutral contribution. Performing density and radial scans experiments should allow to disentangle these two contributions, because of the strong density dependance of the hot CX contribution, and of the radial dependance of the turbulent fluctuation rate.

ACKNOWLEDGMENTS

This work has been supported by the "Spectral Line Shape and Turbulence" (SLST) EURATOM Intra-European Training Fellowship, by a collaboration (LRC DSM 99-14) between the LPIIM and the DRFC, CEA Cadarache, and by the Department of Energy grant DE FGO2ER54677.

REFERENCES

1. Y. Marandet *et al.*, *Europhys. Lett.* (**69**), pp 531–537 (2005).
2. Y. Marandet and J. W. Dufty, *Contrib. Plasma Phys.*, in press.
3. D. Reiter, M. Baelmans and P. Börner, *Fusion Science and Technology* (**47**), pp 172-186 (2005).
4. S. I. Braginskii, *Rev. Plasma Phys.* **1**, p 205 (1965).
5. B. I. Schraiman and E. D. Siggia, *Phys. Rev. E* **49**, pp 2912-2927 (1994).
6. J. Rosato *et al.*, this volume.
7. Special issue on TEXTOR, *Fusion Science and Technology* (**47**)2, pp 138-145 (2005).

ASTROPHYSICS/EARTH ATMOSPHERE

Turbulence Spectra from Doppler-shifted Spectral Lines

A. Lazarian

Department of Astronomy, University of Wisconsin-Madison, lazarian@astro.wisc.edu

Abstract. Turbulence is a key element of the dynamics of astrophysical fluids, including those of interstellar medium, clusters of galaxies and circumstellar regions. Turbulent motions induce Doppler shifts of observable emission and absorption lines. In the review we discuss new techniques that relate the spectra of underlying velocity turbulence and spectra of Doppler-shifted lines. In particular, the Velocity-Channel Analysis (VCA) makes use of the channel maps, while the Velocity Coordinate Spectrum (VCS) utilizes the fluctuations measured along the velocity axis of the Position-Position Velocity (PPV) data cubes. Both techniques have solid foundations based on analytical calculations as well as on numerical testings. Among the two the VCS, which has been developed quite recently, has two advantages. First of all, it is applicable to turbulent volumes that are not spatially resolved. Second, it can be used with absorption lines that do not provide good spatial sampling of different lags over the image of turbulent object. In fact, numerical testing shows that measurements of Doppler shifted absorption lines over a few directions is sufficient for a reliable recovering of the underlying spectrum of the turbulence. Our comparison of the VCA and the VCS with a more traditional technique of Velocity Centroids, shows that the former two techniques recover reliably the spectra of supersonic turbulence, while the Velocity Centroids may have advantages for studying subsonic turbulence. In parallel with theoretical and numerical work on the VCA and the VCS, the techniques have been applied to spectroscopic observations. We discuss results on astrophysical turbulence obtained with the VCA and the VCS.

Keywords: Turbulence, Spectral Lines, MHD, Interstellar Medium, Plasma
PACS: 95.30.Qd, 52.30.Cv, 96.50.Tf, 98.38.-j, 98.38Gt

WHAT CAN TURBULENT SPECTRA TELL US?

As a rule astrophysical fluids are turbulent and the turbulence is magnetized. This ubiquitous turbulence determines the transport properties of interstellar medium (see Elmegreen & Falgarone 1996, Stutzki 2001, Balesteros-Peredes et al. 2006) and intra-cluster medium (see Sunyaev, Norman & Bryan 2003, Ensslin & Vogt 2006, Lazarian 2006), many properties of Solar and stellar winds (see Hartman & McGregor 1980) etc. One may say that to understand heat conduction, propagation of cosmic rays and electromagnetic radiation in different astrophysical environments it is absolutely essential to understand the properties of underlying magnetized turbulence. The fascinating processes of star formation (see McKee & Tan 2002, Elmegreen 2002, Mac Low & Klessen 2004) and interstellar chemistry (see Falgarone et al. 2006 and references therein) are also intimately related to properties of magnetized compressible turbulence (see reviews by Elmegreen & Scalo 2004).

From the point of view of fluid mechanics astrophysical turbulence is characterized by huge Reynolds numbers, *Re*, which is the inverse ratio of the eddy turnover time of a parcel of gas to the time required for viscous forces to slow it appreciably. For

CP874, *Spectral Line Shapes: 18th International Conference,*
edited by E. Oks and M. Pindzola

$Re \gg 100$ we expect gas to be turbulent and this is exactly what we observe in HI (for HI $Re \sim 10^8$). In fact, very high astrophysical Re and its magnetic counterpart magnetic Reynolds number Rm (that can be as high as $Rm \sim 10^{16}$) present a big problem for numerical simulations that cannot possibly get even close to the astrophysically-motivated numbers. The currently available 3D simulations can have Re and Rm up to $\sim 10^4$. Both scale as the size of the box to the first power, while the computational effort increases as the fourth power (3 coordinates + time), so the brute force approach cannot begin to resolve the controversies related to ISM turbulence. This caused serious concerns that while present codes can produce simulations that resemble observations, whether numerical simulations reproduce reality well (see McKee 1999, Shu et al. 2004). We believe that these concerns may be addressed via observational studies of astrophysical turbulence.

Statistical description is a nearly indispensable strategy when dealing with turbulence. The big advantage of statistical techniques is that they extract underlying regularities of the flow and reject incidental details. Kolmogorov description of unmagnetized incompressible turbulence is a statistical one. For instance it predicts that the difference in velocities at different points in turbulent fluid increases on average with the separation between points as a cubic root of the separation, i.e. $|\delta v| \sim l^{1/3}$. In terms of direction-averaged energy spectrum this gives the famous Kolmogorov scaling $E(k) \sim 4\pi k^2 P(\mathbf{k}) \sim k^{5/3}$, where $P(\mathbf{k})$ is a *3D* energy spectrum defined as the Fourier transform of the correlation function of velocity fluctuations $\xi(\mathbf{r}) = \langle \delta v(\mathbf{x}) \delta v(\mathbf{x}+\mathbf{r}) \rangle$. Note that in this paper we use $\langle ... \rangle$ to denote averaging procedure.

The example above shows the advantages of the statistical approach to turbulence. For instance, the energy spectrum $E(k)dk$ characterizes how much energy resides at the interval of scales $k, k+dk$. At large scales l which correspond to small wavenumbers k (i.e. $l \sim 1/k$) one expects to observe features reflecting energy injection. At small scales one should see the scales corresponding to sinks of energy. In general, the shape of the spectrum is determined by a complex process of non-linear energy transfer and dissipation.

In view of the above it is not surprising that attempts to obtain spectra of interstellar turbulence have been numerous since 1950s (see Munch 1958). However, various directions of research achieved various degree of success (see Armstrong, Rickett & Spangler 1995). For instance, studies of turbulence statistics of ionized media were more successful (see Spangler & Gwinn 1990) and provided the information of the statistics of plasma density at scales 10^8-10^{15} cm. However, these sort of measurements provide only the density statistics, which is an indirect measure of turbulence.

Velocity statistics is much more coveted turbulence measure. Although, it is clear that Doppler broadened lines are affected by turbulence, recovering of velocity statistics was extremely challenging without an adequate theoretical insight. Indeed, both velocity and density contribute to fluctuations of the intensity in the Position-Position-Velocity (PPV) space. In what follows we discuss how the observable Doppler-shifted lines can be used to recover a spectrum of turbulent velocity using two new techniques, that, unlike other mostly empirical techniques, have solid theoretical foundations. How to obtain using spectroscopic observations other characteristics of turbulence, e.g. higher order statistics, anisotropies has been reviewed earlier (see Lazarian 2004).

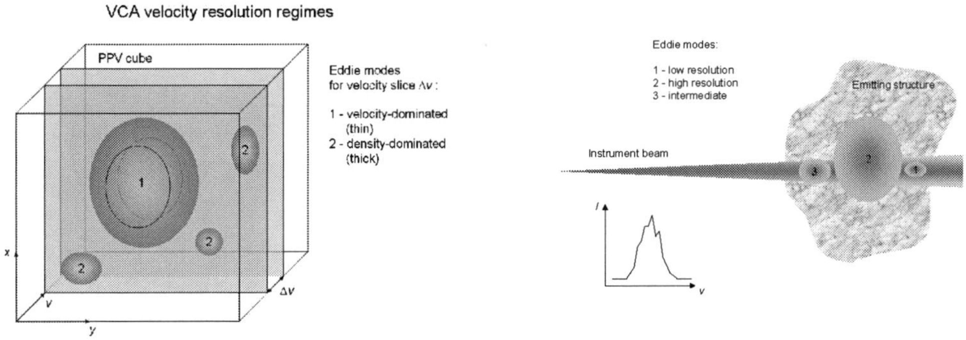

FIGURE 1. *Left Panel*: PPV data cube. Illustration of the concepts of the thick and thin velocity slices. the slices are thin for the PPV images of the large eddies, but thick for the images of small eddies. *Right Panel*: Illustration of the VCS technique. For a given instrument resolution large eddies are in the high resolution limit, while small eddies are in the low resolution limit.

VCA AND VCS: TWO WAYS TO ANALYZE SPECTRAL DATA

Spatial spectra obtained by taking Fourier transform of channel maps had been widely used to study HI before we conducted our theoretical study of what those spectra mean in Lazarian & Pogosyan (2000, henceforth LP00). The channel maps correspond to the velocity slices of PPV cubes as shown in Figure 1 and one may naturally ask a question whether anything depends on the thickness of the channel. It is intuitively clear that if medium is optically thin and the velocity is integrated over the whole spectral line, the fluctuations can depend only on density inhomogeneities. It is also suggestive that the contribution of the velocity fluctuations may depend on whether the images of the eddies under study fit within a velocity slice or their velocity extend is larger than the slice thickness (see Figure 1). According to LP00 this results in the assymptotics that correspond to "thin" and "thick" velocity channels. Note, that these questions were not posed by the earlier research. This resulted in channel map spectra with different spectral indexes the relation of which to the underlying velocity fluctuations was unclear (see Green 1990 and references therein).

Some spectral data, for instance, optically thick CO data were traditionally analyzed in differently (see Falgarone & Puget 1995), namely, spectra of total intensities was studied. The origin of this spectrum and its relation to the underlying velocity and density fluctuations was established in Lazarian & Pogosyan (2004, henceforth LP04). This work also clarified the effects of absorption that were reported for HI data. In terms of the techniques of turbulence study LP04 deals with the VCA, but in the case when absorption is present.

A radically different way of analyzing spectroscopic data is presented in Lazarian & Pogosyan (2006, henceforth LP06) and Chepurnov & Lazarian (2006a). There the spectra along the V-axis of the PPV cube are studied (see Figure 2). The technique was termed Velocity Coordinate Spectrum (VCS) in Lazarian (2004). The formalism can be traced to LP00 work, where the expressions that relate the spectrum of fluctuations along the velocity coordinate and the underlying velocity spectrum were obtained. However, it took some time to understand the advantages that the VCS provides for the practical handling of the emission and absorption data. The first analysis of the data using the VCS is presented in Chepurnov, Lazarian & Stanimirovic (2006). Numerical testing of the technique is provided in Chepurnov & Lazarian (2006b).

BASICS OF THE FORMALISM

Below we provide a brief introduction to the mathematical foundations of the VCA and the VCS (see more in LP00, LP04 and LP06). Our goal is to relate the statistics that can be obtained through spectral line observations, for instance, the structure function of the intensity of emission $I_{\mathbf{X}}(v)$

$$\mathscr{D}(\mathbf{X},v_1,v_2) \equiv \left\langle [I_{\mathbf{X}}(v_1) - I_{\mathbf{X}}(v_2)]^2 \right\rangle \quad , \tag{1}$$

where the z-axis component velocity v is measured by in the direction defined by the two dimensional vector[1] $\mathbf{X}$, to the underlying properties of the turbulent cascade.

The intensities $I_{\mathbf{X}}(v)$ are affected by both the turbulence and the absorption. To quantify these effects we consider the standard equation of radiative transfer $dI_v = -g_v I_v ds + j_v ds$, where $g_v = \alpha(\mathbf{x})\rho(\mathbf{x})\phi_v(\mathbf{x})$, $j_v = \tilde{\varepsilon}\rho(\mathbf{x})\phi_v(\mathbf{x})$, $\mathbf{x}$ is a three dimensional position vector $(\mathbf{X},z)$, $\rho(\mathbf{x})$ is the density and $\phi_v(\mathbf{x})$ is the velocity distribution of the atoms. The turbulent motions affect the velocity distribution. Indeed, for the line-of-sight component, v, of the velocity at the position $\mathbf{x}$, it is a sum of the z-components of the regular gas flow (e.g., due to galactic rotation) $v_{gal}(\mathbf{x})$, the turbulent velocity $u(\mathbf{x})$ and the residual component due to thermal motions. This residual thermal velocity $v - v_{gal}(\mathbf{x}) - u(\mathbf{x})$ has a Maxwellian distribution, so

$$\phi_v(\mathbf{x})\mathrm{d}v = \frac{1}{(2\pi\beta)^{1/2}} \exp\left[-\frac{(v - v_{gal}(\mathbf{x}) - u(\mathbf{x}))^2}{2\beta}\right] \mathrm{d}v \quad , \tag{2}$$

where $\beta = \kappa_B T/m$, m being the mass of atoms. For $T \to 0$ the function ϕ_v tends to a delta-function that depends on regular gas flow and the turbulent velocity u. Taking Fourier transforms we deal with velocity gradients, which are larger for turbulent motions than for large-scale sheer. For instance, the latter for the Galactic rotation is given by the Oort's constant, which is 14 km s^{-1} kpc^{-1}. In comparison, the shear due to typical Kolmogorov-type turbulent motions in the Galaxy with the injection of energy at 10 km s^{-1} at the scale of $L \sim 30$ pc is ~ 300 km s$^{-1}(L/l)^{2/3}$ kpc^{-1}. Thus, in spite of the fact, that regular large-scale galactic shear velocities may be much larger than the

[1] Henceforth we denote by the capital bold letters the two dimensional position-position vectors that specify the line of sight. Small bold letters are reserved to describe the vectors of three dimensional spatial position. The z-axis is chosen to be along the line of sight.

turbulent velocities, they can be neglected for our analysis (LP00 and a numerical study in Esquivel et al. 2003).

For the z-component of the turbulent velocity field (i.e. u) we use the structure function

$$D_z(\mathbf{r}) = \langle (u(\mathbf{x}+\mathbf{r}) - u(\mathbf{x}))^2 \rangle \ , \tag{3}$$

which for a self-similar power-law turbulent motions provide $D_z \sim r^m$, where $m = 1/3$ for Kolmogorov turbulence. These velocity correlations together with the correlations of over-density

$$\xi(r) = \xi(\mathbf{r}) = \langle \rho(\mathbf{x})\rho(\mathbf{x}+\mathbf{r}) \rangle \ , \tag{4}$$

enter the correlation function that can be constructed from the PPV densities ρ_s, which are available through spectroscopic observations. If the gas is confined in an isolated cloud of size S and the galactic shear over this scale is neglected, the zero-temperature correlation function is (see LP06)

$$\xi_s(R,v) \equiv \langle \rho_s(\mathbf{X}_1,v_1)\rho(\mathbf{X}_2,v_2) \rangle \propto \int_{-S}^{S} \mathrm{d}z \left(1 - \frac{|z|}{S}\right) \frac{\xi(r)}{D_z^{1/2}(\mathbf{r})} \exp\left[-\frac{v^2}{2D_z(\mathbf{r})}\right], \tag{5}$$

where the correlation function ξ_s is defined in the PPV space, where R is the spatial separation between points in the plane-of-sky and v is the separation along the V -axis.

TABLE 1. The short-wave asymptotical behavior of power spectra in PPV space. γ is the power-law index of the density correlation function, m is the index of the velocity correlation function. The spectrum of fluctuations in the PPV volume can be presented as the sum of P_ρ, which has both density and velocity contributions and P_v, which has only the contribution arising from turbulent velocity. From LP06.

	1D: $P_s(k_v)$	2D: $P_s(K)$	3D: $P_s(K,k_v)$
	$k_v D_z^{1/2}(S) \gg 1$	$KS \gg 1$	$k_v^2 D_z(S) \gg (kS)^m$
P_ρ :	$(r_0/S)^\gamma \left[k_v D_z^{1/2}(S)\right]^{2(\gamma-1)/m}$	$(r_0/S)^\gamma [KS]^{\gamma+m/2-3}$	$(r_0/S)^\gamma \left[k_v D_z^{1/2}(S)\right]^{-2(3-\gamma)/m}$
P_v :	$\left[k_v D_z^{1/2}(S)\right]^{-2/m}$	$[KS]^{m/2-3}$	$\left[k_v D_z^{1/2}(S)\right]^{-6/m}$

The correlation function of over-density given by Eq. (4) has a constant part that depends on the mean density only and the part that changes with r. For instance, for the power-law density spectrum the correlation functions of over-density take the form (see LP06 for the discussion of cases of $\gamma < 0$ and $\gamma > 0$):

$$\xi(r) = \langle \rho \rangle^2 \left(1 + \left[\frac{r_0}{r}\right]^\gamma\right), \tag{6}$$

where r_0 has the physical meaning of the scale at which fluctuations are of the order of the mean density (see more in LP06). Substituting Eq. (6) in Eq. (5) it is easy to see that

TABLE 2. Conditions for the impact of density inhomogeneities to the PPV statistics exceeds the velocity contribution. γ must be larger than 0.

$m \geq \max\left[\frac{2}{3}, \frac{2}{3}(1-\gamma)\right]$	$v^2 < D_z(S)(r_0/S)^m$
$\frac{2}{3}(1-\gamma) < m < \frac{2}{3}$	$v^2 < D_z(S)(r_0/S)^{\frac{2/3\gamma m}{m-2/3(1-\gamma)}}$
$m \leq \min\left[\frac{2}{3}, \frac{2}{3}(1-\gamma)\right]$	$r_0/S > 1$

TABLE 3. The VCA assymptotics. *Thin* means that the channel width < velocity dispersion at the scale under study; *thick* means that the channel width > velocity dispersion at the scale under study; *very thick* means that a substantial part of the velocity profile is integrated over.

Slice thickness	Shallow 3-D density $P_n \propto k^{-3+\gamma}, \gamma > 0$	Steep 3-D density $P_n \propto k^{-3+\gamma}, \gamma < 0$
2-D intensity spectrum for thin slice	$\propto K^{-3+\gamma+m/2}$	$\propto K^{-3+m/2}$
2-D intensity spectrum for thick slice	$\propto K^{-3+\gamma}$	$\propto K^{-3-m/2}$
2-D intensity spectrum for very thick slice	$\propto K^{-3+\gamma}$	$\propto K^{-3+\gamma}$

the PPV correlation function ξ_s can be presented as a sum of two terms, one of which does depend on the fluctuations of density, the other does not. Taking Fourier transform of ξ_s one gets the PPV spectrum that evidently is also a sum of two terms P_ρ and P_v the assymptotics for which are presented in Table 1.

Expression (5) and its generalizations may be used directly to solve the inverse problem to find the properties of the underlying astrophysical turbulence for an arbitrary spectrum. However, so far, most attention was given to astrophysically important case of power-law turbulence. Note, that using spectra rather than the correlation function has advantages. For instance, the correlations along the V-axis of the PPV cube may be dominated by large-scale gradients, while spectra provide correct result (see explanation in LP04 and LP06). The results for 1, 2 and 3 dimensional spectra are presented in Table 1. In many cases the contributions that depend on density are subdominant. Table 2 presents special cases when density fluctuations are important.

VELOCITY CHANNEL ANALYSIS

The interpretation of the channel maps is the domain of the VCA. Table 3 shows how the power spectrum of the intensity fluctuations depends on the thickness of the velocity channel. Below we provide quantitative discussion of the VCA.

It is easy to see that both steep and shallow underlying density the power law index *steepens* with the increase of velocity slice thickness. In the thickest velocity slices the velocity information is averaged out and we get the density spectral index $-3+\gamma$. The velocity fluctuations dominate in thin slices, and the index m that characterizes the velocity fluctuation can be obtained using thin velocity slices (see Table 1). As we mentioned earlier, the notion of thin and thick slices depends on a turbulence scale under study and the same slice can be thick for small scale turbulent fluctuations and thin for large scale ones (see Figure 1).

One may notice that the spectrum of intensity in a thin slice gets shallower as the underlying velocity get steeper. To understand this effect let us consider turbulence in an incompressible optically thin medium. The intensity of emission in a velocity slice is proportional to the number of atoms per the velocity interval given by the thickness of the slice. Thin slice means that the velocity dispersion at the scale of study is larger than the thickness of a slice. The increase of the velocity dispersion at a particular scales means that less and less energy is being emitted within the velocity interval that defines the slice. As the result the PPV image of the eddy gets fainter. In other words, the larger is the dispersion at the scale of the study the less intensity is registered at this scale within the thin slice of spectral data. This means that steep velocity spectra that correspond to the flow with more energy at large scales should produce intensity distribution within thin slice for which the more brightness will be at small scales. This is exactly what our formulae predict for thin slices (see Table 3).

If density variations are also present they modify the result. For small scale asymptotics of thin slices this happens, however, only when the density spectrum is shallow (i.e. $\gamma > 0$), i.e. dominated by fluctuations at small scales (see Eq. (6)).

VELOCITY COORDINATE SPECTRUM

The VCS is a brand new technique, which, unlike the VCA, was not motivated by the interpretation of the existing observations. In the case of the VCS it were theoretical advances that induced the subsequent data analysis.

Unlike the standard spatial spectra, that are functions of angular wavenumber, the VCS is a function of the wave number $k_v \sim 1/v$, which means that large k_v correspond to small velocity differences, while small k_v correspond to large velocity differences.

A realistic beam has a finite width, ΔB. We remind the reader that we deal with the case in which the turbulent volume extend along the line of sight S is much smaller than the distance to the volume. As the result, the angular extend of the beam is straightforwardly related to the physical scales that we deal with. The VCS near a scale k_v depends on whether the instrument resolves the correspondent spatial scale $\left[k_v^2 D_z(S)\right]^{-1/m} S$. If this scale is resolved then $P_v(k_v) \propto k_v^{-2/m}$ and $P_\rho(k_v) \propto k_v^{-2(1-\gamma)/m}$. If the scale is not resolved then $P_v(k_v) \propto k_v^{-6/m}$ and $P_\rho(k_v) \propto k_v^{-2(3-\gamma)/m}$. These results are presented in a compact form in Table 4. The transition from the low to the high resolution regimes happens as the velocity scale under study gets comparable to the turbulent velocity at the minimal spatially resolved scale. As the change of slope is the velocity-induced effect, it is not surprising that the difference in spectral indexes in the low and high resolution limit is $4/m$ for both P_v and P_ρ terms, i.e it does not depend on the density[2]. This allows for separation of the velocity and density contributions. For instance, Figure 2 illustrates that in the case of shallow density both the density and velocity spectra can be obtained. Potentially, procedures of extracting the information on

[2] In the situation where the available telescope resolution is not sufficient, i.e. in the case of extragalactic turbulence research, the high spatial resolution VCS can be obtained via studies of the absorption lines from point sources.

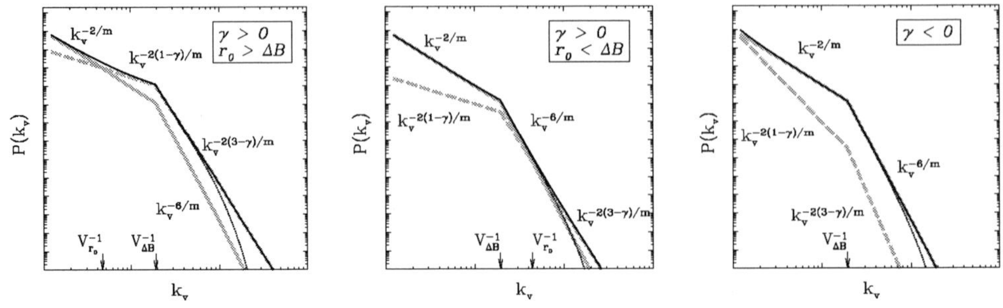

FIGURE 2. Particular cases for the VCS (from LP06). In every panel light lines show contributions from the ρ-term (density modified by velocity, dashed line) and v-term (pure velocity effect, solid line) separately, while the dark solid line shows the combined total VCS power spectrum. Thermal suppression of fluctuations is shown by the dotted line. Labels below the dark solid lines mark the scaling of the subdominant contributions. For the *left* and *middle panels* the density power spectrum is taken to be shallow, $\gamma > 0$. The left panel corresponds to high amplitude of the density correlations, $r_0 > \Delta B$, where density effects become dominant at relatively long wavelengths for which the beam is narrow. In the middle panel, the amplitude of density correlations is low $r_0 < \Delta B$ and they dominate only the smallest scales which results in the intermediate steepening of the VCS scaling. The *right panel* corresponds to the steep density spectrum. In this case the density contribution is always subdominant. In this example the thermal scale is five times shorter than the resolution scale $V_{\Delta B}$.

TABLE 4. Scalings of VCS for shallow and steep densities (from LP06). To the linear scale ΔB corresponds the velocity scale $V_{\Delta B} \equiv \sqrt{D(S)(\Delta B/S)^m}$, equal to the magnitude of turbulent velocities at the separation of a size ΔB. It is not difficult to find that when $k_v^{-1} > V_{\Delta B}$ the beam is narrow, while on smaller scales its width is important.

Spectral term	$\Delta B < S\left[k_v^2 D_z(S)\right]^{-\frac{1}{m}}$	$\Delta B > S\left[k_v^2 D_z(S)\right]^{-\frac{1}{m}}$
$P_\rho(k_v)$	$\propto \left(k_v D_z^{1/2}(S)\right)^{-2(1-\gamma)/m}$	$\propto \left(k_v D_z^{1/2}(S)\right)^{-2(3-\gamma)/m}$
$P_v(k_v)$	$\propto \left(k_v D_z^{1/2}(S)\right)^{-2/m}$	$\propto \left(k_v D_z^{1/2}(S)\right)^{-6/m}$

the 3D turbulent density can be developed for the steep density case as well. However, this requires careful accounting for errors that arise while the major contribution arising from velocity is subtracted from the data[3].

[3] Needless to say, when the turbulent object is resolved, the easiest way to obtain the density spectral index is to study the integrated intensity maps, provided that the absorption is negligible (see criteria for

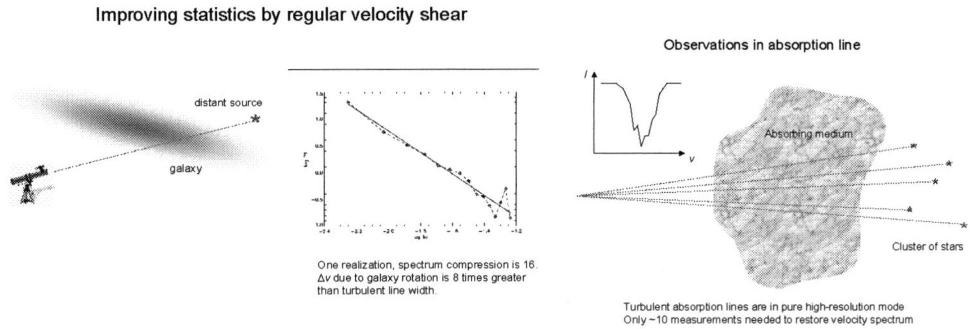

FIGURE 3. Illustration of VCS absorption studies of turbulence. *Left Panel*: In case an external galaxy in which regular shear within the galaxy helps to improve the statistics. *Right Panel*: For a cloud we showed that we can restore the turbulence spectrum with spectroscopic measurements for less than 10 directions. From Chepurnov & Lazarian 2006b

The VCS is applicable to both emission and absorption studies of turbulence. A numerical study in Chepurnov & Lazarian (2006b) shows that having just several lines of sight along which the absorption spectra is measured is sufficient for recovering the underlying turbulent velocity spectrum (see Figure 3). In the case of strong regular shear just one line of sight may be sufficient.

EFFECTS OF ABSORPTION

The issues of absorption were worrisome for the researchers from the very start of the research in the field (see Munch 1958). The erroneous statements about the effects of absorption on the observed turbulence statistics are widely spread in the literature (see discussion in LP04).

Using transitions that are less affected by absorption, e.g. HI, may allow to avoid the problem. However, it looks regretful not to use the wealth of spectroscopic data only because absorption may be present. A study of absorption effects in LP04 and LP06. For the VCA it was found that for sufficiently thin[4] slices the scalings obtained in the absence of absorption still hold provided that the absorption on the scales under study is negligible. The criterion for the absorption to be important is $\alpha^2\langle(\rho_s(\mathbf{X},v_1) - \rho_s(\mathbf{X},v_2))^2\rangle \sim 1$, which for $\gamma < 0$ results in the critical size of the slice thickness V_c given

this in LP04).

[4] The thermal broadening limits to what extend the slice can be thin. This means that in some cases that the actual turbulent velocity spectrum may not be recoverable.

by (LP06)

$$
\begin{aligned}
V_c/D_z(S)^{1/2} &\approx (\alpha\bar{\rho}_s)^{\frac{2m}{m-2}}, \quad & m > 2/3 \\
V_c/D_z(S)^{1/2} &\approx (\alpha\bar{\rho}_s)^{-1}, \quad & m < 2/3,
\end{aligned} \tag{7}
$$

where $\bar{\rho}_s$ is the mean PPV density. The absorption is dominant for the slices thicker than V_c. The difference with the case of $\gamma > 0$ is that in the latter case one should also consider a possibility that density contribution can be important (see Table 2). The criterion (7) coincides with one for the VCS, if we identify the critical k_v with $1/V_c$. If the resolution of the of the telescope is low, another limitation applies. The resolved scale should be less than the critical spatial scale that arises from the condition $\alpha^2\langle(\rho_s(\mathbf{X}_1,v)-\rho_s(\mathbf{X}_2,v))^2\rangle \sim 1$ which for $\gamma < 1$ results in $R_c/S \approx (\alpha\bar{\rho}_s)^{\frac{2}{m-2}}$ (LV06). If only scales larger than R_c are resolved, the information on turbulence is lost.

If integrated intensity of spectral lines is studied in the presence of absorption non-trivial effects emerge. Indeed, for optically thin medium the spectral line integration results in PPV intensity fluctuations that reflect the density statistics. LP04 showed that this may not be any more true for lines affected by absorption. When velocity is dominant a very interesting regime for which intensity fluctuations show universal behavior, i.e. the power spectrum $P(K) \sim K^{-3}$ emerges. When density is dominant (see Table 2), the spectral index of intensity fluctuations in those two situations is the same as in the case an optically thin cloud integrated through its volume. This means that for $\gamma > 0$ in the range of parameter space defined by Table 2 the measurements of intensity fluctuations of the integrated spectral lines reflect the *actual* underlying density spectrum in spite of the absorption effects.

COMPARISON OF TECHNIQUES

Traditionally the techniques to study velocity turbulence, e.g. velocity centroids or VCA, require the observations to spatially resolve the scale of the turbulence under study[5]. This constrains the variety of astrophysical objects where the turbulence can be studied. In this way, the VCS, is a unique tool that allows studies of astrophysical turbulence even when the instrument does not resolve the turbulent fluctuations spatially.

Our study of the effect of finite temperatures for the technique reveals that, unlike the VCA, the temperature broadening does not prevent the turbulence spectrum from being recovered from observations. Indeed, in VCA, gas temperature acts in the same way as the width of a channel. Within the VCS the term with temperature gets factorized and it influences the amplitude of fluctuations (LP06). One can correct for this term[6], which also allows for a new way of estimating the interstellar gas temperature.

---

[5] As it was discussed in LP00, the VCA can be applied directly to the raw interferometric data, rather than to images that require good coverage of all spatial frequencies. However, even with interferometers, the application of the VCA to extragalactic objects is restricted.

[6] To do this, one may attempt to fit for the temperature that would remove the exponential fall off in the spectrum of fluctuations along the velocity coordinate (Chepurnov & Lazarian 2006a).

Another advantage of the VCS compared to the VCA is that it reveals the spectrum of turbulence directly, while within the VCA the slope of the spectrum should be inferred from varying the thickness of the channel. As the thermal line width acts in a similar way as the channel thickness, additional care (see LP04) should be exercised not to confuse the channel that is still thick due to thermal velocity broadening with the channel that shows the thin slice asymptotics. A simultaneous use of the VCA and the VCS makes the turbulence spectrum identification more reliable.

The introduction of absorption in VCS and VCA brings about different results. Within the analysis of velocity slices spectra (VCA) the absorption results in new scalings for slices for which absorption is important. The turbulence spectral indexes can be recovered for the VCA within sufficiently thin slices, provided that the thickness of the slices exceeds the thermal line width. For the VCS when absorption becomes important the spectra get exponentially damped. This simplifies the interpretation of the data.

Both VCA and VCS are applicable to studies of not only emission, but also absorption lines. However, the necessity of using extended emission sources limits the extent of possible VCA studies of turbulence. This is not an issue for the VCS, for which absorption lines from *point sources* can be used (see Figure 3). Interestingly enough, in this case the VCS asymptotics for the high resolution limit should be used irrespectively of the actual beam size of the instrument. Note, that the VCA and the VCS are applicable to lines for which emission depends not only to first but also to the second power of density (see discussions in LP04, LP06).

A more traditional approach to turbulence studies includes velocity centroids, i.e. $S(\mathbf{X}) = \int v_z\ \rho_s(\mathbf{X},v_z)\ dv_z$, where ρ_s is the density of emitters in the PPV space.[7]. Analytical expressions for structure functions[8] of centroids, i.e. $\left\langle [S(\mathbf{X_1}) - S(\mathbf{X_2})]^2 \right\rangle$ were derived in Lazarian & Esquivel (2003). In that paper a necessary criterion for centroids to reflect the statistics of velocity was established. Esquivel & Lazarian (2005) confirmed the utility of the criterion and revealed that for MHD turbulence simulations it holds for subsonic or slightly supersonic turbulence (see also Ossenkopf et al. 2006). This is in contrast to the VCA and the VCS that provide reliable ways to study supersonic turbulence.

VCA is related to the Spectral Correlation Functions (SCF) (see Padoan, Goodman & Juvela 2003 and references therein). The latter technique also deals statistics of velocity slices. However if, ignoring the normalization, we write the informative part of the SCF in our notations as $\langle [\int dv(\rho_s(\mathbf{X_1},v) - \rho_s(\mathbf{X_2},\mathbf{v}))^2]^{1/2} \rangle$ we see that we cannot commute the $()^{1/2}$ and $\langle\rangle$ operations. This is very unfortunate, as if it had been possible, we the SCF would have corresponded to $\xi_s^{1/2}(R,0)$ (see Eq. (5)) for which we have the analytical theory. With all the wealth of data analyzed using the SCF (see Padoan et al. 2003) it is very tempting to investigate what sort of errors this mathematically incorrect

[7] Traditionally the definition of centroids includes a normalization by the integral of ρ_s. This, however does not substantially improve the statistics, but makes the analytical treatment very involved (Lazarian & Esquivel 2003).

[8] Expressions for the correlation functions are straightforwardly related to those of structure functions. The statistics of centroids using correlation functions was used in a later paper by Levier (2004).

commuting entails. Alternatively, it is appealing to reanalyze the data in Padoan et al. 2003 using the VCA. It also seems interesting, although currently not feasible, to find the connections between the VCA and VCS and another statistical tool, namely, Principal Component Analysis (see Heyer & Brunt 2004 and references therein).

In LP00, LP04, LP06 we used HI as an example of species to which the technique is to be applied. Using heavier species that have lower thermal Doppler width of spectral lines allows one to study turbulence up to smaller scales. In addition, we would like to stress that the VCS technique can be used at different wavelength. For instance, the X-ray spectrometers with high spatial resolution can be used to study of turbulence in hot plasma. In particular, the potential of VCS is high for studies of turbulence in clusters of galaxies (cf. Sunyaev et al. 2003 and references therein). A simulated example of such a study with the future mission Constellation X is provided in Lazarian (2006).

Studies of turbulence in objects which are poorly resolved spatially is a natural avenue for the VCS applications. Interestingly enough, in this case one can combine the absorption line studies, which would provide the VCS for the pencil beam, i.e. for the high resolution, with the emission studies that would provide the VCS in the poor resolution limit. Potentially, both velocity and density spectra can be obtained this way.

The importance of this work goes beyond the actual recovery of the particular power-law indexes. First of all, the techniques can be generalized to solve the inverse problem to recover non-power law turbulence spectra. This may be important for studying turbulence at scales at which either injection or dissipation of energy happens. Such studies are important for identifying astrophysical sources and sinks of turbulent energy. Second, studies of the VCS transition from low resolution to high resolution regimes (see Figure 2) allows one to separate thermal and non-thermal contributions to the line-widths as it is discussed in LP06. This could both test the thermal correction that can be applied to extend the power-law into sub-thermal velocity range (see also Chepurnov & Lazarian 2006b) and enable studies of temperature distribution of the gas in atomic clouds (cf. Heiles & Troland, 2003).

TESTING AND APPLYING TO OBSERVATIONS

VCA predictions were tested in Lazarian et al. (2001), Esquivel et al. (2003) and in Chepurnov & Lazarian (2006b) using synthetic maps obtained with simulated power-law data as well as with numerical compressible MHD simulations. Simulated data cubes allowed both density and velocity statistics to be measured directly. Then these data cubes were used to produce synthetic spectra which were analyzed using the VCA. As the result, the velocity and density statistics were successfully recovered[9].

The VCA has been applied to several data sets already. One of the first applications was in Stanimirovic & Lazarian (2001), where the technique was applied to the Small Magellanic Cloud (SMC) data. The analysis revealed spectra of 3D velocity fluctuations roughly consistent with the Kolmogorov scaling (a bit more shallow). LP00 argued

[9] A recent study in Chepurnov & Lazarian (2006b) explains why an attempted testing of the VCA in Miville-Deschenes et al. (2003) had inadequate resolution and therefore brought erroneous results.

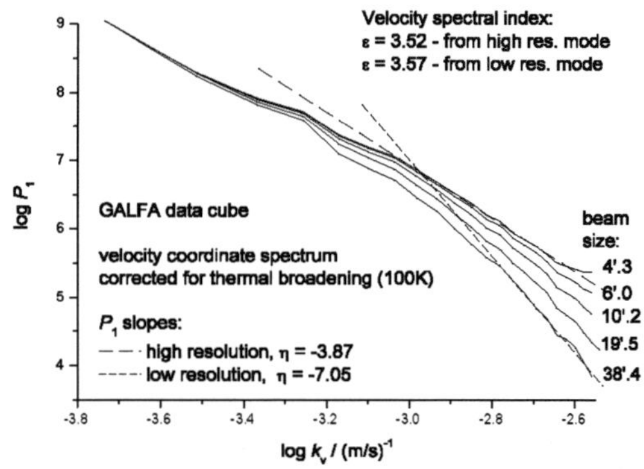

FIGURE 4. Application of the VCS to GALFA high latitude Arecibo data (from Chepurnov, Lazarian & Stanimirovic (2006)). The spatial resolution of the maps was decreased to illustrate the VCS in both high and low resolution regimes. ε here is the power spectrum index, which for the Kolmogorov turbulence is $11/3$. The measured ε is in the range $[3.52;3.57]$.

that the Kolmogorov scaling was expected for the magnetized turbulence appealing to the Goldreich-Shridhar (1995) model. Esquivel et al. (2003) used simulations of MHD turbulent flows to show that in spite of the presence of anisotropy caused by magnetic field the expected scaling of fluctuations is indeed Kolmogorov. Studies by Cho & Lazarian (2003) revealed that the Kolmogorov-type scaling is also expected in the compressible MHD flows. This also supports the conclusion in LP00 that the data in Green (1993) is consistent with MHD turbulence scaling.

Muller et al. (2004) applied VCA to the Magellanic Bridge and discovered two structures with distinctly different properties. If the spectrum of turbulence in the southern part is similar to that in the SMC (velocity is roughly consistent with the Kolmogorov value and the density somewhat shallower the Kolmogorov value), the northern part of the Magellanic Bridge reveals completely different properties. If we interpret the results for the northern part using Table 3 the velocity power spectrum we get is very steep (index ~ 4.2 compared with Kolmogorov ~ 3.7), while the density is shallow (index ~ 2.8). Whether these values should be interpreted in terms of turbulence (some sort of shocks) or some other effects dominate is an open question that requires further studies.

Studies of turbulence are more complicated for the inner parts of the Galaxy, where (a) two distinct regions at different distances from the observer contribute to the emissivity for a given velocity and (b) effects of the absorption are important. However, the analysis in Dickey et al. (2001) showed that some progress may be made even in those unfavorable circumstances. Dickey et al. (2001) found the steepening of the spectral index with the increase of the velocity slice thickness. They also observed the spectral index for strongly absorbing direction approached -3 in accordance with the conclusions in LP04. Note, that the effects of optical depths may explain some other case when the spectral index stayed the same, e.g. -3, while the thickness of the slice was varying (see Kralil et al. 2006). Incidently, this situation can be confused with the situation when the fluctuations arise from density only (see Begum et al. 2006).

21-cm absorption provides another way of probing turbulence on small scales. The absorption depends on the density to temperature ratio ρ/T, rather than to ρ as in the case of emission[10]. However, in terms of the VCA this change is not important and we still expect to see emissivity index steepening as velocity slice thickness increases, provided that velocity effects are present. In view of this, results of Deshpande et al. (2001), who did not see such steepening, can be interpreted as the evidence of the viscous suppression of turbulence on the scales less than 1 pc. The fluctuations in this case should be due to density and their shallow spectrum $\sim k^{-2.8}$ may be related to the damped magnetic structures below the viscous cutoff (see Lazarian, Vishniac, & Cho 2004). This may be also a consequence of the shallow density spectrum in compressible MHD (see Beresnyak, Lazarian & Cho 2005).

Historically, the CO data was analyzed after integration over the entire emission line. Stutzki et al. (1998) presented the power spectra of ^{12}CO and ^{13}CO fluctuations obtained via integrating the intensity over the entire emission line for L1512 molecular cloud. Counter-intuitively, Stutzki et al. (1998) found for both isotopes the power spectrum with a similar spectral index. According to LP04 this may correspond to optically thick assymptotics (i.e. the integration range of velocities is larger than V_c (see Eq. 7)). If the velocity fluctuations dominate, the expected index is universal and equal to -3 (meaning K^{-3}), if the density fluctuations dominate (see Table 2) the expected index is $-3+\gamma$ (meaning $K^{-3+\gamma}$). The index measured in Stutzki et al. (1998) is ~ 2.8, which may either correspond to -3 within the experimental errors, or more likely indicate the that $\gamma \approx 0.2$, i.e. the density spectrum is shallow. The latter possibility is indirectly supported by ^{18}CO measurement for L1551 cloud in Swift (2006), used the VCA (observing the changes of the channel map spectral index while changing the velocity slice thickness) and obtained the shallow density spectrum with $\gamma \approx 0.2$, while his measured velocity spectrum was approximately Kolmogorov (the index is -3.72).

VCS predictions were successfully tested in Chepurnov & Lazarian (2006b). In Chepurnov, Lazarian & Stanimirovic (2006) the technique was applied to the Arecibo high latitude galactic data. Figure 4 illustrates the results obtained for maps with high spectral resolution as well as for the spatially smoothed maps with lower spatial resolution. This allowed to study both the high and low resolution regimes of the VCS. The results obtained for the aforementioned regimes are similar, namely, the power spectrum indexes are somewhat more shallow than the Kolmogorov index.

The practical application of the VCA and the VCS has only started. Nevertheless, it has already provided some intriguing results. While there are many unsolved problems, the techniques has proved to be a promising way of using the wealth of spectroscopic surveys for studies of astrophysical turbulence. We also believe that eventually the techniques will find applications beyond the astrophysical domain, e.g. for plasma turbulence studies.

Acknowledgments I thank Alexey Chepurnov, Dmitry Pogosyan and Snezana Stanimirovic for their input. AL research is supported by by NSF grant AST 0307869 and the

---

[10] In the case of an isobaric medium the product of density and temperature are constant and the problem is similar to studies of transitions for which the emissivity is proportional to ρ^2 that we discussed earlier.

NSF Center for Magnetic Self Organization in Laboratory and Astrophysical Plasmas.

REFERENCES

1. Armstrong, J. W., Rickett, B. J., & Spangler, S. R. *ApJ*, **443**, 209 (1995)
2. Ballesteros-Paredes, J. Klessen, R., Mac Low, M., Vazquez-Semadeni, E. astro-ph/0603357
3. Begum, A., Chengalur, J., & Bharadwaj, S. *MNRAS Letters* in press, astro-ph/0607367 (2006)
4. Beresnyak, A., Lazarian, A., & Cho, J. *ApJ Letters*, **624**, 93 (2005)
5. Chepurnov, A. & Lazarian, A. *ApJ*, submitted (2006a)
6. Chepurnov, A. & Lazarian, A. *ApJ*, submitted (2006b)
7. Chepurnov, A. & Lazarian, A. & Stanimirovic *ApJ*, in preparation (2006)
8. Cho, J., Lazarian, A. *MNRAS*, **345**, 325 (2003)
9. Deshpande, A.A., Dwarakanath, K.S., Goss, W.M., *ApJ*, **543**, 227 (2000)
10. Dickey, J., McClure-Griffiths, N., Stanimirovic, S., Gaensler, B., & Green, A. *ApJ*, **561**, 264 (2001)
11. Falgarone, E, & Puget, J.-L. *A& A*, **293**, 840 (1995)
12. Falgarone, E., Pineau Des Gorets, B., Hily-Blant, P., & Schilke, P. astro-ph/0606222 (2006)
13. Green, D.A. *MNRAS*, **262**, 328 (1993)
14. Goldreich, P., & Sridhar, S. *ApJ*, **438**, 763 (1995)
15. Hartmann, L., & MacGregor, K. B. *ApJ*, **242**, 260 (1980)
16. Heiles, C. & Troland, T. *ApJS*, **151**, 271
17. Heyer, M., & Brunt, C. *ApJ*, **615**, 45 (2004)
18. Elmegreen, B. *ApJ*, **577**, 206 (2002)
19. Elmegreen, B. & Falgarone, E. *ApJ*, 471, 816 (1996)
20. Elmegreen, B. & Scalo, J. *ARA&A*, **42**, 211 (2004)
21. Ensslin, T. & Vogt, C. *A&A*, **453**, 447 (2006)
22. Esquivel, A. & Lazarian, A. *ApJ*, **784**, 320 (2005)
23. Esquivel, A., Lazarian, A., Pogosyan, D., Cho, J. *MNRAS*, **342**, 325 (2003)
24. Khalil, A, Joncas, G., Nekka, F., Kestener, P., & Arneodo, A. *ApJS*, **165**, 512 (2006)
25. Lazarian, A. *ApJ Letters*, **645**, L25 (2006a)
26. Lazarian, A. *AN*, **327**, 609 (2006b)
27. Lazarian, A. *JKAS*, **37**, 563 (2004)
28. Lazarian, A. & Esquivel, A. *ApJ Letters*, **592**, L37 (2003)
29. Lazarian, A. & Pogosyan, D. *ApJ*, **537**, 720 (2000)
30. Lazarian, A. & Pogosyan, D. *ApJ*, **616**, 943 (2004)
31. Lazarian, A. & Pogosyan, D. *ApJ*, in press, astro-ph/0511248 (2006)
32. Lazarian, A., Pogosyan, D., Vazquez-Semadeni, E., & Pichardo, B. *ApJ*, **555**, 130 (2001)
33. Lazarian, A., Vishniac, E., & Cho, J. *ApJ*, **603**, 180 (2004)
34. Levier, F. *A& A*, **421**, 387 (2004)
35. Mac Low, M. & Klessen, R. *RvMP*, **76**, 125 (2004)
36. McKee, C., Tan, J., *Nature*, **416**, 59 (2002)
37. McKee, C. "The Dynamical Structure and Evolution of Giant Molecular Clouds" in *The Origin of Stars and Planetary Systems*, ed. Charles J. Lada and Nikolaos D. Kylafis. Kluwer, 1999 p.29
38. Miville-Deschênes, M., Levrier, F., & Falgarone, E. *A&A*, **593**, 831 (2003)
39. Muller, E., Stanimirovic, S., Rosolowsky, E., Staveley-Smith, L. *ApJ*, **616**, 845
40. Munch, G. *Rev. Mod. Phys.*, **30**, 1035 (1958)
41. Padoan, P., Goodman, A., & Juvela, M. *ApJ*, **588**, 881 (1993)
42. Ossenkopf, V., Esquivel, A., Lazarian, A. & Stutzki, J. *A&A*, **452**, 223
43. Shu, F., Li, Z.-Y., & Allen, A. *ApJ*, **601**, 930 (2004)
44. Stanimirovic, S. & Lazarian, A. *ApJ*, **551**, L53 (2001)
45. Stutzki, J., Bensch, F., Heithausen, A., Ossenkopf, V., & Zielinsky, M. *A&A*, **336**, 697 (1998)
46. Spangler, S., & Gwinn, C. *ApJ*, **353**, L29 (1990)
47. Sunyaev, R.A., Norman, M.L., & Bryan, G.L. *Astronomy Letters*, **29**, 783 (2003)
48. Swift, J. *PhD Thesis*, Univ. of California-Berkeley (2006)

Line Mixing in Water Vapor and Methane

M. A. H. Smith,[1] L. R. Brown,[2] R. A. Toth,[2] V. Malathy Devi[3] and D. Chris Benner[3]

[1] *Science Directorate, NASA Langley Research Center, Mail Stop 401A, Hampton, VA 23681-2199, U. S. A.*

[2] *Jet Propulsion Laboratory, California Institute of Technology, 4800 Oak Grove Dr., Pasadena, CA 91109-8099, U. S. A.*

[3] *Department of Physics, The College of William and Mary, Box 8795, Williamsburg, VA 23187-8795, U. S. A.*

Abstract. A multispectrum fitting algorithm has been used to identify line mixing and determine mixing parameters for infrared transitions of H_2O and CH_4 in the 5-9 μm region. Line mixing parameters at room temperature were determined for two pairs of transitions in the ν_2 fundamental band of $H_2{}^{16}O$, for self-broadening and for broadening by H_2, He, CO_2, N_2, O_2 and air. Line mixing parameters have been determined from air-broadened CH_4 spectra, recorded at temperatures between 210 K and 314 K, in selected R-branch manifolds of the ν_4 band. For both H_2O and CH_4, the inclusion of line mixing was seen to have a greater effect on the retrieved values of the line shifts than on the retrieved values of other parameters.

Keywords: Infrared, line mixing, water vapor, methane, line broadening, line shifts.
PACS: 33.20.-t, 33.20.Ea, 33.70.Jg

INTRODUCTION

A multispectrum fitting algorithm [1], used to analyze pressure-broadening and line shift parameters for transitions of H_2O and CH_4 in the 5-9 μm region, has revealed the characteristic residual patterns due to line mixing between specific transitions within each band. In this paper we will describe observations and quantitative analyses of line mixing in the ν_2 fundamental band of $H_2{}^{16}O$ and in the ν_4 fundamental band of CH_4. These are the first laboratory measurements of line mixing parameters in each band.

WATER VAPOR

For two pairs of transitions in the ν_2 fundamental band of $H_2{}^{16}O$, the characteristic residuals of line mixing were observed very distinctly in least-squares fits of high-resolution FTS spectra of self-broadened H_2O at pressures up to 29 Torr and mixtures of H_2O with H_2, He, CO_2, N_2, O_2 and air at pressures up to 999 Torr [2]. The positions

CP874, *Spectral Line Shapes: 18th International Conference,*
edited by E. Oks and M. Pindzola

and rotational assignments of the four lines (in terms of the rotational quantum numbers J, K_a and K_c) are given as follows:

P branch: (1 0 1)←(2 1 2) at 1539.061 cm^{-1} mixing with (2 1 2)←(3 0 3) at 1540.300 cm^{-1};

R branch: (3 0 3)←(2 1 2) at 1652.400 cm^{-1} mixing with (2 1 2)←(1 0 1) at 1653.267 cm^{-1}.

Note that in each pair of lines, the rotational quantum numbers of the upper state of one transition (2 1 2) are the same as those of the lower state of the other transition.

For each broadening gas, groups of three to seven spectra were analyzed [2] to determine line positions, intensities, halfwidths, pressure-induced shifts and off-diagonal relaxation matrix elements. Each group of spectra was analyzed first using the multispectrum algorithm assuming Voigt line shapes and no line mixing, and then the analysis was repeated with line mixing implemented, allowing the off-diagonal relaxation matrix elements [3] to be determined in the least-squares fit. The observed spectra and residuals from the fits of self-broadened spectra are shown in Fig. 1.

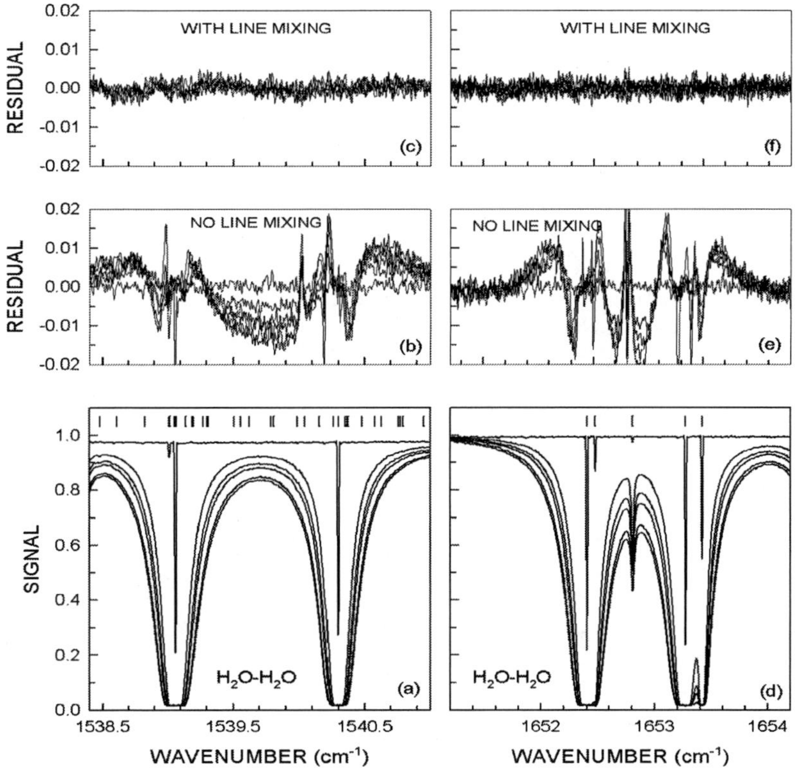

FIGURE 1. Multispectrum fits of mixed pairs of self-broadened H_2O lines in the P branch (left panels) and R branch (right panels) of the ν_2 band. Observed spectra, at six different water pressures ranging from 0.01 to 29 Torr, are plotted in the bottom panels (a, d) along with tick marks indicating the line positions. Middle panels (b, e) show (Obs. – Calc.) residuals for the Voigt line shape *without* line mixing, and the upper panels (c, f) show the corresponding residuals for the fits *with* line mixing [2].

The characteristic residuals of a pair of mixed lines are seen most clearly in the middle left panel of Fig.1, where the calculations without line mixing overestimate the absorption between the two strong P branch lines and underestimate the absorption in the wings of the mixed pair. The zig-zag residuals near the line centers also indicate that the line positions and/or pressure-induced shifts are not well-determined. For the R branch pair, which is overlapped by some other lines, the pattern of residuals is less obvious but still discernable.

Since all of the H_2O spectra had been recorded with samples near room temperature, the temperature dependences of the broadening, shift, and line mixing parameters could not be determined. However, it was possible to compare the mixing parameters retrieved for the various broadening gases. For the two pairs of ν_2 transitions analyzed, the largest mixing effects were observed for self-broadening and H_2-broadening, and the smallest effects were found for He-broadening.

While inclusion of line mixing in the fits did not significantly change the retrieved line intensities and widths (except for self- and H_2-broadening), the retrieved values of the pressure-induced shifts were affected by mixing for all broadeners except He.

Details of this study may be found in the journal article of Brown et al. [2].

METHANE

Multispectrum fits of air-broadened CH_4 spectra, recorded at temperatures between 210 K and 314 K and broadening gas pressures between 50 and 550 Torr, have been carried out for selected R-branch manifolds in the ν_4 band. An example of fits with and without line mixing in the R(11) manifold is shown in Fig. 2. In this figure it is apparent from the residuals that most of the line mixing is occurring between transitions on the low-wavenumber side of the manifold. Examination of the retrieved relaxation matrix elements verifies this initial observation, as the mixing parameters for transitions on the high-wavenumber side of the manifold are effectively zero. Some self-broadened CH_4 spectra were also included in this fit, but the results are very preliminary since only 15 spectra were included out of 64 possible spectra that can be used in the final analysis.

Another interesting spectral region is the R(3) manifold of $^{12}CH_4$, which is overlapped by the R(5) manifold of $^{13}CH_4$. A set of 31 spectra was fit in this region to determine broadening, shift, and line mixing parameters in both manifolds. The observed spectra and residuals from the fit with line mixing are shown in Fig. 3.

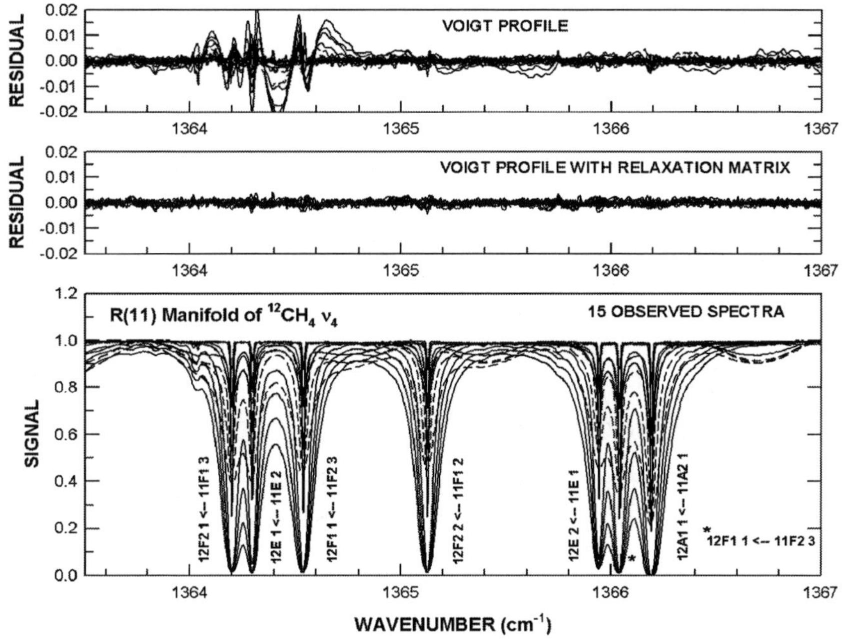

FIGURE 2. Fits of 15 air- and self-broadened room-temperature CH_4 spectra with and without line mixing. Maximum pressure is 550 Torr for air-broadening and 453 Torr for self-broadening.

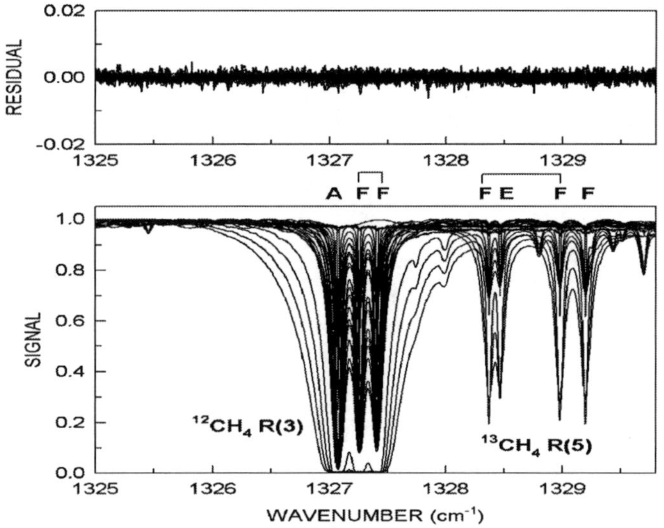

FIGURE 3. Fit, including line mixing, of 31 spectra in the region of the R(3) manifold of $^{12}CH_4$. The data set includes 3 room-temperature spectra of air-broadened $^{13}CH_4$, 26 spectra of air-broadened "natural" CH_4 at 210 K to 314 K, and one spectrum each of low-pressure pure $^{12}CH_4$ and $^{13}CH_4$. Horizontal lines with ticks indicate which transitions are mixing with each other.

The parameters resulting from this fit of the R(3) region are shown in Table 1. The air-broadened widths and temperature dependences, as well as shifts of unmixed lines, retrieved using the multispectrum fits are in excellent agreement with earlier values determined from single-spectrum fits [4, 5]. Furthermore, the parameters determined from multispectrum fits show smaller statistical errors than the corresponding parameters resulting from single-spectrum fits of largely the same data set. Since the $^{13}CH_4$ spectra included in the fit were all recorded at room temperature, no information on the temperature-dependence of parameters for the $^{13}CH_4$ R(5) transitions could be determined, and these were held fixed to average values estimated from corresponding $^{12}CH_4$ R(5) transitions.

TABLE 1. Methane Air-broadening, Shift, and Line Mixing Parameters in the R(3) Region

Position (cm^{-1})	Isotope and HITRAN04 Assignment	Width (cm^{-1} atm^{-1} at 296 K)	n	Shift (cm^{-1} atm^{-1} at 296 K)	δ′ (cm^{-1} atm^{-1} K^{-1})	Mixing (cm^{-1} atm^{-1} at 296 K)	T-Dependence of Mixing
1327.074206(4)	12CH4 4A1 1←3A2 1	0.05836(2) *0.0574(3)*	0.723(2) *0.81(3)*	−0.00164(2) *−0.0018(1)*	0.85(3) $\times 10^{-5}$ *0.7(2)* $\times 10^{-5}$	none	none
1327. 256817(5) *	12CH4 4F1 1←3F2 1	0.06626(5) *0.0667(3)*	0.802(2) *0.64(2)*	−0.00145(3) *+0.0023(3)*	1.17(6) $\times 10^{-5}$ *1.0(3)* $\times 10^{-5}$	0.00608(3)	1.32(2)
1327.409813(8) *	12CH4 4F2 1←3F1 1	0.06689(3) *0.0662(3)*	0.818(2) *0.74(3)*	−0.00157(3) *−0.0050(2)*	0.60(5) $\times 10^{-5}$ *0.5(3)* $\times 10^{-5}$	0.00608(3)	1.32(2)
1328.372213(9) **	13CH4 6F2 1←5F1 2	0.05958(4) *0.0578(2)*	0.72 (fixed)	−0.00190(4) *−0.0016(2)*	0.0 (fixed)	0.00978(27)	1.4 (fixed)
1328.467453(12)	13CH4 6E 1←5E 1	0.05169(5) *0.0516(1)*	0.72 (fixed)	−0.00198(4) *−0.0026(1)*	0.0 (fixed)	none	none
1328.982458(9) **	13CH4 6F1 1←5F2 1	0.06180(4) *0.0611(0)*	0.72 (fixed)	−0.00252(5) *−0.0036(1)*	0.0 (fixed)	0.00978(27)	1.4 (fixed)
1329.200880(9)	13CH4 6F2 2←5F1 1	0.05894(4) *0.0584(1)*	0.72 (fixed)	−0.00193(5) *−0.0029(2)*	0.0 (fixed)	Not measurable	Not measurable

Numbers in parentheses are statistical errors in units of the last digit quoted. Mixing parameter is the larger off-diagonal matrix element of the pair of mixed lines. Mixed pairs are indicated by * or ** in the first column. Results from previous spectrum-by-spectrum analyses [4, 5] are given in italics.

As was the case for H_2O, the inclusion of line mixing was seen to have a greater effect on the retrieved values of the line shifts than on the retrieved values of other parameters. We note that for the transition at 1327.2568 cm^{-1}, the inclusion of line mixing in the fit actually changes the sign of its pressure-induced shift; the magnitude of the shift of its mixing partner at 1327.4098 cm^{-1} also changes significantly.

We have presented here only a sampling of line mixing results for a few transitions of H_2O and CH_4. Work is continuing on analysis of line mixing in these and other bands. As a larger database of experimentally-determined line mixing parameters is accumulated, there will be additional opportunities to compare experimental and theoretical mixing parameters, and to compare mixing parameters for different bands of the same molecule.

ACKNOWLEDGMENTS

The research described in this paper was carried out by the Jet Propulsion Laboratory, California Institute of Technology, Langley Research Center and the College of William and Mary under contracts with the National Aeronautics and Space Administration.

REFERENCES

1. D. Chris Benner, C. P. Rinsland, V. Malathy Devi, M. A. H. Smith, and D. Atkins, *J. Quant. Spectrosc. Radiat. Transfer* **53**, 705-721 (1995).
2. L. R. Brown, D. Chris Benner, V. Malathy Devi, M. A. H. Smith, and R. A. Toth, *J. Mol. Structure* **742**, 111-122 (2005).
3. A. Levy, N. Lacome, and C. Chackerian, Jr., "Collisional Line Mixing" in *Spectroscopy of the Earth's Atmosphere and Interstellar Medium*, edited by K. Narahari Rao and A. Weber, Academic Press, Boston, Massachusetts, 1992, pp. 261-337.
4. M. A. H. Smith, C. P. Rinsland, V. Malathy Devi, and D. Chris Benner, *Spectrochimica Acta* **48A**, 1257-1272 (1992).
5. V. Malathy Devi, C. P. Rinsland, M. A. H. Smith, and D. Chris Benner, *Appl. Opt.* **27**, 2296-2308 (1988).

Alkali Line Shapes for the Spectra of Brown Dwarfs

G. Peach*, S. J. Gibson†, D. F. T. Mullamphy†, V. Venturi† and I. B. Whittingham†

*Department of Physics and Astronomy, University College London, Gower Street, London WC1E 6BT, United Kingdom

†School of Mathematical and Physical Sciences, James Cook University, Townsville, Australia, 4811

Abstract. Highly accurate calculations of the central cores of pressure broadened alkali resonance doublets due to H_2 and He perturbers are needed in order to estimate the effects of dust in brown dwarf atmospheres. We report results for the Lorentzian profiles of the Na I and K I doublets broadened by helium perturbers for temperatures up to 3000 K. They have been obtained from a fully quantum-mechanical close-coupling description of the colliding atoms, the Baranger theory of line shapes and new ab initio potentials for the alkali-rare gas interaction.

Keywords: Brown dwarfs: alkali resonance doublets – Brown dwarfs: synthetic spectra – Spectral line broadening: calculations.
PACS: 32.70.Jz, 34.20.Cf, 95.30.Ky, 97.20.Vs

INTRODUCTION

Accurate pressure broadened profiles of alkali resonance doublets perturbed by H_2 and He are of crucial importance for the modeling of atmospheres of late M, L and T type brown dwarfs and for generating their synthetic spectra in the region 0.6 - 1.1 μm. The dominant lines are the Na I 589.0/589.6 nm and K I 766.5/769.9 nm doublets [1, 2] which occur in the middle of a spectral region where background absorption is particularly small so that both the line centres and wings stand out [3]. There can also be significant contributions from less abundant alkalis such as Li, Rb and Cs, and from subordinate doublets such as Na I 818.3/819.5 nm.

The non-Lorentzian profiles of the strongly broadened alkali doublets have been the subject of several recent studies using models primarily designed to describe the far wings of the profiles. Burrows & Volobuyev [4] used a quasistatic unified Franck- Condon semiclassical model to study the Na I and K I doublets whereas a unified line shape semiclassical model has been used by Allard et al. [5] for the Na I and K I doublets and, more recently, for the Rb I and Cs I doublets [6]. Alternatively, quantum mechanical single-channel models based upon bound-free and free-free transitions in the transient molecules formed during collision have been used

CP874, *Spectral Line Shapes: 18th International Conference,*
edited by E. Oks and M. Pindzola

to calculate the emission and absorption spectra for the wings of the Li I [7, 8], Na I [7, 9] and K I [9] resonance lines in the presence of He perturbers. Despite these studies, highly accurate calculations of the central Lorentzian cores are still needed [10] in order to estimate the effects of dust in brown dwarf atmospheres.

We report results for the Lorentzian widths over the temperature range $70 \leq T \leq 3000$ K of the Na I and K I doublets broadened by helium perturbers. The calculations extend our previous study [11] of pressure broadening of the Na I doublets by He at laboratory temperatures up to 500 K and are based on a fully quantum-mechanical close-coupling description of the colliding atoms and the Baranger [12] theory of line shapes. As input for our calculations we have computed new *ab initio* potentials for the alkali- rare gas interaction. Preliminary results for the widths of two Na I lines for temperatures up to 2000 K were given in [13].

INTERATOMIC POTENTIALS

The adiabatic molecular potentials ${}^{2S+1}V_\Lambda(R)$ for the X^*–He system (where X = Li, Na or K) have been obtained by using a three-body model in which the alkali X^* is treated as an X^+ ion plus an active electron and the perturbing He atom is represented by a polarizable atomic core. Model potentials are used to represent the electron-atom and electron-atomic ion interactions and the basic methods adopted for obtaining these model potentials are discussed by Peach [14]. These potentials reproduce not only the correct energies of the bound states of X^* but also generate wavefunctions for both the bound and scattering states of $X+e^-$ and $He+e^-$ that contain the correct number of nodes. The model potentials also support unphysical bound states corresponding to the presence of closed shells in the X^+ ion and in He. This effect is taken into account in the calculation of the molecular potential by including the unphysical states in the atomic basis used in the diagonalization of the Hamiltonian for the three-body model.

The most important recent change from the earlier approach is that the core-core interaction is itself calculated directly using a three-body model composed of X^+ and He^+ cores plus one electron. A more detailed discussion of this development is given for the Li-He case by Behmenburg et al. [15]. Further numerical refinements have been incorporated into the computer program since then and new calculations carried out for the three alkali-helium systems considered in our study.

SPECTRAL LINE PROFILES

The width and shift of the spectral lines have been calculated using the quantum- mechanical impact theory of Baranger [12] for non-overlapping spectral lines in which the profile of each isolated line is a Lorentzian. The half half-width w and shift d of the Lorentzian spectral profile for a line involving initial and final emitter states $|j_i\rangle$ and $|j_f\rangle$ respectively of total electronic angular momentum $\mathbf{j} = \mathbf{L} + \mathbf{S}$ (we suppress the quantum numbers (n, L, S) for convenience) is given by [11]

$$w + id = N\int_0^\infty f(E)S(E)dE \tag{1}$$

where $f(E)$ is the normalized Maxwellian perturber energy distribution:

$$f(E) = 2\pi\left(\pi k_B T\right)^{-3/2}\sqrt{E}\exp\left(-\frac{E}{k_B T}\right), \tag{2}$$

N is the perturber number density and

$$\begin{aligned} S(E) = \frac{\hbar^2\pi}{M^2}\sqrt{\frac{M}{2E}}\sum_{l,l'}\sum_{J_i,J_f}(2J_i+1)(2J_f+1)(-1)^{l+l'}\begin{Bmatrix} J_f & J_i & 1 \\ j_i & j_f & l \end{Bmatrix} \\ \times\begin{Bmatrix} J_f & J_i & 1 \\ j_i & j_f & l' \end{Bmatrix}\left\{\delta_{l,l'} - \langle j_i l' J_i|S|j_i l J_i\rangle\langle j_f l' J_f|S|j_f l J_f\rangle^*\right\}. \end{aligned} \tag{3}$$

Here l and l' are the values of the relative emitter-perturber angular momentum $\mathbf{L}_R$ before and after the collision, $\mathbf{J} = \mathbf{L}_R + \mathbf{j}$ is the total angular momentum of the emitter-perturber system, $\langle j'l'J|S|j\,l\,J\rangle$ is the interaction picture scattering matrix in the coupled $\{|j\,l\,J\rangle\}$ representation, $\begin{Bmatrix} a & b & c \\ d & e & f \end{Bmatrix}$ is the $6-j$ symbol and M is the reduced mass of the emitter-perturber system.

The scattering matrix elements in (3) are determined from the asymptotic behaviour of the radial functions $G^J_{jl}(R)$ for each scattering channel $(j,\ l,\ J)$ which satisfy the coupled equations [11]

$$\left[\frac{\partial^2}{\partial R^2} - \frac{l(l+1)}{R^2} + k_j^2\right]G^J_{jl,j''l''}(R) = \frac{2M}{\hbar^2}\sum_{j',l'}V^J_{jl,j'l'}(R)G^J_{j'l',j''l''}(R) \tag{4}$$

where $(j'',\ l'')$ labels the linearly independent solutions of (4) and the Born-Oppenheimer coupling terms have been neglected. The parameter

$$k_j^2 = 2M\left[E - \left(E^{(L,S)} + \varepsilon_j(\infty)\right)\right]/\hbar^2 \tag{5}$$

is positive for open scattering channels and negative for closed channels. Here E is the total energy of the emitter-perturber system, $E^{(L,S)}$ is the energy of the state of the separated atoms to which the molecular state dissociates adiabatically and the fine structure parameter $\varepsilon_j(R)$ has been assumed to have its asymptotic value $\varepsilon_j(\infty)$.

The interaction potential matrix elements $V^J_{jl,j'l'}(R)$ [11] are, for the ^2S_j level,

$$V^J_{jl,j'l'}(R) = \delta_{j,j'}\delta_{l,l'}\,{}^2V_\Sigma(R), \tag{6}$$

for the ^2P_j levels,

$$V^J_{jl,j'l'}(R) = \delta_{j,j'}\delta_{l,l'}\,{}^2V_\Pi(R) + C^{(1)}_{jl,j'l'}\left[{}^2V_\Sigma(R) - {}^2V_\Pi(R)\right] \tag{7}$$

and, for the ^2D_j levels,

$$\begin{aligned} V^J_{jl,j'l'}(R) = {} & \delta_{j,j'}\delta_{l,l'}\,{}^2V_\Delta(R) + B_{jl,j'l'}(-1)^{j'-j}\left[1+(-1)^{l'+l}\right]\left[{}^2V_\Pi(R) - {}^2V_\Delta(R)\right] \\ & + C^{(2)}_{jl,j'l'}\left[{}^2V_\Sigma(R) - {}^2V_\Delta(R)\right] \end{aligned} \tag{8}$$

where the coefficients

$$\begin{aligned} C^{(n)}_{jl,j'l'} = {} & \sum_\Omega (-1)^{j'-j}\, C(J\,j\,l;-\Omega\Omega 0)\, C(J\,j'\,l';-\Omega\Omega 0) \\ & \times C\left(n\,\frac{1}{2}\,j;0\,\Omega\,\Omega\right) C\left(n\,\frac{1}{2}\,j';0\,\Omega\,\Omega\right) \end{aligned} \tag{9}$$

and

$$\begin{aligned} B_{jl,j'l'} = {} & \sum_\Omega C(J\,j\,l;-\Omega\,\Omega\,0)\, C(J\,j'\,l';-\Omega\,\Omega\,0) \\ & \times C\left(2\,\frac{1}{2}\,j;1\;\Omega-1\;\Omega\right) C\left(2\,\frac{1}{2}\,j';1\;\Omega-1\;\Omega\right) \end{aligned} \tag{10}$$

are symmetric with respect to the interchange $(j,\,l) \leftrightarrow (j',\,l')$. Here $C(j_1\,j_2\,j;\,m_1\,m_2\,m)$ is the Clebsch-Gordan coefficients, Ω denotes the projection of an angular momentum onto the internuclear axis $\mathbf{R}$ and $^{2S+1}V_\Lambda(R)$, where $\Lambda \equiv |\Omega_L|$, are the adiabatic molecular potentials.

The equations (4) decouple into two sets of opposite parity $(-1)^{J\pm 1/2}$. Thus the $^2S_{1/2}$, $^2P_{3/2,1/2}$ and $^2D_{3/2}$ alkali states are represented by one, three and five coupled differential equations for each parity respectively.

The close-coupled equations (4) have been solved used using the $\mathcal{R}$-matrix packages RPROP2 and FARM [17], the latter as modified by [18], and the solutions fitted to free-field boundary conditions to extract the scattering matrix elements. $\mathcal{R}$-matrix techniques have high numerical stability and computational efficiency, especially where large numbers of scattering energies are needed. The RPROP2 package uses the solution following propagator developed by Burke, Baluja and Morgan (BBM) whereas the FARM package incorporates a combination of the BBM solution following propagator at small distances where $V(R)$ is rapidly varying and the potential following Light-Walker propagator at larger distances where $V(R)$ is slowly varying. FARM also includes an accelerated Gailitis expansion to minimize the distance to which the matching procedure is undertaken.

RESULTS

Calculations have been completed for the Na I doublet 3p $^2P_{3/2,1/2} \rightarrow$ 3s $^2S_{1/2}$ and subordinate doublet 3d $^2D_{3/2} \rightarrow$ 3p $^2P_{3/2,1/2}$ and the K I doublet 4p $^2P_{3/2,1/2} \rightarrow$ 4s $^2S_{1/2}$. The main computational issues associated with extending our previous calculations up to the higher temperatures (3000 K) of astrophysical interest are convergence of the sum over partial waves in (3) and of the integration (1) over perturber energies. The present calculations required up to 1000 partial waves and energy nodes extending to 5000 K to obtain satisfactory convergence.

The temperature dependence of the computed half half-widths for the Na I and K I lines are shown in Figure 1 and Figure 2 respectively. All results can be conveniently represented by the fit

$$w(T) = w_0 + w_1\hat{T} + w_2\hat{T}^2 + w_3\hat{T}^3 + w_4\hat{T}^4 \tag{11}$$

where $\hat{T} \equiv T/1000$, with the fit coefficients w_i given in Table 1.

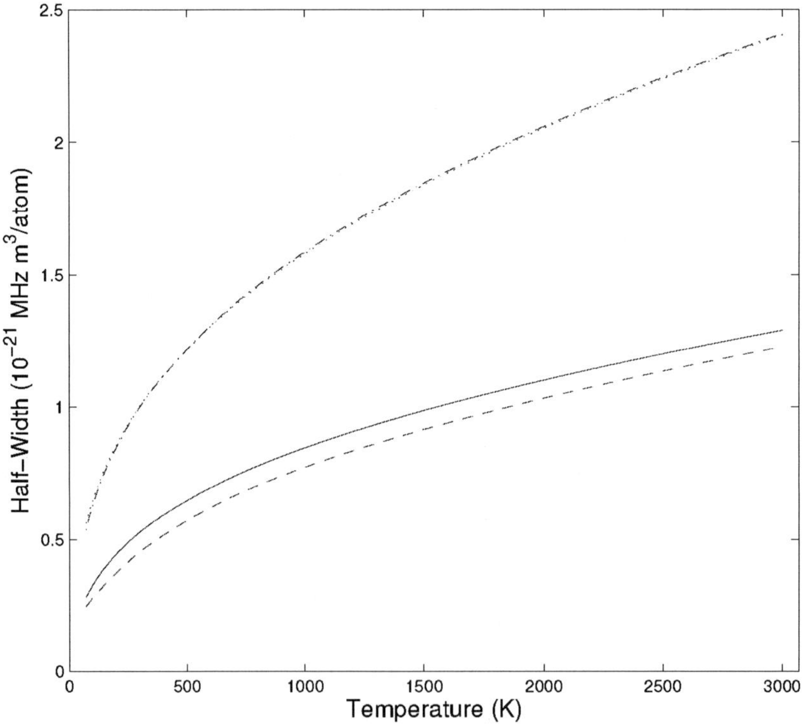

FIGURE 1. Temperature dependence of the Lorentzian half half-widths for the Na I doublet and subordinate doublet (in units of 10^{-21} MHz m^3/atom) pressure broadened by He perturbers. The results

shown are for the 589.0 (solid line), 589.6 (dashed line), 818.3 (dotted line) and 819.5 (dash-dotted line) nm spectral lines. Note that, on the scale shown, the 818.3 and 819.5 nm results can not be distinguished.

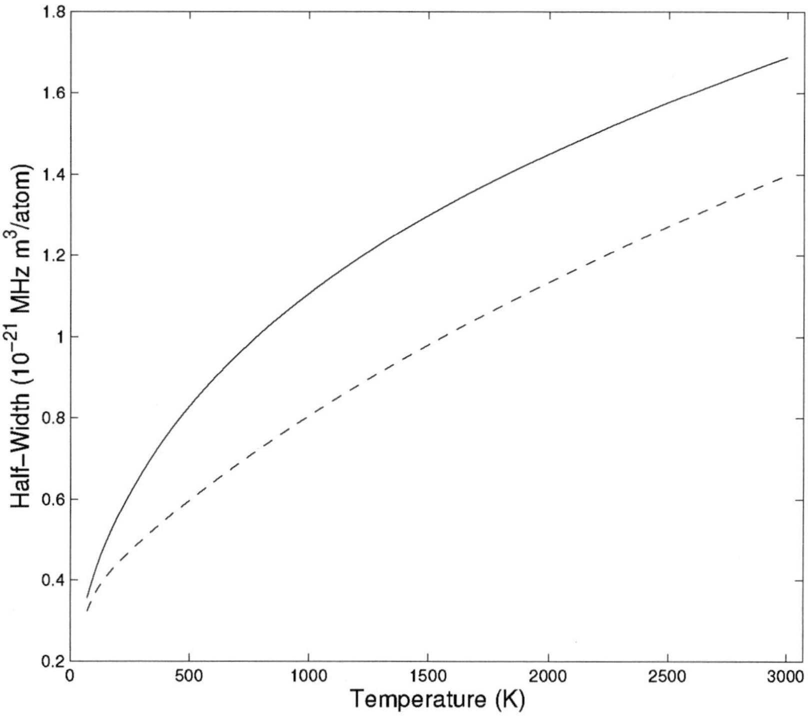

FIGURE 2. Temperature dependence of the Lorentzian half half-widths for the K I doublet (in units of 10^{-21} MHz m^3/atom) pressure broadened by He perturbers. The results shown are for the 766.5 (solid line), 769.9 (dashed line) nm spectral lines.

TABLE 1). Coefficients w_i for the temperature dependence of the Lorentzian half half-widths (in units of 10^{-21} MHz m^3/atom) for alkali doublets perturbed by He for temperatures between 70 and 3000 K.

Alkali	Transition	λ nm	w_0	w_1	w_2	w_3	w_4
Na	$3p\ ^2P_{3/2} \rightarrow 3s\ ^2S_{1/2}$	589.0	0.255	1.039	-0.632	0.218	-0.028
	$3p\ ^2P_{1/2} \rightarrow 3s\ ^2S_{1/2}$	589.6	0.203	0.957	-0.541	0.181	-0.023
	$3d\ ^2D_{3/2} \rightarrow 3p\ ^2P_{1/2}$	818.3	0.518	1.841	-1.101	0.381	-0.050
	$3d\ ^2D_{3/2} \rightarrow 3p\ ^2P_{3/2}$	819.5	0.498	1.910	-1.161	0.403	-0.053
K	$4p\ ^2P_{3/2} \rightarrow 4s\ ^2S_{1/2}$	766.5	0.310	1.353	-0.775	0.261	-0.033
	$4p\ ^2P_{1/2} \rightarrow 4s\ ^2S_{1/2}$	769.9	0.310	0.677	-0.243	0.074	-0.009

CONCLUSIONS AND FUTURE WORK

The present calculations have demonstrated that the formalism, calculational techniques and potentials can be successfully applied to the computation of pressure broadening of alkali doublet lines by He perturbers at temperatures pertaining to the atmospheres of brown dwarfs. Extensions of the calculations to the Li I doublet lines will be completed shortly.

Future work will involve extension to the line wings as the present calculations are only valid for the central regions of the profiles. The line wing will be treated using the one-perturber approximation theory of line broadening [19] which employs the full X^*-He wavefunctions and not just the asymptotic form as required by the impact theory.

REFERENCES

1. Y. Pavlenko, M. R. Zapatero Osorio and R. Rebolo, Astron. and Astrophys., **355**, 245–255 (2000)
2. A. Burrows, W. B. Hubbard, J. I. Lunine, and J. Liebert, Rev. Mod. Phys., **73**, 719–765 (2001)
3. C. Johnas, These proceedings
4. A. Burrows and M. Volobuyev, Astrophys. J., **583**, 985–995 (2003)
5. N. F. Allard, F. Allard, P. H. Hauschildt, J. F. Kielkopf, and L. Machin, Astron. and Astrophys., **411**, L473–L476 (2003)
6. N. F. Allard and F. Spiegelman, Astron. and Astrophys., **452**, 351–356 (2006)
7. C. R. Mason, PhD thesis, Uni. London (1991)
8. C. Zhu, J. F. Babb, and A. Dalgarno, Phys. Rev. A, **71**, 052710 (2005)
9. C. Zhu, J. F. Babb, and A. Dalgarno, Phys. Rev. A, **73**, 012506 (2005)
10. Y. Pavlenko, Private communication (2004)
11. P. J. Leo, G. Peach and I. B. Whittingham, J. Phys. B: At. Mol. Opt. Phys., **33**, 4779–4797 (2000)
12. M. Baranger, Phys. Rev., **112**, 855–865 (1958)
13. G. Peach, D. F. T. Mullamphy, V. Venturi and I. B. Whittingham, Mem. S. A. It, **7**, 145–148 (2005)
14. G. Peach, Comment. At. Mol. Phys., **11**, 101–118 (1982)
15. W. Behmenburg, A. Makonnen, A. Kaiser, F. Rebentrost, V. Staemmler, M. Jungen, G. Peach, A. Devdariani, S. Tserkovnyi, A. Zagrebin, and E. Czuchaj, J. Phys. B: At. Mol. Opt. Phys., **29**, 3891–3910 (1996)
16. L. A. Morgan, Comp. Phys. Commun., **31**, 419–422 (1984)
17. V. M. Burke and C. J. Noble, Comp. Phys. Commun., **85**, 471–500 (1995)
18. V. Venturi, I. B. Whittingham, P. J. Leo, and G. Peach, Phys. Rev. A, **60**, 4635–4646 (1999)
19. G. Peach, Adv. Phys., **30**, 367–474 (1981)

Line-Mixing Effects in the Microwave Spectrum of O_2

P. W. Rosenkranz

Massachusetts Institute of Technology, Cambridge, MA 02139-4307 USA

Abstract. The microwave spectrum of molecular oxygen consists of magnetic-dipole transitions in which the unpaired electronic spin changes orientation with respect to the rotational angular momentum N. These transitions form a band at 60 GHz and a single line at 119 GHz. The selection rules for magnetic dipoles also allow diagonal transitions, or nonresonant absorption, which are equivalent to a band at zero frequency. The presence of a magnetic field such as that of the earth splits the lines into Zeeman components, which are polarized. Pressure broadening produces a hierarchy of line-mixing effects among the Zeeman components, among the spin-rotation lines, and between the resonant and nonresonant bands. Measurements that quantify the magnitude of these effects are reviewed. In comparison to the mixing between spin-rotation lines within the 60-GHz band, mixing between Zeeman components within a line increases from none in the first rotational level to approximately one-fourth as large an effect at N=7 to 11, and possibly more at higher N.

Keywords: line mixing, oxygen microwave absorption, Zeeman effect
PACS: 33.20.Bx, 33.70.Jg, 33.55.Be

INTRODUCTION

The ground electronic state ($X^3\Sigma_g^-$ in spectroscopic notation) of the oxygen molecule has electronic spin of 1 due to two unpaired electron spins. The spin vector couples to rotational angular momentum, with quantum number N, to form a triplet of levels with total angular momentum quantum number J=N, N−1, and N+1 for each rotational state [1]. J=N is always the highest in energy of the three. For the most abundant isotopomer $^{16}O_2$, symmetry permits only odd values of N [2]. Transitions between J=N−1 and J=N (designated N^-) and between J=N+1 and J=N (designated N^+) lie at frequencies between roughly 50 and 70 GHz, with one significant exception (1^-) at 118.75 GHz. The selection rules for the magnetic dipole moment of O_2 also permit diagonal elements of the dipole-moment operator; these produce a nonresonant absorption which can be represented as zero-frequency transitions. The above transitions constitute the microwave spin-rotation spectrum of oxygen. Transitions between different values of N also are permitted [3], but these occur at sub-mm wavelengths, and will not be part of the present discussion.

In the presence of a magnetic field, the oxygen magnetic-dipole transitions are split

CP874, *Spectral Line Shapes: 18th International Conference,*
edited by E. Oks and M. Pindzola

by the Zeeman effect. The Zeeman components of each line can be identified by the orientational quantum number M in the J=N level and by its change ΔM in the transition; ΔM=0, ±1 are permitted. The earth's magnetic field ranges from 0.3 to 0.7 gauss, and 0.5 gauss produces a total Zeeman splitting of each oxygen line of approximately ±1 MHz [4,5]. Although the number of components increases with J, they are more closely spaced in frequency, yielding approximately the same maximum splitting. The ΔM=–1, 0, or +1 types each have different characteristic polarizations which depend on the angle between the static magnetic field and the direction of propagation. That leads to significant polarization dependence of radiometric measurements for the purpose of atmospheric temperature profiling at altitudes above ~50 km [4,6,7], where pressure broadening is comparable to or less than Zeeman splitting. (Doppler-broadening half-width of the transitions is ~10^{-6} times their frequency.) Even for measurements made in atmospheric windows, there is a very small residual polarization of the atmospheric emission, which can be a source of confusion for ground-based measurement of the cosmic-background polarization [8].

Oxygen also interacts with radiant energy by means of collision-induced dipoles [9]. This absorption is very small at 60 GHz, although it becomes significant at sub-mm wavelengths.

LINE MIXING

The theory of line mixing originated with work by Baranger [10] and Kolb and Griem [11] to describe hydrogen lines in plasmas. Further developments were made by Fano [12], Ben-Reuven [13,14], and others. The absorption coefficient at frequency ν is

$$\alpha(\nu) = \Gamma \ \mathrm{Im}[\, \mu' (\nu \mathbf{I} - \nu_L - i\,\mathbf{W})^{-1} \rho\, \mu], \tag{1}$$

where μ is a column vector in the line-space notation, containing the dipole-moment matrix elements for the radiating transitions, μ' = transpose μ, **I** is the identity matrix, ν_L is a diagonal matrix containing the transition frequencies (including zero-frequency and negative-frequency lines, the latter corresponding to emission transitions), and **ρ** is the density matrix containing the Boltzmann factors, divided by the partition function, on its diagonal. The relaxation matrix **W** embodies the effect of collisions on the spectrum and each of its two indices corresponds to a radiating transition. For temperature T and frequencies such that $h\nu \ll kT$,

$$\Gamma = \frac{8\pi^2 \nu^2 n_a}{3ckT}, \tag{2}$$

in which n_a is the absorbing-gas number density.

All of the spin-rotation lines of oxygen lie within the first 5 cm^{-1} of the spectrum, so duration-of-collision effects can safely be neglected here. Assuming also binary collisions, **W** can be represented as [15]

$$\mathbf{W} = 2\pi n_p \int_0^\infty b \int_0^\infty v\, F(v)\, (\mathbf{I} - \mathbf{S}_{\text{final}} \otimes \mathbf{S}^*_{\text{initial}})\, dv\, db, \tag{3}$$

where n_p is the perturbing-gas number density, b is the collision impact parameter, F(v) is a relative-speed distribution, and $\mathbf{S}_{\text{final}}$ and $\mathbf{S}_{\text{initial}}$ are Hilbert-space scattering matrices for the initial and final states of the radiating transitions. **W** does not contain off-diagonal elements mixing Zeeman components with different ΔM [13], but line mixing can occur between Zeeman components with the same value of ΔM.

The matrix inverse in eq.(1) can always be evaluated by diagonalization [10,11] with a transformation **X** such that

$$\mathbf{X}^{-1}\, (\boldsymbol{\nu}_L + i\, \mathbf{W})\, \mathbf{X} = \boldsymbol{\Lambda} \text{ (diagonal)}. \tag{4}$$

Then

$$\alpha(\nu) = \Gamma \sum_j \frac{\operatorname{Im}\Lambda_j \operatorname{Re}G_{jj} + (\nu - \operatorname{Re}\Lambda_j)\operatorname{Im}G_{jj}}{(\nu - \operatorname{Re}\Lambda_j)^2 + (\operatorname{Im}\Lambda_j)^2}, \tag{5}$$

where

$$\mathbf{G} = \mathbf{X}^{-1}\, \boldsymbol{\rho}\, \mu\, \mu'\, \mathbf{X}. \tag{6}$$

Approximations and Limiting cases

To first order in **W** (i.e. first-order in the gas density or pressure) [16],

$$\Lambda_j = \nu_j + i\, W_{jj}, \tag{7}$$

where ν_j is a diagonal element of $\boldsymbol{\nu}_L$, and

$$G_{jj} = \mu_j^2\, \rho_j + i\, 2\, \mu_j\, \rho_j \sum_{k \neq j} \frac{\mu_k W_{kj}}{\nu_j - \nu_k}. \tag{8}$$

In this limit, the diagonal elements of **W** are the half-widths of the resonant terms in eq. (5), and the effect of off-diagonal elements on the spectrum is characterized by one additional pressure-proportional parameter for each line:

$$Y_j = \operatorname{Im} G_{jj} / \operatorname{Re} G_{jj} = \frac{2}{\mu_j} \sum_{k \neq j} \frac{\mu_k W_{kj}}{\nu_j - \nu_k}. \tag{9}$$

When a number of transitions form an indistinguishable group, either because the pressure causes them to merge completely or because the observed frequency is on their far wings, the size of **W** can be reduced. In the latter case, we can substitute eqs. (7,8) into (5) and assume that $(\nu-\nu_j)^2 >> W_{jj}^2$ for $a\leq j\leq b$; then, making use of the symmetry of the product **W**ρ [14], the submatrix with $a\leq j\leq b$ and $a\leq k\leq b$ makes a contribution

$$\alpha_a^b(\nu) = \Gamma\sum_{j=a}^{b}\sum_{k=a}^{b}\frac{\mu_j\mu_k W_{kj}\rho_j}{(\nu-\nu_j)(\nu-\nu_k)} \tag{10}$$

to the sum in eq.(5). Hence, for frequencies lying far outside the group, with $(\nu-\nu_j)\cong(\nu-\nu_k)$, this submatrix acts like resonances

$$\alpha_a^b(\nu) = \Gamma\sum_{j=a}^{b}\frac{\mu_j^2\rho_j}{(\nu-\nu_j)^2}\sum_{k=a}^{b}\frac{\mu_k W_{kj}}{\mu_j} \tag{11}$$

with reduced broadening parameters given by the second summation in eq. (11), because off-diagonal elements of **W** are typically opposite in sign to the diagonal ones due to the presence of the identity matrix in eq.(3). A result equivalent to eq.(10) was obtained in [10] by starting from eq.(1) instead of the perturbation equations (7,8). Thus, although low pressure (roughly, such that $|Y_j|<1$ for all j) is a sufficient condition for applicability of eqs.(7-9), it is not a necessary one in the far wings. But within the resonant region, the perturbation solution eventually becomes invalid with increasing pressure.

The linearity of eqs.(7-11) in **W** suggests two ways to compute absorption from a group of lines: (a) sum over each transition using the half-widths W_{jj} and mixing parameters Y_j as defined above, or (b) use the reduced half-widths corresponding to the merged transitions and mixing parameters corresponding to mixing with lines *outside* the group. These two formulations will yield the same result on the far wings, but within the resonant region, one or the other will be more accurate depending on the pressure regime. Computations of atmospheric brightness temperature [6,8] employed formulation (b) with respect to the Zeeman components within each oxygen line, since the half-widths and mixing coefficients were determined from laboratory measurements at zero magnetic field.

Determination of W and/or Y

Direct calculations of oxygen line mixing using scattering theory have been done in refs. [15,17-19], often showing good agreement with measurements. Experimental determination of line mixing parameters can be done either by constructing a model for the **W** matrix with a small number of free parameters [16,20], or by directly fitting the Y-parameters to measured spectra [21-24]. In the latter case some type of smoothness constraint is usually necessary; consequently, the fitted Y's constitute a set which minimizes a "cost function" on the spectrum rather than being individually determined parameters like the half-widths (which are measured at much lower pressures). Figure 1 shows an example of measurements [23,24] that permit a comparison between pure oxygen and air. Line shifts were found to be undetectably small, in accord with the assumption being made here that **W** is real. The line half-widths and mixing coefficients for different O_2 concentrations diverge somewhat at higher rotational levels due to different intermolecular potentials for O_2-O_2 vs. O_2-N_2 collision partners. The observational signature of these differences is a ratio of absorption in pure O_2 vs. air that is greater on the wings of the 60-GHz band than close to the center. The line-mixing parameters were determined at one atmosphere; at that pressure, higher-order terms not included in eqs.(7,8) were a source of error that had to be taken into account along with measurement error in the procedure of fitting the Y's to the measured spectra [24].

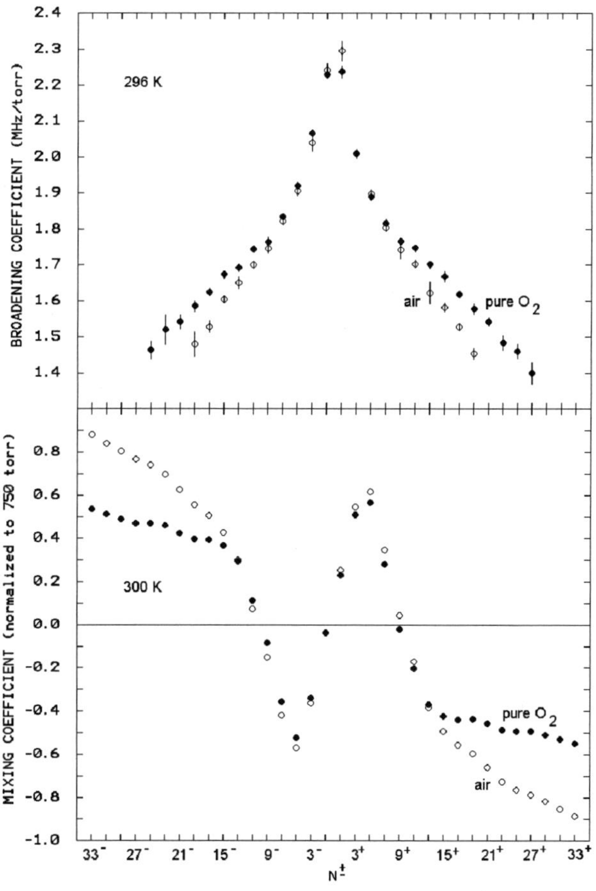

FIGURE 1. Line-broadening and mixing coefficients in pure oxygen (filled circles) and air (open circles) [23,24].

By treating all of the positive-frequency lines as a single resonance, Gordon [25] reduced the **W** matrix to a 3x3 form with indices corresponding to positive-, zero- and negative-frequency resonances:

$$\mathbf{W} = \begin{bmatrix} W_b & -W_b/2 & -W_b/2 \\ -W_b/2 & W_b & -W_b/2 \\ -W_b/2 & -W_b/2 & W_b \end{bmatrix}. \tag{12}$$

The parameter W_b was related to the collisional-average of $\sin^2\alpha$ in a semiclassical picture where α is the angle by which the rotational angular momentum is reoriented with respect to the spin vector. Although this treatment ignores the separation of the 119-GHz line from the rest of the spin-rotation lines, good agreement with measured absorption in pure O_2 on the low-frequency side of the 60-GHz band was obtained using W_b = 0.54 GHz/atm = 0.71 MHz/torr. The difference between this effective broadening parameter and the half-widths of the resolved lines (see Fig. 1) is a measure of the combined contributions of off-diagonal elements in **W** that mix lines within the 60-GHz band.

Figure 2 compares half-widths of transitions measured in a strong magnetic field [26] with half-widths in zero magnetic field [27,28]. (The measurements of [24] could also have been plotted in Fig. 2, but they differ from those of [27] by less than 3%). For the N=1 level, there are no significant differences between the broadening of the resolved Zeeman transitions and that of the zero-field lines. That is because the 1^- and 1^+ lines have just three Zeeman components of different ΔM, which do not mix [13]. (In the case of the 1^+ line, transition frequencies with different initial values of M are degenerate, so there are only the three distinguishable components.) For N>1, the zero-field half-widths decrease with N more rapidly than the resolved Zeeman transitions, differing by ~0.3 MHz/torr at N=11.

The experimental determinations of pressure broadening for $1 \leq N \leq 11$ are summarized in Fig. 3, showing the broadening observed when all of the quantum numbers are resolved, when M is unresolved, and when neither M nor N is resolved. As an example, consider the 7^+ line; the difference between zero-field half-width and Zeeman half-widths is ~0.3 MHz/torr, or 1/4 of the difference between zero-field half-width and W_b. However, in a typical terrestrial magnetic field of 0.5 gauss, the spacing of the Zeeman components for each ΔM-type within the 7^+ line is 0.15 MHz, while the spacing of lines in the N^+ branch is roughly 5000 times larger. Referring to eq.(9), one can estimate the maximum magnitude of Y-coefficients for the resolved Zeeman components as ~10^3 times as large as for the spin-rotation lines in zero field. That implies that a first-order treatment of Zeeman-component mixing by method (a) would be valid within the core of the line only for pressures less than 1 torr. Fortunately, that is the pressure range where Zeeman splitting is of most significance for atmospheric measurements.

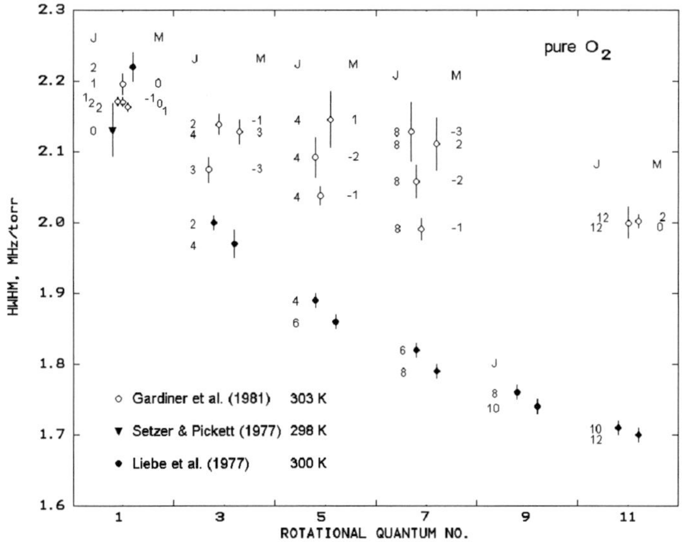

FIGURE 2. Half-widths (HWHM) of oxygen transitions in a strong magnetic field (open circles) [26] and in zero magnetic field (filled circles and triangle) [27,28]. Quantum numbers of the lower energy level are written next to each data point.

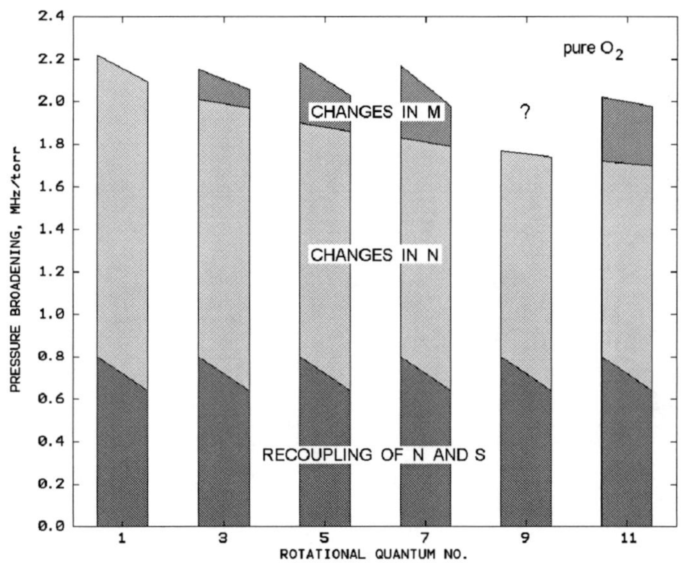

FIGURE 3. Allocation of pressure-broadening to types of motion during collisions. The slanted tops of the bars represent the range of experimental data.

CONCLUSIONS AND SUGGESTIONS FOR FURTHER WORK

Published measurements of absorption in oxygen gas suggest that mixing between Zeeman components could have an effect on measurements of the atmosphere at high altitudes. Whether brightness temperatures calculated by the two methods (a) and (b) differ significantly may depend on the line being observed, on the observing instrument's passbands, on the observation geometry, and on the temperature profile of the atmosphere. Further development of models for the **W** matrix could make use of the scaling relations among its elements that arise from the properties of angular momentum [29,30]. Although lower-altitude measurements are in reasonable agreement with existing models [31] there may be room for improvement there as well.

The discussion here (as well as elsewhere) has not considered mixing of the sub-millimeter rotational transitions, and mixing between those lines and the microwave spin-rotation lines. Such effects would be small due to the large line spacing, but not necessarily undetectable, as shown by the example of the single line at 119 GHz [23].

Experimental evidence indicates that first-order line shifts are negligible in the O_2 microwave spectrum, but this may not be the case for other molecules. Implications of a complex **W** have been discussed in [32].

ACKNOWLEDGMENTS

This research was supported by NASA under contract NNG04HZ51C.

REFERENCES

1. G. Herzberg, *Spectra of Diatomic Molecules*, Van Nostrand Reinhold, New York 1950, pp. 221-224.
2. R. R. Gamache, A. Goldman, and L. S. Rothman, *J. Quant. Spectrosc. Rad. Transf.* **59**, 495-509 (1998).
3. M. Tinkham and M. W. P. Strandberg, *Phys. Rev.* **97**, 937-951 (1955).
4. W. B. Lenoir, *J. Geophys. Res.* **73**, 361-376 (1968).
5. H. J. Liebe, *Radio Science* **16**,1183-1199 (1981).
6. P. W. Rosenkranz and D. H. Staelin, *Radio Science* **23**, 721-729 (1988).
7. A. Stogryn, *IEEE Trans. Geosci. Remote Sensing* **27**, 279-289 (1989).
8. S. Hanany and P. Rosenkranz, *New Astron. Rev.* **47**, 1159-1165 (2003).
9. J. Boissoles, C. Boulet, R. H. Tipping, A. Brown and Q. Ma, *J. Quant. Spectrosc. Rad. Transf.* **82**, 505-516 (2003).
10. M. Baranger, *Phys. Rev.* **111**, 494-504 (1958).
11. A. C. Kolb and H. Griem, *Phys. Rev.* **111**, 514-521 (1958).
12. U. Fano, *Phys. Rev.* **131**, 259-268 (1963).
13. A. Ben-Reuven, *Phys. Rev.* **141**, 34-40 (1966)

14. A. Ben-Reuven, *Phys. Rev.* **145**, 7-22 (1966).

15. K. S. Lam, *J. Quant. Spectrosc. Rad. Transf.* **17**, 351-383 (1977).

16. P. W. Rosenkranz, *IEEE Trans. Ant. and Prop.* **AP-23**, 498-506 (1975).

17. U. Mingelgrin, R. G. Gordon, L. Frenkel and T. E. Sullivan, *J. Chem. Phys.* **57**, 2923-2931 (1972).

18. U. Mingelgrin and R. G. Gordon, *J. Chem. Phys.* **70**, 3828-3839 (1979).

19. E. W. Smith, *J. Chem. Phys.* **74**, 6658-6673 (1981).

20. H. Tran, C. Boulet and J.-M. Hartmann, *J. Geophys. Res.* **111**, (in press 2006).

21. P. W. Rosenkranz, *J. Quant. Spectrosc. Rad. Transf.* **39**, 287-297 (1988).

22. H. J. Liebe, P. W. Rosenkranz and G. A. Hufford, *J. Quant. Spectrosc. Rad. Transf.* **48**, 629-643 (1992).

23. M. Yu. Tretyakov, G. Yu. Golubiatnikov, V.V. Parshin, M. A. Koshelev, S. E. Myasnikova, A. F. Krupnof, and P. W. Rosenkranz, *J. Molec. Spectros.* **223**, 31-38 (2004).

24. M. Yu. Tretyakov, M. A. Koshelev, V. V. Dorovskikh, D. S. Makarov, and P. W. Rosenkranz, *J. Molec. Spectros.* **231**, 1-14 (2005).

25. R. G. Gordon, *J. Chem. Phys.* **46,** 448-455 (1967).

26. W. C. Gardiner, H. M. Pickett and M. H. Proffitt, *J. Chem. Phys.* **74**, 6037-6043 (1981).

27. H. J. Liebe, G. G. Gimmestad and J. D. Hopponen, *IEEE Trans. Ant. and Prop.* **AP-25**, 327-335 (1977).

28. B. J. Setzer and H. M. Pickett, *J. Chem. Phys.* **67**, 340-343 (1977).

29. G. C. Corey and F. R. McCourt, *J. Chem. Phys.* **81**, 3892-3907 (1984).

30. H. M. Pickett, *J. Chem. Phys.* **61**, 1923-1933 (1974); **63**, 2153-2161 (1975).

31. P. W. Rosenkranz, *IEEE Trans. Geosci. Remote Sensing* **41**, 362-368 (2003).

32. J. Boissoles, C. Boulet, D. Robert and S. Green, *J. Chem. Phys.* **90,** 5392-5398 (1989).

Recent Progress In Molecular Line Shape Theory For Lorentzian Parameters: Application to Atmospheric Systems

R. H. Tipping*, Q. Ma[†], and C. Boulet[‡]

*Department of Physics and Astronomy, University of Alabama, Tuscaloosa, AL 35487, USA
[†]Department of Applied Physics, Columbia University, and Institute for Space Studies, Goddard Space Flight Center, 2880 Broadway, New York, NY, 10025, USA
[‡]Laboratoire de Photophysique Moléculaire, UPR 3361 CNRS, Université Paris-Sud, Centre d'Orsay, France

Abstract. We have made two modifications to the Robert-Bonamy (RB) formalism that has been widely used for calculating Lorentzian spectral line half-widths and shifts for decades. The first one comes from a correction to their derivation of the expressions for the widths and shifts. Numerical calculations show the new half-width and shift values differ from the previous ones, and the stronger the interaction between two molecules is, the larger these differences are. Numerical results will be presented to illustrate the magnitudes of the changes. The second correction is to the expression for the first-order term in the expansion of the S matrix that is essential in calculating vibration-rotation pressure shifts. They made an incorrect assumption that the trajectories of interest are vibrationally independent. As a result, their formalism is not applicable for calculating shifts for molecular lines involving vibrational transitions. In comparison with the original expression, the new formula contains extra terms that represent the contributions from vibration-dependent trajectories. Recently, we have reformulated their (modified) theory in terms of the coordinate representation. This allows us to avoid truncation errors that exist in many previous calculations, and to obtain converged results no matter how complicated the potential models are or what kind of trajectories are chosen. Further refinements that are needed to increase the accuracy will also be discussed briefly.

INTRODUCTION

The accurate specification of the spectroscopy of atmospheric molecules, both the resonant (lines) and non-resonant (continuum) absorptions are essential to the accuracy of satellite-based retrievals used to study the Earth's energy budget, the abundance of greenhouse molecules, etc. This applies to microwave sounding instruments, where the minimum retrieval altitude for tropospheric water vapor is determined from the baseline that is left after the sharp line contributions are removed

CP874, *Spectral Line Shapes: 18th International Conference*, edited by E. Oks and M. Pindzola

and to IR and visible instruments as well. Because the background is the atmospheric continuum, which depends on moisture, temperature, and pressure, accurate retrievals depend on accurate continua.

The strongest continua are due to absorptions by pairs of colliding molecules. These continua arise from two different physical mechanisms: the far wings of allowed dipole lines, such as the self (H_2O-H_2O) and foreign (H_2O-N_2, H_2O-O_2) water continua, and collision-induced absorption (N_2-N_2, N_2-O_2, O_2-O_2) resulting from transient dipoles that only exist during binary collisions. The self and foreign water continua are important in all spectral regions, particularly in the atmospheric window between 800 – 1200 cm^{-1}. Because of this there exist much more experimental data than in other regions. During the past 15 years, we have developed a comprehensive theory for both the self and foreign water vapor continuum, based on first-principle calculations of the far wings of water lines broadened by various molecules.[1,2] The accuracy of the theoretical absorption depends on the accuracy of the interaction potential; this in turn can be refined by comparisons with accurate laboratory data for the absorption. Except for some uncertainty in the interaction potential, all the physics is treated correctly in our theory, which is outlined in the following section. During the same period, we have presented new experimental measurements and theoretical analysis for the collision-induced absorptions of the pure rotational and fundamental vibrational bands.[3,4] Recently, we have extended our work to study the collision-induced electronic transitions in O_2-CO_2, O_2-N_2 and O_2-O_2 in the near IR and visible regions. [5]

In addition to the far-wing absorption, one needs to know the parameters describing the line shapes near the line centers. These include the pressure-induced Lorentzian half-widths and shifts and their temperature indices. A considerable literature exists for these data for atmospheric molecules as compiled in the HITRAN database.[6] Because it is not always possible or practical to measure the Lorentzian parameters for the ambient conditions in the Earth's or planetary atmospheres, one has to rely on accurate theoretical calculations. Most of the theoretical values that appear in the HITRAN database were calculated using the theoretical expressions formulated almost 30 years ago by Robert and Bonamy.[7] While there have been improvements in the sophistication of the interaction potential and the trajectories describing the translational motion, no modifications of the starting expressions for the widths and shifts have been made until recently [8,9], as we will describe later.

Because the Lorentzian parameters (like the far-wing continua) depend critically on the interaction potential, comparisons between theory and experimental results will allow us to refine the potential. Previously, we have constrained the potential by requiring that it also give good results for both the temperature dependence of the second virial coefficient and scattering cross sections. These constraints, however, are less stringent than that imposed by line shapes. Our ultimate goal is to be able to interpolate between these two limits(near line center and far wings) in order to obtain results for the line shapes in the near wings for which no accurate theory exist for the molecular systems of atmospheric interest.

FAR-WING LINE SHAPES

The theoretical calculation of the far-wing line shapes for H_2O-H_2O and H_2O-N_2 [1, 2] were carried out assuming only binary collisions and the quasistatic approximation that is valid in the far wings. In this approximation, the translational motion is treated classically, while the internal motion is treated quantum mechanically. When one introduces a sophisticated short-range anisotropic interaction potential such as

$$V_{atom-atom}(t) = \sum_{l_1 l_2 l} \sum_{m_1 m_2 m} \sum_{n_{\{ij\}}} \sum_{wq} \frac{U(l_1 l_2 l, n_{\{ij\}}, wq)}{R^{l_1+l_2+q+2w}(t)} \times C(l_1 l_2 l, m_1 m_2 m) Y_{l_1 m_1}(\Omega_a) Y_{l_2 m_2}(\Omega_b) Y^*_{lm}(\omega(t)), \quad (1)$$

where $C(l_1l_2l,m_1m_2m)$ are Clebsch-Gordan coefficients, $n_{\{ij\}}$ runs over all pairs of atom, q = 6 or 12, w is an integer index from 0 to infinity, and definitions for $U(l_1l_2l, n_{\{ij\}}, wq)$ can be found in the literature [10], the mathematical calculations become intractable when one carries out the calculations in terms of the internal state functions for linear or for asymmetric top molecules. Because one can express the line shapes and the absorption coefficient in terms of a trace, and this is independent of the basis set, we formulated the theory in terms of the coordinate representation in which the orientations of the system are taken to be products of delta functions that describe the angular orientations of the colliding pair of molecules. This has the advantage that the anisotropic interaction is diagonal and is simply a function not a tensor operator. One can thus include as many states as desired in the calculation. However, the density matrix is off-diagonal but its calculation is straightforward. Thus, the sums over the state functions become multidimensional integrals over the initial and final orientations. For systems consisting of two linear molecules, one linear and one asymmetric top molecule, or two asymmetric molecules, the dimensionality is seven, nine, and eleven, respectively. Fortunately, one can calculate these integrals using the Monte Carlo method in reasonable cpu times on a workstation. [1,2]

The resulting line shapes for H_2O-H_2O and H_2O-N_2 are super-Lorentzian in the near wings but decrease exponentially with the frequency displacement from the line center in the far wings as expected. The resulting absorption from the far wings in the atmospheric windows between the vibrational bands is found to be in agreement with measured values. [1,2] Because of the large number of potential parameters and the absence of accurate ab initio interactions, especially the short-range anisotropic part that is instrumental in determining the far-wing line shape, we wanted to test the accuracy of the potentials obtained by calculating the near line center shape, where the impact approximation is valid for moderate number densities, and the lines are well represented by the Lorentzian half-widths and shifts. This imposes more constraints than the calculation of other properties, such as virial data or transport properties.

THE ROBERT-BONAMY FORMALISM FOR LORENTZIAN WIDTHS AND SHIFTS

The Robert-Bonamy (RB) formalism was developed more than two decades ago and has been widely used in calculating the Lorentzian half-widths and shifts for spectral lines for many molecular systems of importance. This is an improvement over the older Anderson-Tsao-Curnutte (ACT) theory. This theory originally developed for linear molecules, has been extended to other types of molecules, and successfully applied to infrared and Raman spectra. The main features of this theory are the non-perturbative treatment of the scattering matrix S through the use of the linked cluster theorem and a convenient description of classical trajectories for large impact parameters as well as for the closest approach. Some progress has been made recently to improve the RB original results, but most of these efforts have been related to the extension to more complicated interaction potentials, vibrational dependences, more accurate trajectories, etc. None has focused on making further improvements to the basic theory itself. As a result, the calculations of half-widths and shifts using the RB results have become routine, and the results widely applied to the self- and foreign-broadened widths and shifts for atmospheric molecular systems, such as those compiled in the HITRAN database. In cases where the calculated values did not match the experimental data well, the strategy used for improvements was to use more complicated potential models and higher potential expansion cut-offs, to take into account more sophisticated vibrational dependences, and to consider the translational motion moving along “exact” trajectories rather than to scrutinize the RB theory in detail. Recently, we have checked these authors’ early papers and found there was an invalid assumption made in their derivation of the starting formulas. More specifically, the problem occurred when Robert and Bonamy used Bloch’s linked-cluster theorem

$$\begin{aligned} U(t) &= U_L(t) < 0 \,|\, U(t) \,|\, 0 > \\ &= U_L(t)\, e^{<0|U(t)|0>^{(C)}}, \end{aligned} \tag{2}$$

where |0> is the non-degenerate ground state of the unperturbed Hamiltonian and U(t) is the time evolution operator, to evaluate the matrix elements <<j2’i2’|$\hat{S}$|j2i2>>. Here j and i are simple notations to represent states of the absorber molecule and 2 and 2’ for the states of the bath molecule. We have recently shown how one can correct this problem and derive modified expressions for the starting point for the calculation of Lorentzian widths and shifts.

For the half-widths, the RB formula and the new one are given by

$$\gamma_{RB} = \frac{n_b}{2\pi c} \int_0^{+\infty} v f(v) dv \int_0^{+\infty} 2\pi b db < 1 - \cos(S_1 + \mathrm{Im}(S_2)) e^{-\mathrm{Re}(S_2)} >_{j_2}, \tag{3}$$

$$\gamma_{new} = \frac{n_b}{2\pi c}\int_0^{+\infty} vf(v)dv\int_0^{+\infty} 2\pi bdb[1-\cos(<S_1>_{j_2} + \mathrm{Im}(<S_2>_{j_2}))e^{-\mathrm{Re}(<S_2>_{j2})}], \quad (4)$$

respectively, where f(v) is the Maxwell-Boltzmann distribution function. Meanwhile, for the shifts, the RB formula and the new one are given by,

$$\delta_{RB} = \frac{n_b}{2\pi c}\int_0^{+\infty} vf(v)dv\int_0^{+\infty} 2\pi bdb < \sin(S_1 + \mathrm{Im}(S_2))e^{-\mathrm{Re}(S_2)} >_{j_2}, \quad (5)$$

$$\delta_{new} = \frac{n_b}{2\pi c}\int_0^{+\infty} vf(v)dv\int_0^{+\infty} 2\pi bdb \ \sin(<S_1>_{j_2} + \mathrm{Im}(<S_2>_{j_2}))\ e^{-\mathrm{Re}(<S_2>_{j2})}. \quad (6)$$

In contrast with the RB formulas Eqs. (3) and (5) where the summation over j_2 is outside of the cumulant expansion and the other summation over j_2' is inside, in the new Eqs. (4) and (6), both the summations over j_2 and j_2' are inside and are treated on an equal footing. The new treatment is consistent with how one should treat the average of a Liouville operator over the internal degrees of the bath molecule mentioned above, and thus this treatment is more physically realistic. Finally, we would like to emphasize that these different orders of averaging over bath degrees of freedom are not arbitrary. It is this subtle error in applying the cumulant expansion in the RB formalism compared to as done in the present study that leads to the differences.

To investigate the order of magnitudes of the differences, we have carried out numerical comparisons for the HF-N_2 and HF-HF systems. For illustration purposes,

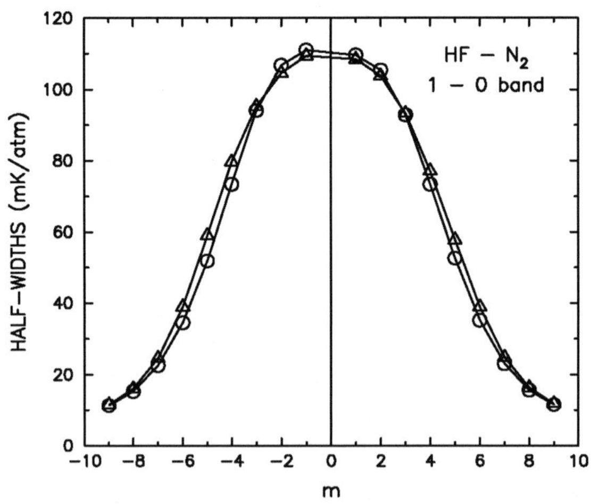

FIGURE 1. Calculated half-widths of the 0-1 band of HF broadened by N_2 at 296 K. Values derived from the RB and the new formulas are denoted by ○ and Δ, respectively.

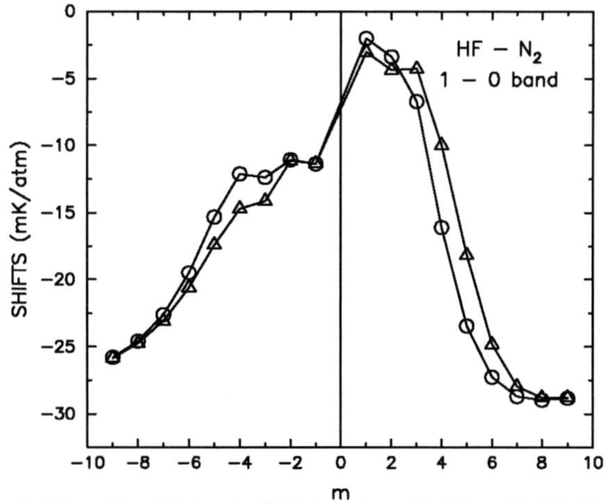

FIGURE 2. Calculated shifts of the 0-1 band of HF broadened by N_2 at 296 K. Values derived from the RB and the new formulas are denoted by ○ and Δ, respectively.

we use a simple anisotropic potential model that consists of the dipole-dipole, dipole-quadrupole, quadrupole-dipole, and quadrupole-quadrupole interactions. With respect to the isotropic potential, we use the Lennard-Jones model with vibrational dependence. The translation motion is assumed to follow a parabolic trajectory. As shown in Figs. 1 and 2, for the half-widths and shifts, respectively, for the fundamental band of HF perturbed by N_2, the values obtained from these two formulas differ slightly; the shifts differ more than the widths. This behavior is expected because the dipole moment of N_2 is zero and the interaction between HF and N_2 is relatively weak.

We next consider a pair of HF-HF molecules whose interaction is very much

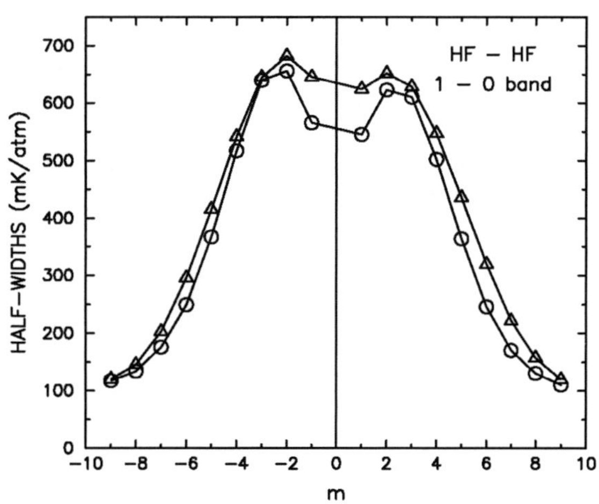

FIGURE 3. The same as Fig. 1 except for HF broadened by HF.

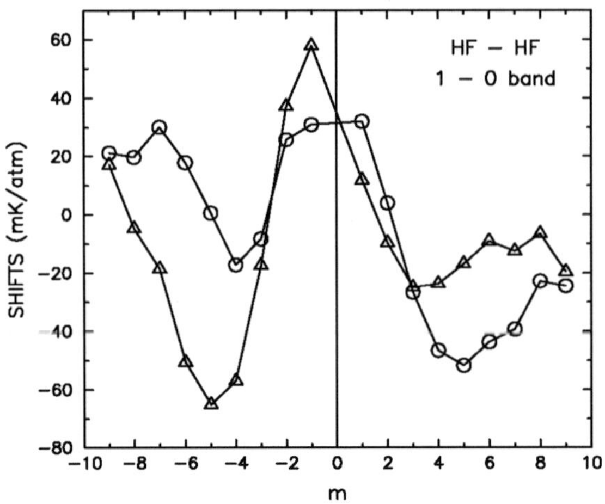

FIGURE 4. The same as Fig. 2 except for HF broadened by HF.

stronger than that for HF-N_2. We see from Fig. 3 that, in general, values of the half-width obtained from the new expression are larger than those found from the RB results. On the other hand, one can see from Fig. 4 that in comparison with the shifts based on the RB theory, the new calculated values change dramatically.

VIBRATION-DEPENDENT TRAJECTORIES AND DEPHASING

In order to calculate accurate widths and shifts for vibration-rotation spectral lines, one has to use more realistic trajectories to describe the translational motion. In the ACT theory, one assumed straight-line motion at a constant speed. In the RB formalism, several different trajectory models have been proposed: parabolic trajectories and the "exact" trajectories, assuming only the isotropic part of the interaction potential. It is well known that in a lot of molecular collision theories, the Born-Oppenheimer approximation is introduced to simplify calculations. With this approximation, the total wavefunction is written as a product of two components associated with the translational and internal motions, respectively, and the original Schrödinger equation is separated into two equations governed by two Hamiltonians. In addition, the Hamiltonian determining the translational motion consists of a kinetic operator K and a potential operator V_0. The latter can be expressed as

$$V_0 = <\alpha|V|\alpha> + \sum_{\alpha'\neq\alpha} \frac{<\alpha|V|\alpha'><\alpha'|V|\alpha>}{E_\alpha - E_{\alpha'}} + \cdots, \tag{7}$$

where V is the total potential between the two molecules, α is a simple notation used to denote all internal quantum numbers, and E_α corresponds to its eigenvalue. It is worth mentioning that no matter if the translational motion is treated quantum mechanically or classically, V_0 depends on the internal degrees of freedom.

In semi-classical theories, the translational motion is treated classically. In addition, if one assumes V_0 is not sensitive to the rotational quantum numbers, one can conclude that the trajectories are vibrationally dependent. We note that at present for most of theoretical half-width and shift calculations, the trajectory is governed by the isotropic potential only. In the following, we provided details to show how to derive formulas for vibration-dependent trajectories and their effects on vibrational dephasing.

In calculating molecular line half-widths and shifts using straight-line trajectories, the vibrational dephasing makes no contributions to S_1, the first-order term in the expansion of the S matrix, defined by the following equation

$$S_1 = \frac{1}{\hbar}\int_{-\infty}^{+\infty} dt\,[< f\,|V_0|\,f > - < i\,|V_0|\,i >]. \tag{8}$$

Here i and f denote the initial and final internal states. Usually, by introducing the normalized vibrational coordinate ξ defined by $\xi \equiv (r - r_e)/r_e$, where r is the vibrational displacement and r_e the equilibrium displacement, the isotropic potential V_0 is expressed as an expansion in ξ by

$$V_0(R(t),\xi) = C_{12}R(t)^{-12}(1 + s_1\xi + t_1\xi^2) - C_6R(t)^{-6}(1 + s_2\xi + t_2\xi^2). \tag{9}$$

In the ATC theory, the trajectories are only characterized by the impact parameter, independent of V_0. The translational motion is given by

$$R(t) = \sqrt{b^2 + (vt)^2}. \tag{10}$$

By introducing a dimensionless parameter $z = vt/b$ and assuming a Lennard-Jones form for the isotropic potential, one can evaluate S_1 analytically

$$\begin{aligned}
S_1 &= \frac{1}{\hbar}\int_{-\infty}^{+\infty} dt[< f\,|V_0(R(t),\xi)|\,f > - < i\,|V_0(R(t),\xi)|\,i >] \\
&= \frac{2C_{12}}{\hbar v b^{11}}\int_0^{+\infty} dz \frac{1}{(1+z^2)^6}\times[< f\,|1 + s_1\xi + t_1\xi^2|\,f > - < i\,|1 + s_1\xi + t_1\xi^2|\,i >] \\
&\quad - \frac{2C_6}{\hbar v b^5}\int_0^{+\infty} dz \frac{1}{(1+z^2)^3}\times[< f\,|1 + s_2\xi + t_2\xi^2|\,f > - < i\,|1 + s_2\xi + t_2\xi^2|\,i >] \\
&= \frac{3\pi\varepsilon\sigma}{2\hbar v}\{\frac{21}{32}\cdot(\frac{\sigma}{b})^{11}\Delta_{fi}^{(1)}(\xi) - (\frac{\sigma}{b})^5\Delta_{fi}^{(2)}(\xi)\},
\end{aligned} \tag{11}$$

where simple notations $\xi_{(f)} = < f \mid \xi \mid f >$, $\xi^2_{(f)} = < f \mid \xi^2 \mid f >$, $\xi_{(i)} = < i \mid \xi \mid i >$, and $\xi^2_{(i)} = i \mid \xi^2 \mid i >$ have been used; $\Delta_{fi}^{(1)}(\xi) = s_1(\xi_{(f)} - \xi_{(f)}) + t_1(\xi^2_{(f)} - \xi^2_{(f)})$; and where $y = \Delta_{fi}^{(1)}(\xi) / \Delta_{fi}^{(2)}(\xi)$ and $a_{0,1} = \Delta_{fi}^{(2)}(\xi)$.

Parabolic Trajectories in the RB Theory

In the Robert-Bonamy (RB) formalism, the straight-line trajectory model has been refined by a "parabolic" trajectory model in which effects from V_0 have been taken into account. The translational motion is described by movements with an "apparent" velocity v_c' along rectilinear paths or curves such that R(t) is given by

$$R(t) = r_c \sqrt{1 + z^2}, \tag{12}$$

where z is a dimensionless parameter defined by $z = v_c' t / r_c$. In the above expression, r_c is the distance of closest approach defined in terms of the impact parameter b by

$$\frac{b^2}{r_c^2} = 1 - \frac{8\varepsilon}{mv^2}[(\frac{\sigma}{r_c})^{12} - (\frac{\sigma}{r_c})^6]. \tag{13}$$

Meanwhile, v_c' is defined by

$$v_c' = v\{1 + \frac{8\varepsilon}{mv^2}[5(\frac{\sigma}{r_c})^{12} - 2(\frac{\sigma}{r_c})^6]\}^{1/2}. \tag{14}$$

By following the same derivation as above, one can easily obtain an expression for S_1 applicable for the "parabolic" trajectory model:

$$S_1 = \frac{3\pi\varepsilon\sigma}{2\hbar v_c'}\{\frac{21}{32} \cdot (\frac{\sigma}{r_c})^{11} \Delta_{fi}^{(1)}(\xi) - (\frac{\sigma}{r_c})^5 \Delta_{fi}^{(2)}(\xi)\}. \tag{15}$$

On comparing the above expression with Eq. (11), it is obvious that it is the same except for replacing v by v_c' and b by r_c.

Refinement of Parabolic Trajectories

In general, the current expression for S_1 in the RB formalism is applicable in calculating line half-widths and shifts not involving vibrational transitions. However, for lines of interest where vibrational transitions take place, one must be careful to take the vibrational dependence of the translational motion into account correctly. As a result, two integrations in the expression for S_1 should be treated separately

$$
\begin{aligned}
S_1 &= \frac{1}{\hbar}\int_{-\infty}^{+\infty} dt[< f\,|V_0(R(t),\xi)|\,f > - < i\,|V_0(R(t),\xi)|\,i >] \\
&= \frac{2r_c}{\hbar v_{c,(f)}{}'}\int_0^{+\infty} dz < f\,|V_0(R(z),\xi)|\,f > - \frac{2r_c}{\hbar v_{c,(i)}{}'}\int_0^{+\infty} dz < i\,|V_0(R(z),\xi)|\,i >,
\end{aligned} \tag{16}
$$

where $v_{c,(f)}'$ and $v_{c,(i)}'$ differ from each other and they also differ from v_c' defined by Eq. (14) because the isotropic potential has a vibrational dependence.

According to Eq. (14), the values of v_c' are determined by the LJ parameters ε and σ associated with $V_0(R, \xi = 0)$. In order to find $v_{c,(f)}'$ we need to know $< f \,|\, V_0(R, \xi)|\, f >$ given in terms of a LJ model with two new parameters labeled as $\varepsilon_{(f)}$ and $\sigma_{(f)}$. We rewrite $< f \,|\, V_0(R, \xi)|\, f >$ as

$$
\begin{aligned}
< f\,|V_0(R,\xi)|\,f > &= < f\,|C_{12}R^{-12}(1+s_1\xi+t_1\xi^2) - C_6R^{-6}(1+s_2\xi+t_2\xi^2)|\,f > \\
&= C_{12}(1+s_1\xi_{(f)}+t_1\xi^2_{(f)})R^{-12} - C_6(1+s_2\xi_{(f)}+t_2\xi^2_{(f)})R^{-6} \\
&\equiv C_{12,(f)}R^{-12} - C_{6,(f)}R^{-6}.
\end{aligned} \tag{17}
$$

The definitions of $C_{12,(f)}$ and $C_{6,(f)}$ are given in the above equation. Furthermore, with these two vibrational dependent coefficients $C_{12,(f)}$ and $C_{6,(f)}$, we can find two corresponding parameters $\sigma_{(f)}$ and $\varepsilon_{(f)}$,

$$
\begin{aligned}
\sigma_{(f)} &= [\frac{C_{12,(f)}}{C_{6,(f)}}]^{1/6} = [\frac{C_{12}(1+s_1\xi_{(f)}+t_1\xi^2_{(f)})}{C_6(1+s_2\xi_{(f)}+t_2\xi^2_{(f)})}]^{1/6} \\
&= \sigma[(1+(s_1-s_2)\xi_{(f)}+(t_1-t_2)\xi^2_{(f)}+\cdots]^{1/6} \\
&\approx \sigma[1+\frac{1}{6}(s_1-s_2)\xi_{(f)}+\frac{1}{6}(t_1-t_2)\xi^2_{(f)}],
\end{aligned} \tag{18}
$$

and

$$
\begin{aligned}
\varepsilon_{(f)} &= \frac{[C_{6,(f)}]^2}{4C_{12,(f)}} = \frac{C_6{}^2(1+s_2\xi_{(f)}+t_2\xi^2_{(f)})^2}{4C_{12}(1+s_1\xi_{(f)}+t_1\xi^2_{(f)})} \\
&= \varepsilon\,(1+2s_2\xi_{(f)}+2t_2\xi^2_{(f)}+\cdots)(1-s_1\xi_{(f)}-t_1\xi^2_{(f)}+\cdots) \\
&\approx \varepsilon(1+(2s_2-s_1)\xi_{(f)}+(2t_2-t_1)\xi^2_{(f)}).
\end{aligned} \tag{19}
$$

In deriving the above expressions, we have taken into account the fact that magnitudes of $\xi_{(f)}$ and $\xi^2_{(f)}$ are of the same order. Similarly, we are able to obtain expressions for $\sigma_{(i)}$ and $\varepsilon_{(i)}$ which are the same as except for replacing the subscript (f) by (i). It appears that these ε's depend on the vibrational states more sensitively than the σ's. As a next step, by performing the integrations analytically in, we find

$$
\begin{aligned}
S_1 &= \frac{2r_c}{\hbar v_{c,(f)}'}\int_0^{+\infty} dz < f\,|V_0(R(z),\xi)|\,f > - \frac{2r_c}{\hbar v_{c,(i)}'}\int_0^{+\infty} dz < i\,|V_0(R(z),\xi)|\,i > \\
&= \frac{2r_c}{\hbar v_{c,(f)}'}\Big[\frac{C_{12,(f)}}{r_c^{12}}\int_0^{+\infty} dz\frac{1}{(1+z^2)^6} - \frac{C_{6,(f)}}{r_c^{6}}\int_0^{+\infty} dz\frac{1}{(1+z^2)^3}\Big] \\
&\quad - \frac{2r_c}{\hbar v_{c,(i)}'}\Big[\frac{C_{12,(i)}}{r_c^{12}}\int_0^{+\infty} dz\frac{1}{(1+z^2)^6} - \frac{C_{6,(i)}}{r_c^{6}}\int_0^{+\infty} dz\frac{1}{(1+z^2)^3}\Big] \\
&= \frac{2r_c}{\hbar v_{c,(f)}'}\times\frac{3\pi}{16}\Big[\frac{21}{32}\cdot\frac{C_{12,(f)}}{r_c^{12}} - \frac{C_{6,(f)}}{r_c^{6}}\Big] - \frac{2r_c}{\hbar v_{c,(i)}'}\times\frac{3\pi}{16}\Big[\frac{C_{12,(i)}}{r_c^{12}}\cdot\frac{21}{32} - \frac{C_{6,(i)}}{r_c^{6}}\Big] \\
&= \frac{3\pi\varepsilon_{(f)}\sigma_{(f)}}{2\hbar v_{c,(f)}'}\Big[\frac{21}{32}\cdot\Big(\frac{\sigma_{(f)}}{r_c}\Big)^{11} - \Big(\frac{\sigma_{(f)}}{r_c}\Big)^{5}\Big] - \frac{3\pi\varepsilon_{(i)}\sigma_{(i)}}{2\hbar v_{c,(i)}'}\Big[\frac{21}{32}\cdot\Big(\frac{\sigma_{(i)}}{r_c}\Big)^{11} - \Big(\frac{\sigma_{(i)}}{r_c}\Big)^{5}\Big].
\end{aligned}
\tag{20}
$$

By comparing Eq. (20) with Eq. (15), one can see that there are additional terms and these will affect the vibrational dephasing, thus giving different shifts than those obtained in the usual RB formalism. The magnitudes of the differences depend on the system considered. A similar analysis can be carried out for the "exact" trajectory model. [11]

CONCLUSIONS

There are several conclusions that can be drawn concerning the accuracy of the theoretical results calculated using the standard Robert-Bonamy theory. First, because of the error made in applying the linked cluster theorem in the original derivation of the theoretical expressions for the widths and shifts, there are other, tacit approximations beyond those of the binary collision and impact approximations have been introduced. As shown above, for strongly interacting systems, the differences can be significant, more so for the shifts than the half-widths. For weakly interacting systems, the differences are smaller. Second, when applying the modified RB expressions for the half-widths and shifts to vibration-rotational transitions when using more physically realistic trajectory models, one has to be sure to include all the terms arising from the vibrational-dependence of the trajectories in S_1. Third, because of the necessary cut-offs in the usual application of the RB formalism [11], one has to be sure that the results are converged. We have found that this is not always the case for theoretical values published in the literature. To avoid this last problem, we have reformulated the theory for the calculation of Lorentzian parameters in terms of the coordinate representation, and in this formulation, all couplings between interaction potential terms having the same angular dependence are included and all convergence problems are eliminated. [11]

Finally, only after the refinements discussed above have been made, can one use the comparisons between experiment and theory to determine if further refinements such as translation-rotation coupling, line mixing, third-order terms in the scattering matrix, etc. are necessary. Ultimately, the theoretical line shape depends on the interaction potential, and until ab initio results and close-coupling calculations for more complicated systems become feasible, one must rely on mathematically tractable theories to obtain the results required for practical applications. As described above, one stringent test of the accuracy of the interaction potentials would be to obtain accurate results for both the far-wing and near-line-center shapes using the same potential. This would then allow one to obtain more accurate approximations for the near-wing line shapes that can play an important role in applications to molecular systems of atmospheric importance, and to assess the need for further refinements.

ACKNOWLEDGMENTS

Two of authors (Q. Ma and R. H. Tipping) acknowledge financial support from NASA under grants NAG5-13337, NNG06GB23G, and FCCS-547. Q. Ma wishes to acknowledge financial support from the Biological and Environmental Research Program (BER), U.S. Department of Energy, Interagency Agreement No. DE-AI02-93ER61744.

REFERENCES

1. Q. Ma and R. H. Tipping, J. Chem. Phys. **111**, 5909 (1999).
2. Q. Ma and R. H. Tipping, J. Chem. Phys. **116**, 4102 (2002).
3. G. Moreau, J. Boissoles, C. Boulet, R. H. Tipping and Q. Ma, J. Quant. Spec. Rad. Transfer **64**, 87 (2000).
4. G. Moreau, J. Boissoles, R. Le Doucen, C. Boulet, R. H. Tipping and Q. Ma, J. Quant. Spec. Rad. Transfer **69**, 245 (2001).
5. R. H. Tipping, Q. Ma, C. Boulet, and J.-M. Hartmann, J. Mol. Structure **742**, 83 (2005).
6. L. S. Rothman, et al., J. Quant. Spec. Rad. Transfer **96**, 139 (2005).
7. D. Robert and J. Bonamy, J Phys **40**, 923 (1979).
8. R. R. Gamache, R. Lynch, and S. P. Neshyba, J. Quant. Spec. Rad. Transfer **59**, 19 (1998).
9. J.Buldyreva, J. Bonamy, and D. Robert, J. Quant. Spec. Rad. Transfer **62**, 321 (2000).
10. C. G. Gray and K. E. Gubbins, *Theory of Molecular Liquids*, Clarendon, Oxford, 1984.
11. Q. Ma, R. H. Tipping, and C. Boulet, J. Chem. Phys, **124**, 014109 (2006).

Collision Induced Absorption and Line-Mixing Effects in the O_2 A-Band (Abstract)

H. Tran, C. Boulet, and J.-M. Hartmann

Laboratoire Interuniversitaire des Systèmes Atmosphériques (LISA)
Université Paris XII, Faculté des Sciences et Technologie
61 avenue du Général de Gaulle, 94010 CRETEIL Cedex, France

The absorption by oxygen in the region of the O_2 A-band near 760 nm has been measured in the laboratory under various conditions of pressure (20-200 atm) and temperature (200-300 K) for both pure O_2 and O_2-N_2 mixtures. In order to calculate the contribution of the "allowed" A-band transitions, Lorentzian profiles and a model accounting for line-mixing (LM) effects using the Energy Corrected Sudden (ECS) approximation have been used. The differences between computed spectra and measured values enable extraction of the collision induced absorption (CIA) contribution. It is shown that neglecting line-mixing overestimates absorption in the wings and underestimates absorption at the P and R branch peaks, whereas the CIA extracted by the line-mixing approach shows the "smooth" profile expected. Applying this approach to our spectra enables determination of the CIA and allowed contributions for both O_2-O_2 and O_2-N_2 collisions versus temperature and pressure. The resulting model and data are then used to build a database and some software suitable for the calculation of oxygen (in air) atmospheric absorption and for easy inclusion in radiative transfer codes (available upon request). These tools are then applied to a theoretical study of the influences of both line-mixing and collision induced processes on atmospheric photon path escape factors and on cloud-top altitude retrievals. It is shown LM and CIA make significant contributions and explain a large part of the discrepancies between measured and calculated atmospheric absorption observed recently.

CP874, *Spectral Line Shapes: 18th International Conference*,
edited by E. Oks and M. Pindzola

The Roles of the S_1 and S_2 Scattering Matrix Terms on Half-Widths and Their Temperature Dependence for the Water Vapor-Nitrogen System

R. R. Gamache[*], B. K. Antony[*] and P. R. Gamache[*] and Jean-Michel Hartmann[†]

[*]*Department of Environmental, Earth & Atmospheric Sciences, Intercampus Graduate School of Marine Sciences and Technology University of Massachusetts Lowell, 265 Riverside Street, Lowell, MA 01854-5045 USA*

[†]*Laboratoire Inter-Universitaire des Systèmes Atmosphériques (LISA) CNRS, Université Paris 12 et Paris 7, 61 ave Général de Gaulle 94010 Créteil Cédex France*

Abstract. The effect of the first (S_1) and second (S_2) order successive approximations to the S matrix on the half-width and the temperature dependence of the half-width are studied for the water vapor-nitrogen system. The calculations are done via the Robert-Bonamy semi-classical formalism. It is found that the influence of the S_1 and S_2 terms depends on the rotational states of the transition. For the H_2O-N_2 system, half-widths for H_2O transitions with low J quantum numbers are dominated by the S_2 terms. Transitions with high J values the half-width is determined largely by the S_1 term. For transitions with intermediate values of J both the S_1 and S_2 terms contribute. The temperature dependence of the half-width also depends on whether it is the S_2, S_1 or S_2+S_1 that are contributing to the half-width. It is found that the standard power law temperature dependence of the half-width is only valid for transitions where Re(S_2) dominates. When S_1 dominates the temperature dependence is negative and when both terms contribute the temperature dependence is not modeled by the power law.

Keywords: half-width, temperature dependence of the half-width, H_2O-N_2.
PACS: 30

INTRODUCTION

The complex the Robert-Bonamy (CRB) formalism [1] to calculate the half-width and line shift for a ro-vibrational transition is a semi-classical method based on a second order successive expansion of the S matrix. This approach yields a first order term that is purely isotropic and imaginary, S_1, and a complex second order term, S_2. Depending on the collision interaction system each term can contribute more or less. In this work it is found that for the H_2O-N_2 collision system the contributions of the S_1 and S_2 terms changes from transition to transition. The mechanism for this is detailed below. The CRB calculations used a potential which consists of the leading

CP874, *Spectral Line Shapes: 18th International Conference,* edited by E. Oks and M. Pindzola electrostatic components for the H_2O-N_2 (the dipole and quadrupole moments of H_2O with the quadrupole moment of N_2), atom-atom interactions, and isotropic induction and London dispersion interactions in the S_1 term. Here the atom-atom part of the potential is expanded to 8th order to insure convergence of this part of the potential [2]. The isotropic component of the atom-atom potential are fit to an effective isotropic (or "hetero-molecular") Lennard-Jones 6-12 potential which is used to define the trajectory of the collisions within the semi-classical model of Robert and Bonamy [1]. All parameters used in the calculations are the best available from the literature and can be found in Ref. [3].

RESULTS AND DISCUSSION

The calculation of the half-width and line shift for a ro-vibrational transition $f \leftarrow i$ is made by averaging over all collisionally induced transitions. In the collision the absorbing molecule is driven from state i to state i' (or f to state f') with an energy gap $\Delta E_{i \to i'}$ in the same collision the perturbing molecule goes from state J_2 to $J_{2'}$ with an energy gap $\Delta E_{J2 \to J2'}$. If the energy gaps match each other the collision is said to be on resonance and the collisional contribution to the half-width will be large. These components are in the $Re(S_2)$ term. For conditions where the energy gaps do not match the collision is said to be off resonance and the collisional contribution is small. In such cases it is the S_1 term and the $Im(S_2)$ term that contribute to the half-width What is observed for the H_2O-N_2 system is that for low J states of water the collisions are on resonance and the half-width is determined mostly from collisional contributions, the $Re(S_2)$ term. For high J states most of the collisions are off resonance. In such cases the half-width is determined from vibrational contributions (S_1 and $Im(S_2)$).. These tend to be narrow lines.

The temperature dependence of the half-width was worked out by Birnbaum [4] assuming on resonance collisions and a single term in the intermolecular potential. The result is the standard power law seen in the literature, for example see Wagner et al. [5]. We have observed that for low J lines, which are determined mostly by the real part of S_2, the power law model works well. (Note, all figures are for the ν_2 band of water.) Figure 1a shows the fit for the $2_{0\,2} \leftarrow 3_{1\,3}$ transition. The slope (temperature exponent) is 0.74 and the correlation coefficient is 1 to four significant figures indication the power law model is nearly perfect for this line. For the high J lines, where the S_1 and $Re(S_2)$ terms determine the half-width, the temperature exponents can be negative and the power law is approximate, see Fig. 1b. What is interesting is the intermediate J region where both the S_1 and S_2 parts contribute to the half-width. Figure 1c is a plot for the $9_{0\,9} \leftarrow 8_{1\,8}$ transition. The temperature dependence of these lines is not modeled by the power law.

Caution should be used when modeling the temperature dependence of the half-width of N_2-broadened water vapor transitions. When the collisions are on resonance the usual power law model correctly describes the temperature dependence. For high J lines where the collisions are far off resonance the power law approximately models the temperature behavior. However for intermediate J transitions the power law fails to model the observed temperature dependence of the half-width. Similar results are found for O_2- and air-broadening of water vapor transitions. This could have large effects in modeling radiative transfer in mediums where the temperature is changing, i.e. Earth's atmosphere, combustion media, etc. For such cases a more sophisticated model is needed.

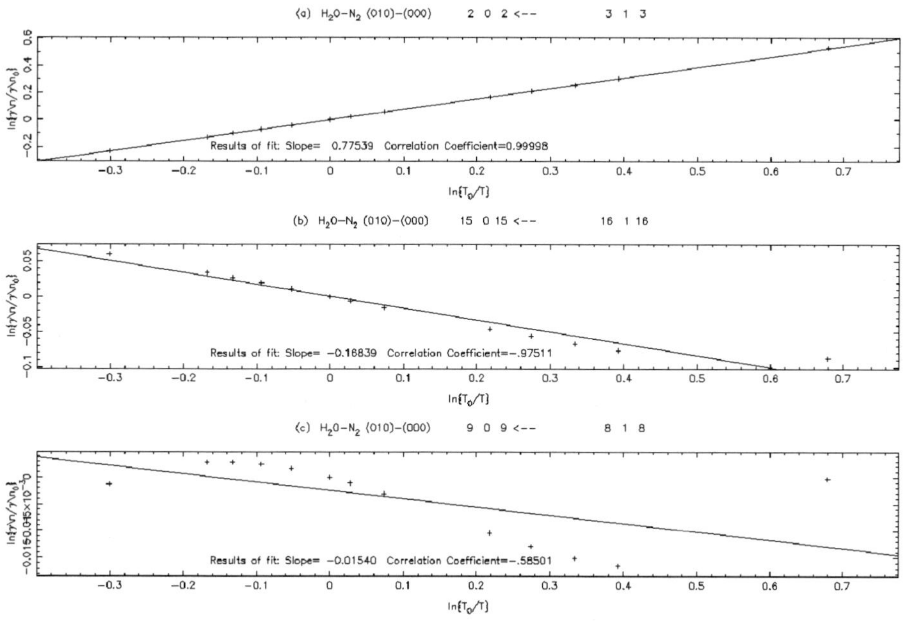

FIGURE 1. Plots of $\ln\{\gamma(T)/\gamma(T_0)\}$ versus $\ln\{T_0/T\}$ to determine the temperature exponent in the temperature range 150-400 K. (a) $2_{0\,2} \leftarrow 3_{1\,3}$ ν_2 band transition, (b) $15_{0\,15} \leftarrow 16_{1\,16}$ ν_2 band transition, (c) $9_{0\,9} \leftarrow 8_{1\,8}$ ν_2 band transition.

ACKNOWLEDGMENTS

The authors are pleased to acknowledge support of this research by the National Aeronautics and Space Administration (NASA) through Grant No. NAG5-11064 and by the National Science Foundation (NSF) through Grant No. ATM-0242537.

Any opinions, findings, and conclusions or recommendations expressed in this material are those of the author(s) and do not necessarily reflect the views of NASA or NSF.

REFERENCES

1 D. Robert and J. Bonamy, J. Phys. Paris. 40, 923 (1979).
2 S.P. Neshyba and R.R. Gamache, J. Quant. Spectrosc. Radiat. Transfer **50**, 443 (1993).
3 G. Durry, V. Zeninari, B. Parvitte et al., J. Quant. Spectrosc. Radiat. Transfer **94**, 387 (2005).
4 George Birnbaum, Intermolecular Forces **12**, 487 (1967).
5 Georg Wagner, Manfred Birk, Robert. R. Gamache et al., J. Quant. Spectrosc. Radiat. Transfer **92**, 211 (2004).

Alkali Line Profiles for Brown Dwarfs and their Application

Christine M. S. Johnas*, Nicole F. Allard†, Derek Homeier**, France Allard‡,† and Peter H. Hauschildt*

*Hamburger Sternwarte, Univ. of Hamburg, Gojenbergweg 112, 21029 Hamburg, Germany
†IAP, Univ. of Pierre & Marie Curie, 98bis, bd Arago, 75014 Paris, France
**Inst. für Astrophysik, Univ. of Göttingen, Friedrich-Hund-Platz 1, 37077 Göttingen, Germany
‡CRAL-ENS, 46 Alleé d'Italie, 69364 Lyon Cedex 07, France

Abstract. Since 1995, there have been great advances and extremely exciting results in the search for brown dwarfs and extrasolar giant planets. In the formation of their spectra, the far wings of the alkali resonance lines, especially Na I and K I, play an extraordinary role. Model atmospheres are necessary in order to calculate synthetic spectra and colors and to derive reliable parameters and the atmospheric chemical composition for such objects. Using alkali resonance line profiles [1] and [2], the differences in the synthetic spectra compared to earlier results [3], [4] will be outlined. Besides the line formation process dependent on the effective temperature and the chemistry in the atmosphere will be discussed. For the synthetic spectra, the general purpose stellar atmosphere code PHOENIX [5] is used.

Keywords: Stars: atmospheres Line: profiles Stars: low-mass, brown dwarfs
PACS: 95.30.Jx, 97.10.Ex, 97.20.Vs

THEORETICAL BACKGROUND

In the atmospheres of brown dwarfs, pressure broadening is prevalent. With a semi-classical approach, it is possible to compute the complete profile function of an alkali perturbed by H_2 or He. The alkali absorption line profiles are represented by two terms, one describing the line core with the impact approximation [6], [7] and the other describing the line wing with the one-perturber approximation [7]. For a detailed derivation see Allard and Kielkopf [6] and Allard et al. [7]. More detailed calculations of the perturbed alkali absorption doublets of Na I, Li I, K I, and Rb I have been supplied [1], [2].

Using the multi-purpose stellar atmosphere code PHOENIX, atmosphere models have been constructed for a range of effective temperatures and surface gravities typical for low-mass stars and brown dwarfs, tracing the behavior of the alkali absorption profiles under the low temperatures and high gas pressures prevalent in these atmospheres.

CP874, *Spectral Line Shapes: 18th International Conference,*
edited by E. Oks and M. Pindzola

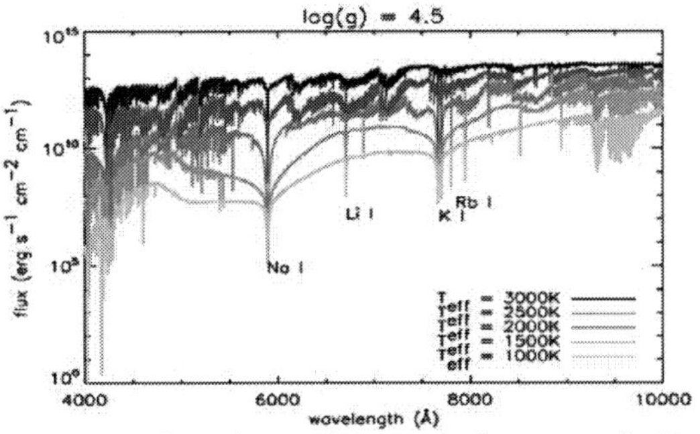

FIGURE 1. Effective temperature dependence of the synthetic spectra for T_{eff} from 1000 K to 3000 K.

MODELS AND SYNTHETIC SPECTRA

When comparing typical synthetic spectra calculated with PHOENIX (AMES-Cond, [3]) and different approximations of the pressure broadened profiles of the alkalies, differences are noticeable. The synthetic spectra calculated with the impact approximated van der Waals broadened profiles [3] underestimate observed line strengths out to ~ 9000 Å (for the K I lines), but produce too strong absorption at larger distances from the line core. The synthetic spectra calculated with the 2nd generation of accurately calculated alkali profiles provide more opacity in the intermediate region, but then drop off rapidly as seen in the observations. The wings of the two most abundant alkalies, Na I and K I, form the pseudo continuum of Li I and Rb I and thus also affect line formation analyzes of these species.

EFFECTIVE TEMPERATURE DEPENDENCE OF THE LINE WINGS

Figure 1 shows the effective temperature dependence of the synthetic spectra for $T_{\rm eff}$ = 1000 K to 3000 K. Towards lower effective temperatures, the alkali absorption lines and especially the far wings of the most abundant alkalies dominate increasingly the spectrum. This behavior is contrary to what might be expected from the profiles introduced in [1], [2] and [8], in which the HWHM and the contributions of the far wing profile increase towards higher gas temperatures. However, analyzing the pressure relations in the different model atmospheres, one finds a decreasing gas pressure in the line forming regions towards higher effective temperatures. Consequently, the observed line broadening decreases with higher effective temperatures. This effect is augmented by increasing background opacity from molecular band absorption, mostly due to TiO, VO, CaH and FeH, which masks much of the alkali line wings at $T_{\rm eff}$ = 3000 K.

Furthermore, when studying the relative abundances of the three most abundant species (H_2, He, and H I) in the synthetic spectra, as depicted in Fig. 2, a change of the

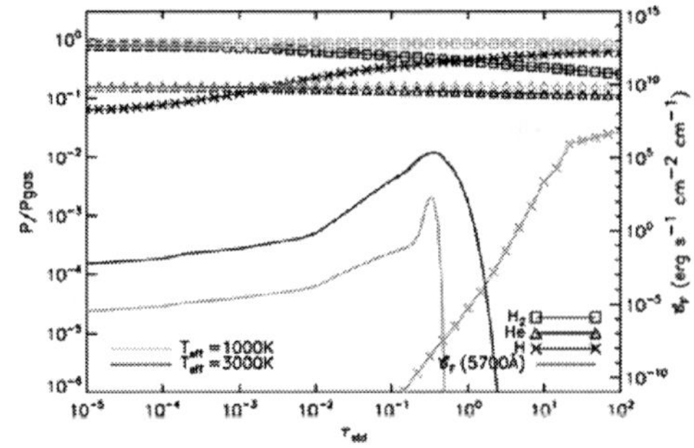

FIGURE 2. Relative abundance of H2, He and H I in the atmosphere at $T_{\rm eff}$ = 1000 K and 3000 K with log(g) = 4.5 (left ordinate). Besides the flux contribution function at a wavelength of 5700 Å is depicted (right ordinate)

relative abundance is visible. While He remains at its cosmic abundance in all parts of the atmosphere, almost all hydrogen is distributed between H_2 and H I according to dissociation equilibrium. At low effective temperatures, H_2 thus is the most abundant species, followed by He. Only in the innermost part of the atmosphere, H I forms in appreciable amounts. However, towards higher effective temperatures, the abundance of H I increases drastically. At $T_{\rm eff}$ = 3000 K, H I represents the most abundant species followed by H_2 from $|_{\rm std} = 1.0$ on inwards. Keeping this in mind and taking a look at the flux contribution function C_F [9], [10], studies show, that the wings of spectral lines form farther inside the atmosphere than the cores of the line. For a more detailed study of the formation of the line wing of Na I D_2, we compare the run of C_F for $T_{\rm eff}$ = 1000 K and 3000 K at a wavelength of 5700 Å in Fig. 2. Its maximum marks the location of the formation of the line in the atmosphere. For $T_{\rm eff}$ = 1000 K the maximum of C_F is not located close to regions of significant H I partial pressure. This relation changes towards higher effective temperatures. At $T_{\rm ef}$ = 3000 K, the locations in the atmosphere where H I becomes dominant and the line wing forms are very close. Hence, at higher effective temperatures, it will be more important to consider also collisional broadening due to H I in order to obtain a more accurate line profile.

REFERENCES

1. N. F. Allard, F. Allard, and J. F. Kielkopf, *A&A* **440,** 11951201 (2005).
2. N. F. Allard, and F. Spiegelmann, *A&A*, accepted (2005).
3. F. Allard, P. H. Hauschildt, D. R. Alexander, A. Tamanai, and A. Schweitzer, ApJ 556, 357372 (2001).
4. N. F. Allard, F. Allard, P. H. Hauschildt, J. F. Kielkopf, and L. Machin, *A&A* **411**, L473L476 (2003).
5. P. H. Hauschildt, and E. Baron, **JCAM** *109*, 4163 (1999).
6. M. Baranger, *Phys. Rev.* **111**, 494-504 (1958).
7. A. Royer, *Phys. Rev..A* **3**, 2044-2049 (1971).
8. N. F. Allard, and J. Kielkopf, *Rev. Mod. Phys.* **54**, 11031182 (1982).
9. N. F. Allard, R. A., J. F. Kielkopf, and N. Feautrier, *Phys. Rev. A* **60**, 10211033 (1999).
10. N. F. Allard, F. Allard, C. M. S. Johnas and J. Kielkopf, *A&A, submitted* (2006).
11. P. Magain, *A&A* **163**, 135139 (1986).
12. B. Fuhrmeister, C. Short, and P. H. Hauschildt, *A&A*, accepted (2006).

CONFERENCE PARTICIPANTS

Name	E-mail addresses	Country
Abdel-Naby, Shahin	abdelsh@auburn.edu	USA
Benner, Chris	dcbenn@wm.edu	USA
Dalimier, Elisabeth	elisabeth.dalimier@upmc.fr	France
Demura, Alexander V.	Alexander.Demura@hepti.kiae.ru	Russia
Ducloy, Martial	ducloy@galilee.univ-paris13.fr	France
Fontes, Christopher	CJF@lanl.gov	USA
Forrest, Phil	forrest@physics.auburn.edu	USA
Gamache, Robert	Robert-Gamache@uml.edu	USA
Gavrilenko, Valeriy P.	gavrilen@fpl.pgi.ru	Russia
Gonzalez, Manuel A	manuelgd@termo.uva.es	Spain
Griem, Hans	griem@glue.umd.edu	USA
Green, Brian	greenbg@auburn.edu	USA
Hartmann, Jean-Michel	hartmann@lisa.univ-paris12.fr	France
Helbig, Volkmar	v_helbig@web.de	Germany
Herman, Roger	rmh@phys.psu.edu	USA
Hernandez, Jesus	hernajv.@auburn.edu	USA
Hodges, Joseph T.	joseph.hodges@nist.gov	USA
Iglesias, Carlos A	iglesias1@llnl.gov	USA
Johnas, Christine	cjohnas@hs.uni-hamburg.de	Germany
Landers, Allen	landers@physics.auburn.edu	USA
Lazarian, Alex	alazarian@facstaff.wisc.edu	USA
Lewis, John K.C.	court@physics.mun.ca	Canada
Li, Zhiling	zhiling.li@grenoble.cnrs.fr	UK
Lisgo, Steven	steve.lisgo@ukaea.org.uk	UK
Lisitsa, Valery	lisitsa@nfi.kiae.ru	Russia
Loch, Stuart	loch@physics.auburn.edu	USA
Mancini, Roberto	rcman@physics.unr.edu	USA
Marandet, Yannick	yannick.marandet@piim.up.univ-mrs.fr	France
Maslowski, Piotr	pima@phys.uni.torun.pl	Poland
Niemax, Kay	niemax@ansci.de	Germany
Oks, Eugene	goks@physics.auburn.edu	USA
Parigger, Christian	cparigge@utsi.edu	USA
Peach, Gillian	g.peach@ucl.ac.uk	UK
Pichler, Marin	marin.pichler@goucher.edu	USA
Pindzola, Michael	pindzola@physics.auburn.edu	USA
Robicheaux, Francis	francisr@physics.auburn.edu	USA
Rosato, Joel	jrosato@piima1.univ-mrs.fr	France
Rosenkranz, Philip	pwr@mit.edu	USA
Rosmej, Frank	frank.rosmej@piim.up.univ-mrs.fr	France
Rowan, William (Bill)	w.l.rowan@mail.utexas.edu	USA
Schott, Romain	romain.schott@upmc.fr	France
Smith, Maryann H.	Mary.Ann.H.Smith@nasa.gov	USA
Stamm, Roland	roland.stamm@piim.up.univ-mrs.fr	France

Tabisz, George C.	tabisz@shaw.ca	Canada
Thibault, Franck	Franc.Thibault@univ.rennes1.fr	France
Tipping, Richard	rtipping@bama.ua.edu	USA
Topcu, Turker	topcut1@auburn.edu	USA
Wehr, Rick	rick@atmosp.physics.utoronto.ca	Canada
Woods, Melanie	woodsme@auburn.edu	USA
Woolsey, Nigel	ncw2@york.ac.uk	UK
Zeitoun, Philippe	philippe.zeitoun@ensta.fr	France

Author Index

A

B

C

D

F

G

H

I

J

K

L

M

N

O

P

Q

R

S

T

U

V

W